Concrete and Sustainability

Concrete and Sustainability

Per Jahren
Tongbo Sui

CRC Press
Taylor & Francis Group
Boca Raton London New York

CRC Press is an imprint of the
Taylor & Francis Group, an **Informa** business

A SPON PRESS BOOK

 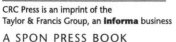 Chemical Industry Press

Cover design: "Strength through Humans" by Professor Boge Berg, Asker, Norway.

CRC Press
Taylor & Francis Group
6000 Broken Sound Parkway NW, Suite 300
Boca Raton, FL 33487-2742

First issued in paperback 2017

© 2014 by Chemical Industry Press
CRC Press is an imprint of Taylor & Francis Group, an Informa business

No claim to original U.S. Government works

Version Date: 20130621

ISBN 13: 978-1-4665-9249-0 (hbk)
ISBN 13: 978-1-138-07350-0 (pbk)

Library of Congress Cataloging-in-Publication Data

Jahren, Per.
    Concrete and sustainability / Per Jahren, Tongbo Sui.
        pages cm
    Includes bibliographical references and index.
    ISBN 978-1-4665-9249-0 (hardback)
    1. Concrete construction. 2. Concrete--Environmental aspects. 3. Sustainable construction. I. Sui, Tongbo. II. Title.

TA681.J35 2013
624.1'8340286--dc23   2013004843

Visit the Taylor & Francis Web site at
http://www.taylorandfrancis.com

and the CRC Press Web site at
http://www.crcpress.com

Contents

Foreword

The International Panel on Climate Change (IPCC), made up of 600 scientists from 40 countries, in a landmark report in February 2007, laid to rest any doubts about global warming. According to the IPCC, global warming is here, is primarily due to human activity, and is irreversible. In fact, global warming is the most serious sustainability issue confronting the global community today. The principal sources of global warming are CO_2 emissions, and if no action is taken to curb CO_2 emissions, it is most likely that the earth's average temperature could increase by 4°C by the end of this century, greatly impacting food production, water availability, and extreme weather events worldwide. The recent Super Storm Sandy in the United States is one such example. The damage caused by this storm is estimated to exceed 50 billion USD. The principal CO_2-emitting sectors of the economy include coal-powered plants, transportation, and major industries such as steel, cement, and concrete.

The production of 1 tonne of Portland cement clinker, the main component of modern hydraulic cements, is accompanied by the direct release of approximately 1 tonne of CO_2. Therefore, considerable CO_2 emissions are attributable to the global production of more than 3 billion tonnes of cement, the binder component of concrete.

There are several publications available that discuss the sustainability issues related to the production of Portland cement, but this is the first book that covers comprehensively the sustainable issues concerning the aggregate cement and concrete industry.

The authors, Mr. Per Jahren of Norway and Professor Sui Tongbo of China, are world-renowned engineers and concrete technologists. Their vast experience in cement and concrete is evident in their research of the sustainable issues concerning the concrete industry. They have skillfully addressed the issues of recycling of concrete and water shortages, which are the next crises on the horizon. Such distant topics as radiation and building materials are also discussed. The wealth of information on sustainability and the concrete industry brought together in this book is remarkable, and the authors' efforts are to be commended. All those interested in the concrete industry and sustainability will find this book very informative and of immense value.

Dr. V. Mohan Malhotra
Recipient of the 2012 Sustainability Award of the
American Concrete Institute
Recipient of the 2007 Sustainability Award of the
Norwegian Concrete Institute
Recipient of the 2006 Lifetime Achievement Award of the
Coal Combustion Products Partnership/EPA, United States
Honorary member of the American Concrete Institute
Honorary member of the Concrete Society, United Kingdom
Honorary member of the Brazilian Concrete Institute
Fellow, ASTM, United States
Fellow, Canadian Society of Civil Engineers
Fellow, Engineering Institute of Canada

Foreword

Sustainability has increasingly become the global solution in the 21st century to combating the challenges of climate change and natural resources depletion the world is facing. The cement industry, as one of the key fundamental materials industries, on the one hand plays a very important role in social and economic development, and, on the other, may also impose great challenge in terms of its large consumption of natural resources and energy and the emissions of greenhouse gases. This issue is of great importance, especially for China, as the world's largest cement and concrete producer and consumer, accounting for more than half of the world's cement production.

Concrete and Sustainability presents to the readers a holistic view and approach as well as worldwide thinking and methodologies covering the levels of safety, durability, functionality and economical feasibility, environmental compatibility, and social responsibility to addressing the sustainability issues. The wide spectrum of possible solutions given in this book provides a way to understand and deal with these global issues. In this context, there is no doubt that this is the first book of its kind, introducing not only technological solutions, but also methodologies and a way of thinking as a whole.

The author, Mr. Per Jahren, is a famous concrete specialist, and has enthusiastically participated in many cement and concrete conferences and seminars in China within the last 10 years, sharing his experience and skills, which are of great value to his Chinese colleagues. His coauthor, Dr. Sui Tongbo, is a leading scientist in the chemistry of cement and concrete, and is dedicated to the R&D of low energy and low CO_2 emission and advanced cement-based materials, with remarkable gains achieved in the development and application of belite-based cements.

This highly informative book, based on the knowledge and experience of the authors, will hopefully be of great benefit to researchers, engineers, architects, university students, and decision makers.

Dr. Sun Wei
Member of the Chinese Academy of Engineering
Professor of Southeast University, China

Preface

In light of the development of world concrete and construction, we see an evolution in the focus and direction of

Safety → Durability → Serviceability/Functionality → Sustainability

It is important in this context to learn at least two things:

- All the components in the evolution process are closely linked to each other and function upon need instead of occurring and existing independently or replacing one by another.
- The latest developed component, sustainability, has not only evolved from the previous components, but works as a function of them as well.

We therefore believe that sustainability is not only an environmental issue, but is indeed a holistic thinking/approach that can be considered the function of safety, durability, functionality and economical feasibility, environmental compatibility, and social responsibility. The level and magnitude of each component to sustainability varies depending on the specific requirement of the target and local boundary conditions.

We have, over recent decades, seen a growing worldwide concern and understanding of sustainability issues, not only in society in general, but also in the cement and concrete industry. The increased focus on climate change has definitely been an important catalyst in this process. There is hardly any doubt about the importance of greenhouse gas emission and the negative effects of climate change. However, sustainable development and environmental issues are much more than this.

This book tries to provide readers with the widest possible views on the sustainability issues of concrete, the world's largest construction material, the complexity of the challenges from different angles, and the versatility and possibilities of solutions to address these challenges. It is a methodologically technical book on concrete and sustainability, rather than only a concrete scientific book, and encourages the readers to think, understand, and reconsider the discussed topic and solutions that might be taken in a holistic way in terms of resource availability, technical viability, economical feasibility, and environmental compatibility. The intention has not been to give detailed technical advice or recommendations on sustainable issues, but to show some of the manifold versatility in the sustainability efforts that have taken place in the concrete environment around the world. Hopefully, we have been able to offer some ideas to concrete technologists or enthusiasts in their efforts to find sustainable solutions.

In Chapter 1, we explore the sustainability issue of concrete from a social and economic perspective. Chapter 2 gives an introduction into the various rules and regulations that the concrete environment is facing in society. Chapters 2 through 4 are about the various environmental challenges that the cement and concrete industry is facing. As emissions, absorptions, and recycling have been the most central elements in discussions in the cement and concrete environment so far, these topics have their own chapters. All the other issues are treated in alphabetical order in Chapter 5. When we say "all the other issues," we are aware that other listings are possible. We have chosen to use the issues that have been used in the Norwegian environmental database since 2002. Experience has shown that the importance of the various issues changes with the various platforms in the industry and society. We therefore treat the issues in alphabetical order instead of importance or significant sequence. In Chapters 6 and 7, we have provided some comments about future development. Finally, a comprehensive reference list, with some 500 references given in the book, might be of help to readers who want to obtain more technical details.

Per Jahren
President of P.J. Consult AS
Former president and honor member
of the Norwegian Concrete Society

Tongbo Sui
Ph D, Professor
Director General of Sinoma Research Institute
Vice President of China Sinoma
International Engineering Co. Ltd.

Acknowledgements

We offer our sincere thanks to the many writers who are the foundation of this publication.

Our grateful thanks to friends around the world who share our interest in the subject and who have sent materials we did not have previously.

Our heartfelt thanks are extended to Per Brevik, Norcem, Norway, for interesting discussions and information about carbon capture and storage (CSS); Inge Richard Eeg, Norbetong, Norway, for pictures from Oreid; Jan Eldegard and the website, www.byggutengrenser.no, Norway, for various building photos; Christian J. Engelsen, Norway, for leaching information; Svein B. Eriksson, Norcem, Norway, for information about utilisation of cement kiln dust and soil stabilisation; Kristin S. Kvisvik, Norcem, Norway, for various pictures; Naomi Matea and the Maeda Concrete Industry, Japan, for various illustrations; Gordana Petkovic, Norwegian Public Road Administration, for recycling pictures; John Erik Reiersen and the Norwegian Precast Concrete Association, for use of a picture; Terje Reiersen, Basal, Norway, for pipe illustrations; Jan-Erik Solberg and Sevaagbygg, Norway, for illustrations from Fossum Terrasse; Nils Chr. Thrane of Thrane & Thrane Teknikk AS, Asker, Norway, and Moscow, Russia, for information and an illustration regarding washing equipment.

Our sincere thanks also to Zhou Chunying, CBMA, for her kind help in collecting useful information; Li Zengkuan, CTC, regarding the radiation in building materials in China; Wang Zhihong, Beijing University of Technology, for providing information on ISO methodology and tools for environment performance evaluation; Li Wenwei, China TGP Co., for providing the picture of the Three Gorges Project; and our colleagues at the Sinoma Research Institute, Zhang Lijun, Chen Zhifeng, Zhou Jian, and Li Jing, for their kind support and cooperation.

Special thanks to Qin Shengyi, Renchsand Sci&Tech, for providing information on an innovative water management system; and Zhao Hongyi, Shandong Hongyi Sci&Tech, and Tan Guoren, Yunnan Yongbao Special Cement Co., for providing useful information on energy savings of the cement sector in China. Gratitude is extended as well to them for providing funding for covering the increased cost of the initial color printing of this book in China.

Appreciation also goes to the support of MOST, China, and the International Collaborative Project (2012DFA70870), and the State Pillar Plan Project (SQ2011BAJY3505).

The authors

Per Jahren is semiretired and has, in later years, concentrated most of his time on writing.

In 2011, the book *Concrete—History and Accounts* was published in Norwegian and English, and, in 2012, the book *Concrete—Manifold and Possibilities* in Norwegian.

Per Jahren was born in Oslo, Norway in 1939, and, after primary school and college, served his military service in the engineer core. His civil engineering education was in England and Norway, receiving a master's degree in civil engineering with concrete as a major in 1965, from the Norwegian Technical University in Trondheim. He later earned a business degree from Norway and Stanford University in the United States.

Through his nearly 50-year professional career he has been involved in cement and concrete—design, production, marketing, and development work in consulting companies, precast concrete production, cement production, and admixture and materials production and marketing. Since 1986, he has been involved in his own company, P.J. Consult AS in Asker, Norway.

Per Jahren became interested in environmental issues in the cement and concrete industry in his career and gave his first paper on the topic well over 30 years ago.

He was editor of the Norwegian environmental database for concrete, www.miljobasen.no, and chairman of the environmental committee of the Norwegian Concrete Society from 2002 to 2011. He is a former president and an honorary member of the Norwegian Concrete Society. He is a longtime member of ACI and a member of several ACI committees.

Per Jahren has written more than 150 papers, articles, and books on concrete, and has given lectures on concrete, and concrete and sustainability topics in particular, in 30 different countries.

Sui Tongbo, PhD, professor, is the founder and director general of Sinoma Research Institute and vice president of China Sinoma International Engineering Co. Ltd (Sinoma International) in Beijing. He joined Sinoma International in May 2010. Currently, as a visiting professor at Jinan University and Wuhan University, his focus is in the R&D of low energy and low CO_2 emission cements.

Before joining Sinoma International, he had been engaged in the R&D of special cements and new cementitious materials since 1991 in the China Building Materials Academy (CBMA), working as research engineer, senior engineer, chief of special cements division, and director of the Research Institute of Cement and New Building Materials under CBMA until February 2006. From March 2006 to April 2010, he was a professor and vice president of CBMA.

Dr. Tongbo has presided over 10 China state-supported key research projects in the cement and concrete area, as well as over 10 international cooperative projects with partners from Australia, France, Norway, and the United States. He is the recipient of various awards, including the national expert for excellent achievement in research on cement-based materials awarded by the State Council in 2005, and the second-class national prize for technological invention on low energy and low emission high belite cement awarded by the State Council in 2006. He also received an international award for his outstanding contribution to technology for cement and concrete sustainability at Seville, Spain in 2009 by the Organization of International Conferences for the Advances of Concrete Technology for Sustainable Development, formerly the CANMET/ACI Council.

He is the main founder of the International Center for Materials Technology Promotion under the United Nations Industrial Development Organization (UNIDO-ICM), located in Beijing since 2002. He had worked as director and international coordinator of UNIDO-ICM (2003–2007) for promoting technical diffusion and transfer in the building materials field and affordable housing technology for South-South cooperation. He is also the main organizer of many international symposiums and workshops in China in the cement and concrete sector.

He has been the co-chair of the World Business Council for Sustainable Development–Cement Sustainability Initiative (WBCSD-CSI) Task Force 1 for Climate Protection since 2011. Dr. Tongbo was a key member of the Cement Task Force–China under the seven-country Asian Pacific Partnership for Climate and Clean Mechanism (APP-CTF, 2006–2011). Additionally, he has held many memberships in professional associations, including deputy director of the Science and Education Committee of the China Building Materials Industry Federation, deputy director of the Cement Branch of the Chinese Ceramic Society, and member of ASTM C01.

Dr. Tongbo was born in 1965 in Shandong, China, and graduated from the Department of Applied Chemistry in the Shandong Building Materials Institute with a BS degree in 1988, and from the Department of Inorganic Materials Science and Engineering in the China Building Materials Academy with an MS degree (1991) and a PhD degree (2001). He was a visiting researcher in 1996 in the Department of Civil Engineering and Environment at the University of Illinois, Urbana-Champaign, supported by the World Bank.

Chapter 1

Introduction

There is no universally accepted definition for sustainable development. The most commonly referred to definition is from the Bruntland Commission: "Sustainable development is development that meets the needs of the present without compromising the ability of future generations to meet their own needs." What normally is referred to as the Bruntland Commission is the United Nations World Commission on Environment and Development (WCED). The commission was headed by Gro Harlem Bruntland. (She was the first Norwegian minister of the environment, 1974–1979, and the first female Norwegian prime minister, 1986–1989 and February–October 1991.[1]) The commission presented its report at the World Earth Summit in Rio de Janeiro, Brazil, in 1992. The work was followed by a new world summit in Kyoto, Japan, also known as the Kyoto Protocol of December 11, 1997.

Sustainability is also referred to as the intersection among economy, social development, and environmental challenges, as shown in Figure 1.1. Sustainable development can be understood as an optimal balance between the developments of these three sectors.

We have seen publications using similar expressions, like *social responsibility* and *economic prosperity*. Probably we could just as well have used *social prosperity* and *economic responsibility*.

However, there are many definitions of the term sustainable development. Malhotra,[2] in 2003, referred to the Massachusetts Institute of Technology (MIT), Cambridge, Massachusetts report that had recorded as many as 57 competing definitions of the term. Today the number is probably higher.

Concrete is the second biggest commodity in the world, after water, and consequently concrete has a formidable impact on all many sectors. The main focus in this book is to look at concrete and the environmental challenges, but before we do that, we will take a brief look into the effect of concrete in other sectors.

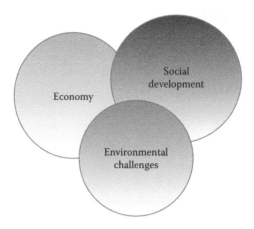

Figure 1.1 Sustainability.

1.1 THE ECONOMICAL IMPACT OF CONCRETE

Annually the amount of concrete cast in the world is similar to a mountain with an area of 1 × 1 km, and a height comparable to that of Mt. Everest. This volume is more than twice that of all the other building materials together. Over the last 10 to 15 years the centre for the use of concrete has shifted from Europe and North America to Asia. China is now the largest concrete producer and consumer, using more than half of the cement in the world, or twice as much as Europe and the United States together. The consumption per capita in China is more than three times that of the United States, for example (Figure 1.2).

In 2008, India passed the United States as the second largest production and consumption country of cement. While growth in the consumption in China is expected to slow down and probably culminate in a few years, the consumption growth in India is formidable. In 2010, India had a consumption of 216 billion tonnes, which is expected to grow to 425 billion tonnes in 2020 and 860 billion tonnes in 2030.[4]

Kulkarni,[4] with reference to the planning commission of the government of India, expects a growth in GDP in India of 9% by 2011–2012 (Figures 1.3 and 1.4).

There is no or moderate growth in the consumption of concrete in the typical industrial countries, and the consumption is typically between 200 and 600 kg of cement per capita. There are many prognoses for the possible growth in cement consumption in the world.

Here is a simple estimate: It is expected that the population in the world will culminate sometime between 2050 and 2070, and will be 8 to 9 billion people. With consumption per capita of 500 kg, this is pointing

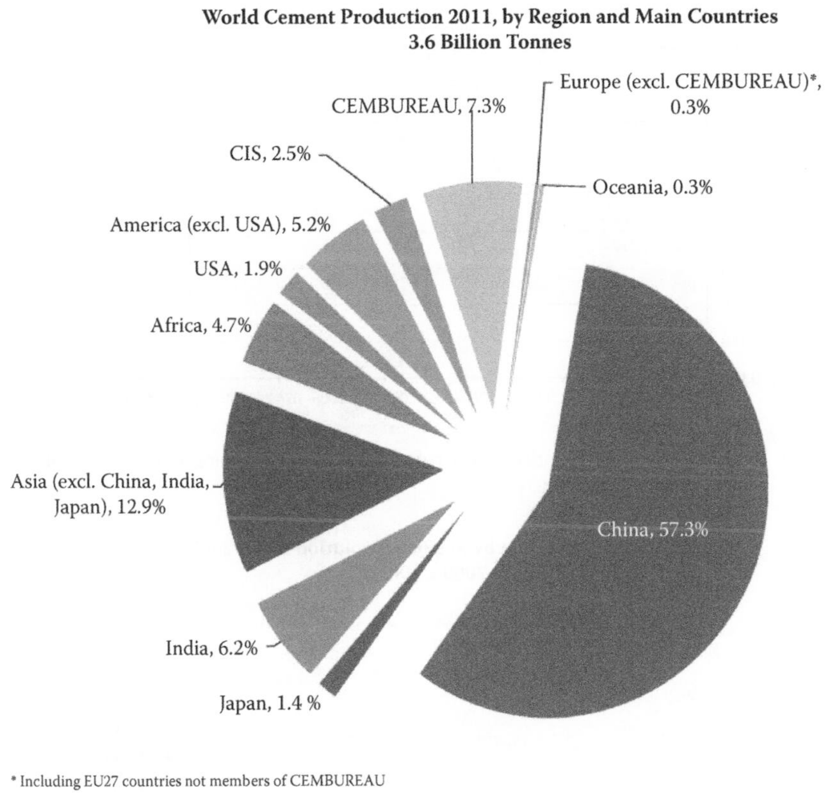

World Cement Production 2011, by Region and Main Countries
3.6 Billion Tonnes

* Including EU27 countries not members of CEMBUREAU

Figure 1.2 World cement production 2011.[3]

at a culmination of the consumption of cement, sometime in the future, of about 4 billion tonnes.

For comparison, it should be mentioned that the consumption of cement in 1900 was about 10 million tonnes. It passed 100 million tonnes in 1948, 0.5 billion tonne in 1968, and 1 billion tonne in 1989.

What does this enormous quantity of cement mean in economic terms?

The prices of cement and concrete vary considerably from one country to another. Here are some typical average prices (Figure 1.5):

United States (August 2011)
    Cement: 3 to 5 USD per 80 lb, i.e., about 110 USD per tonne (110 USD per tonne in New York in 2007[5])
    Concrete: 100 USD per yard,[3] i.e., about 130 USD per m^3
China (August 2011)
    Cement: 300 RMB per tonne, i.e., about 53 USD per tonne
    Concrete: 350 to 375 RMB per m^3 (C30), i.e., about 65 USD per m^3

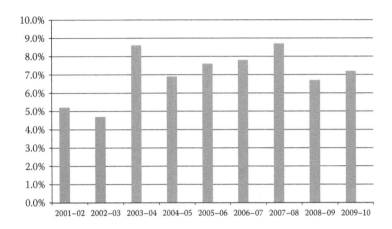

Figure 1.3 Growth in GDP, India,[4] with reference to the planning commission of the government of India.

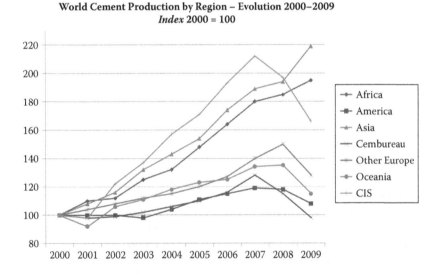

Figure 1.4 Growth is big in Asia and Africa, but seems to have culminated in other places in the world.[3]

Norway (August 2011)
    Cement: 800 NOK per tonne, i.e., about 146 USD per tonne
    Concrete: 890 NOK per m³, i.e., about 163 USD per m³

As China consumes about half the concrete in the world, its price is obviously a major influence on the total economic impact in the world.

Relative Prices

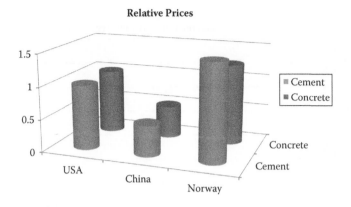

Figure 1.5 Relative prices.

As a conservative weighted average we have used the following rough averages in further evaluation:

Cement: $53 \times 0.60 + 110 \times 0.25 + 146 \times 0.15 = 81.2$ (using 80) USD per tonne

Concrete: $65 \times 0.60 + 130 \times 0.25 + 163 \times 0.15 = 96$ (using 90) USD per m^3

We further know that the total consumption of cement in the world is about 3 billion tonnes, which points toward a total concrete volume of 8 to 10 billion m^3.

As a rough average, the basic price of concrete is about one-third of the total price of the raw construction. The other two-thirds comes from reinforcement, formwork, labour cost, and rigging.

Consequently, the value of the concrete cast in the world each year amounts to roughly $90 \times 3 \times 9$ billion USD, which is about 2,400 billion USD.

The sales value of the cement produced is about $80 \times 3 = 240$ billion USD, or about 10% of the price of the final constructions.

As a comparison, looking at the list of the biggest economies in the world (GDP 2010, according to the International Monetary Fund) (Figure 1.6):[6]

The world	62,909 billion USD
5 France	2582 billion USD
6 United Kingdom	2247 billion USD
7 Brazil	2090 billion USD
8 Italy	2055 billion USD
9 Canada	1574 billion USD
10 India	1537 billion USD

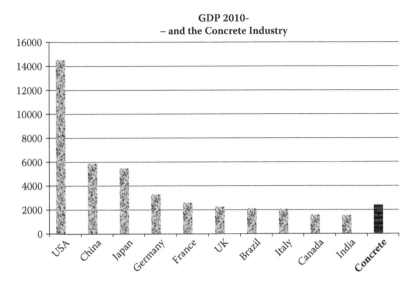

Figure 1.6 Concrete used in countries with the largest GDPs.

This should mean that the importance of the concrete industry in the world is nearly 4% of the total economy, or about the size of the economy in countries like France or the United Kingdom.

The above are very rough, easily understood estimates. We have done a few control evaluations.

In June 2011 a group of economists presented an analysis of the economic importance of the Norwegian concrete industry.[7]

Using figures from their analysis, we find:

- The total sales of the companies in the concrete industry in Norway (contractors not included) are about 4.1% of the Norwegian GDP.
- The cement industry represents 7.3% of this value.
- The added value from the concrete industry production represents 1.1% of the total GDP.
- The 10 largest companies represent 42% of the added value figures and 39% of the total sales value.

Comparing with China:

- The sales value of cement is approximately 85 billion USD.
- The sales value of the concrete produced in China is nearly 1000 billion USD.
- With a Chinese GDP of 5,878 billion USD, the concrete production is comparable to 16% of the GDP, and the cement industry represents

8.5% of this. It is important to remember that China today produces about three times as much cement and concrete per capita as the United States and typical countries in Europe.

The analysis indicates that the previous rough estimates are in the right order. The above simple evaluation should give a reasonable indication of the importance of the concrete industry.

For comparison, we refer to some data and statements from a UN Environment Programme publication:[8]

The building and construction sector typically provides 5–10% of employment at the national level and normally generates 5–15% of the GDP. It literally builds the foundations for sustainable development, including housing, workplace, public buildings and services, communications, energy, water and sanitary infrastructures, and provides the context for social interactions as well as economic development at the micro-level. Numerous studies have also proven the relationship between the built environment and the public health.

1.2 CONCRETE AND SOCIAL PROGRESS

Concrete is both a very old and a new material with constant new developments that have had and will continue to have a significant importance for social development in the world.

A quick look at concrete history is now given.[9] Concrete was probably invented simultaneously in various places in the world several thousand years ago. We have many examples:

- Mortar, Core Turkey, 2600 BC
- Concrete floor, Southern Galilee, 7000 BC
- Cabin floor, Lipenski Vir near Danube, 5000 BC
- Baihuimian mortar, China 5–3000 BC
- Floor, Dadinsan, China, 3000 BC
- Concrete in Middle and South America, 2500 BC
- Pyramids in Egypt? 2500 BC
- Concrete in Mesopotamia, 2500 BC
- Egyptian illustrations, 1950 BC
- "Yellow mud," China, 1600 BC
- Lightweight concrete, Mexico, 1100–800 BC
- King Salomo's water channels, 1000 BC

With the Roman Empire, about 2,000 years ago, came a technological breakthrough, by the use of pozzolan mortar, that lead to architectural

Figure 1.7 Aqueduct, Segovia, Spain.

possibilities and use in a much wider span of utilisation. The Romans developed use of concrete in their special walls, arches, and water channels, dams, and roads (Figure 1.7).

With the fall of the Roman Empire, concrete, to a great degree, went out of use for nearly 1500 years. The knowledge and technology, however, seem to have been kept secret in religious institutions.

Toward the end of the 1700s, the use of concrete increased in popularity again, and a technological breakthrough came in 1824 with the invention of Portland cement. Toward the middle and end of the century, the use of reinforcement was developed, and concrete increased in popularity and was more and more sophisticated in practical utilisation.

Now, more than 100 years later, concrete is a tool used to solve challenges in most areas of our social life.

One of the first globally known concrete technologists advocating for a more sustainable and holistic attitude in the development of concrete technology was P. Kumar Mehta. Since the first international discussions and seminars on the topic a few decades ago, much has happened; new and younger people have picked up the gauntlet, and new thoughts have come forward. However, much of Mehta's thinking is still valid today. In the social challenge context, we have chosen to cite a quote from a paper

Mehta presented at one of the first international concrete conferences on the sustainability topic, in 1998:[10]

> Another issue of considerable importance for the future is the large disparity in the standard of living in different parts of the world. In fact, it seems that we live in two worlds which exist side by side, and both are intent upon exploiting the earth's natural resources. The first world consists of the people in North America, western Europe, and Japan, which comprise 12% of the world's population but accounts for two-thirds of the total energy consumption and enjoy a high standard of living. Their economies are driven by consumerism and generate considerable waste and pollution on a per capita basis. The people in Asia (except Japan), Africa and South America, numbering almost 5 billion, from the second world. For a variety of reasons, including high rates of population growth, these countries are less developed industrially and the people, in general, have much lower standard of living. From the number of variety of ongoing or recently completed large infrastructure projects in western Europe, North America, and Japan, it is obvious that the industrially-advanced world is not slowing down in the use of the earth's natural resources. At the same time, in pursuit of a better life for the poor billions, many less developed countries have greatly accelerated the pace of industrialization. It is not difficult to imagine the end result of a process which encourages the continuation of high rates of consumption of natural resources of the earth on the one hand, and correspondingly high rates of environmental loading by a variety of polluting materials on the other hand. Clearly, a global environmental disaster would be unavoidable. It is in this context that both the industrially-rich and the industrially-poor worlds share equal responsibility to search and adopt the technologies for sustainable development.

Since Mehta wrote his paper 14 years ago, much has happened. Among other things, China and India have increased their cement production sixfold, and several years ago took over the positions of numbers one and two in the world in cement and concrete production. At the same time, the consumption in the countries that Mehta named the first world stagnated in their consumption. However, that does not change his key message about increased awareness of a sustainable development.

Somewhat different from one part of the world to the other, the use of concrete in buildings is from one-half to three-quarters of the total concrete consumption.

Buildings alone represent a majestic testimony of the versatility of concrete, from Dante Bini's creative shell structures based on a pneumatic flexible formwork technology, to new Chinese apartment buildings in a massive

Figure 1.8 Binishell structure, Narrabeen Public School, built 1973. Picture taken in 2011.[11]

Figure 1.9 Glimpse of Chinese apartment buildings, eastern suburb of Beijing, through a car window, August 2011.

effort to improve the social standard in the world's largest population, to churches and television towers (Figures 1.8 to 1.11).

But, the influence of concrete in our lives is much more than buildings. As a reminder, we have, as examples, picked out 25 other areas of use, mentioned in alphabetic order, not in order of importance:

Agricultural concrete products: A number of special concrete products have been developed for modern agricultural production. An example is slotted cattle floors for cattle barns (Figure 1.12).

Block work: Concrete blocks were probably the first precast product ever produced, several thousand years ago, and are probably the most important single concrete product worldwide—and of extreme

Figure 1.10 Polar Sea Cathedral, Tromsø, Norway.

importance in social development in the nonindustrial countries (Figure 1.13).

Bridges (Figure 1.14)

Canoes: Concrete has been used in a number of floating structures since the first concrete rowing boat was built by Jean-Louis Lambot in France in 1846. The first seagoing concrete ship, Namsenfjord, was built by Nic. Fougner in Norway in 1917. Since that time, thousands of ships, barges, floating docks, floating lighthouses, sailing yachts, etc., have been built. Canoes are a special chapter in the history of floating concrete structures.

In June, the annual National Concrete Canoe Competition, arranged by the American Society of Civil Engineers (ASCE), is held between American universities in concrete canoes. The competition started in the 1960s but was extended in 1988 when BASF came in as a sponsor. Now, over 200 university teams compete in the national competition, through qualifications in 18 regional races. From these regional competitions, the two best together with the five best from the previous year take part in the final in June (Figure 1.15).

Dams (Figure 1.16): Dams are important, but concrete is also used downstream from the production of the electricity, for large as well as modest structures.

Erosion protection (Figure 1.17)

Everyday articles (Figure 1.18)

Figure 1.11 KL Tower, Kuala Lumpur, Malaysia.

Furniture (Figure 1.19)

Garden and park products (Figure 1.20)

Harbours and quays: For more than 2000 years, concrete has been an important tool in securing safe transport on the seas. Julius Caesar used pozzolan concrete when building the Caligula wharf in Pozzuoli, Italy, in 44 BC.[9]

When John Smeaton built the Eddystone Lighthouse in concrete in 1724, a new era started in the shipping history, with more lighthouses that reduced the number of shipwrecks considerably.

Already more than 100 years ago, casting of underwater concrete was part of civil engineering education (Figure 1.21).

Today we find use of concrete in all types of harbours (Figures 1.22 and 1.23).

(A)

(B)

Figure 1.12 Precast slotted cattle floors, Norway. (Photo by John-Erik Reiersen.)

Figure 1.13 Block work wall bicycle workshop, Lhasa, Tibet, China.

Locks: Ship transport along the inland waterways in the world gives a very important contribution to sustainable development. To overcome height differences and connect one waterway to another, locks are important tools. Heavy concrete structures are important in lock design, sometimes combined with dams, and sometimes as independent structures (Figure 1.24).

NOx absorbers: By adding photocatalytic TiO_2 powder to the surface layer in concrete, the concrete products will get the ability to absorb NOx (NO and NO_2) from the air. The titanium dioxide powder acts as a catalyst, and the NOx will be absorbed as NO_3 ions in the concrete. Later the NO_3 ions will be washed out with water (Figure 1.25).

Offshore platforms (Figure 1.26)

Piles and poles: Piles and poles have a lot in common production-wise, as they often are produced locally with similar technology. But the production technology and shapes vary considerably in different parts of the world (Figure 1.27).

Pipes: Concrete pipes are produced in a number of dimensions and shapes, with various production methods, and for many different water and sewage transport purposes (Figure 1.28).

(A)

(B)

Figure 1.14 (A) The Glenfinnan Viaduct, England, built from July 1897 to October 1898, with 21 spans and length of 300 m, was a breakthrough in the use of concrete in bridges. (B) The elegant Salganitobel Bridge, near Schiers in Switzerland, designed by Robert Maillart and built in 1929–1930, was a breakthrough in modern concrete bridge design.

(A)

(B)

Figure 1.15 Canoes from the National Concrete Canoe Competition, USA. (Photos courtesy of BASF.)

Concrete has for thousands of years been a natural choice in managing transport and storage of water, from the nomads and Greeks, who secured their wells with concrete, to the Roman water transport systems, to modern water towers, ground-level reservoirs, pumping systems, river abstractions, and desalination plants. The concrete pipes are only a small but important piece in the water management sector, and unlike most of the other structures, are prefabricated in standardised lengths and diameters in concrete plants all over the world.

Railway sleepers/ties (Figure 1.29)

Reefs: Artificial fish reefs have been utilised for more than 200 years, and an investigation from 1998 found reef structures in more than 62 countries (Figure 1.30). The first concrete reefs were made in Japan in 1954.[13] The various reef structures differ considerably depending on the motives for their use. Some uses include:

- Gathering fish to promote more cost-efficient and resource-efficient fishing
- Breeding/increasing biomass production
- Bringing life to the sea bottom deserts
- Purification of water
- Utilisation of waste nutrition
- Refinement for growth of valuable fish food
- Securing areas for coastal population
- Tourist attraction

Sculptures: Concrete seems to be a wonder material for artists. The sculpturing of concrete material has over the years taken many shapes and directions, from small statuettes to gigantic sculptures, from nonfigurative creative wonders to mass production of animals, etc., to a place in the gardens.

One of the most famous artists who made a number of famous concrete sculptures in a number of countries is Pablo Picasso, through cooperation with the Norwegian artist Carl Nesjar. Instead of showing one of their sculptures, we are showing their fascinating sculpturing relief at the end wall of the Norwegian government building in Oslo (Figure 1.31). The picture was taken only 3 weeks before the terrorist attack on the building, July 22, 2011.

Snow shelters (Figure 1.32)

Stairs and stairways: There are numerous alternatives in the use of concrete in stairs and stairways. Concrete stairways are fundamentally used in buildings throughout the world. We can only give a few examples (Figure 1.33).

(A)

(B)

Figure 1.16 (A) The Three Gorges Dam, China, was built from 1993 to 2009. The dam is going to produce 84.68 BkWh. For the dam, 28 million m³ of concrete was used, where 16 million m³ was used in the dam construction itself. In comparison, this is the same as five times the annual concrete consumption in Norway.[12] (B) The dam and power station in Uglich, Russia, with the administration building to the right. The dam was built in the 1940s and is damming up the Volga River just before the Rubinsky reservoir.

Tanks: Concrete has been used to ease the storage of water for thousands of years. An Arabic nomadic tribe, the Nabataesans, used pozzolan-based concrete to secure their secret wells in the desert 300 years BC, and the Greeks used concrete based on Santorin to secure wells before the Romans. About 2000 years ago, the Romans used pozzolan-based concrete in many of their impressive hydraulic structures. Later came the tanks based on concrete in many versions (Figure 1.34).

However, concrete has also been utilised to store other liquids.

Tree repair (Figure 1.35)

(C)

Figure 1.16 (Continued) (C) Small transformer box in concrete, Strømstad, Sweden.

Walls: One of the first items that concrete was used for was walls; walls mark a territory or frontier. Walls are made to keep someone or something out, or to keep them inside.

We know that concrete was used in the mortar in the Great Wall in China. Concrete from the Berlin Wall is now mostly kept as souvenirs around the world. But, we also have attractive walls as sound barriers and to shape gardens and parks (Figure 1.36).

Water channels (Figure 1.37)

Windmills (Figure 1.38)

We limited the versatile list of concrete to 25 areas. A complete list would have been much longer. Just think about all the various sport stadiums and structures, the various use of concrete in foundations, from small signs to massive structures, and all the floating structures, from sailing boats to barges and gigantic oil terminals and airports.

Concrete is, in more than one meaning of the word, one of the foundations for social developments!

(A)

(B)

Figure 1.17 (A) Precast concrete solution in Japan. (B) Along the Moscow river canal in Russia.

(C)

Figure 1.17 (Continued) (C) Sprayed concrete along a main road, Sandvika, Norway.

(A)

Figure 1.18 (A) Garbage container, top of Mt. Tai, China's most holy mountain, Shandong Province, China.

(B)

(C)

Figure 1.18 (Continued) (B) Clock designed by Gil Tsafrir, United States (www.zulasurfing. etsy.com). (C) Flower vase by Maeta Concrete Ltd., Sakkata, Japan.

(A)

(B)

Figure 1.19 Outdoor furniture, national tourist roads, Norway. (A) Strømbu main stop at National Road 27 in Alndalen, Rondane. Architect: Carl-Viggo Hølmebakk. (Photo courtesy Vegard Moen/Statens vegvesen.) (B) Liasanden stop in Leirdalen, Lom, Sognefjellet. Architect: Jensen & Skodvin. (Photo courtesy Jiri Havran/Statens vegvesen.)

(C)

(D)

Figure 1.19 (Continued) (C) Cafeteria furniture, Bird Park, Kuala Lumpur, Malaysia. (D) Elegant chair in high-performance concrete by Omer Arbel, Canada.

(A)

(B)

Figure 1.20 Some examples of garden and park products from Asak Miljøstein, Norway.

(C)

(D)

Figure 1.20 (Continued)

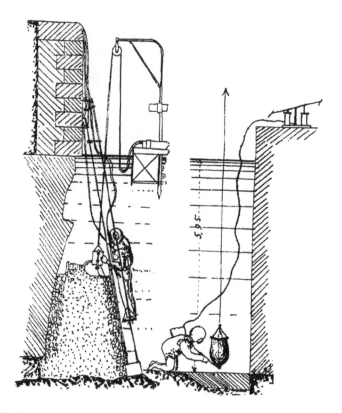

Figure 1.21 Illustration in textbook in Finland 1895.[9]

Figure 1.22 Floating concrete components in a small-boat harbour in Strømstad, Sweden. The photo was taken in September. The summer season is over. In July this is a very busy harbour.

Figure 1.23 One of the world's largest container terminals, Osaka, Japan.

Figure 1.24 A tanker on its way out of the lock on the Svir River in Northeast Russia. This is one of the 19 locks on the waterway between St. Petersburg and Moscow.

Figure 1.25 Pedestrian areas with NOx-reducing pavers, Omiya, Japan.

(A)

(B)

Figure 1.26 (A) Towing out the troll platform to the North Sea in Norway. Troll was the largest concrete structure ever moved. (B) The Arco Barge, built by Concrete Technology Corp., Tacoma, Washington, in 1974, and towed 16,000 km across the Pacific Ocean to the Java Sea to function as a storage facility for 36,000 tonnes of propane, and to be the living quarters for a crew of 50.

(A)

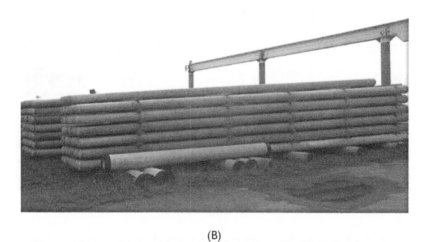

(B)

Figure 1.27 Piling with (A) square-shaped piles in Norway and (B) hollow, circular piles in stock in Japan.

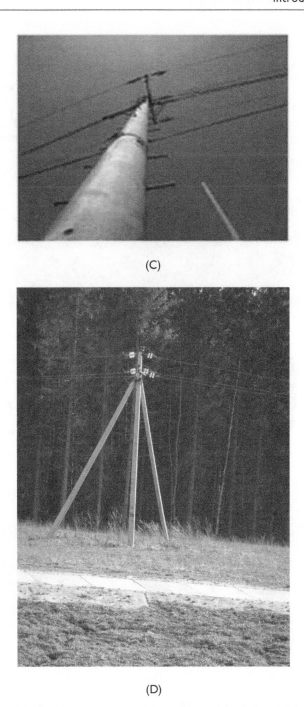

(C)

(D)

Figure 1.27 (Continued) Concrete poles in (C) Japan and (D) Russia.

(A)

(B)

Figure 1.28 Circular pipes for (A) main sewage lines, Norway; (B) oval pipes, Finland.

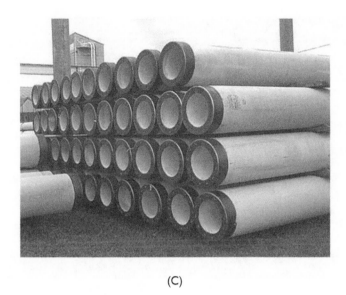

(C)

Figure 1.28 (Continued) (C) prestressed pressure pipes, Japan.

(A)

(B)

Figure 1.29 (A) Railway sleepers in production at Spenncon's factory at Hønefoss, Norway. (B) Railway sleepers in use in front of the Airport Express Train at Asker Station, Norway.

(C)

(D)

Figure 1.29 (Continued) (C and D) Sleepers for the new Chinese high-speed train.

(A)

(B)

Figure 1.30 In the years 2008–2013, a new extension of the "harbour city" in Oslo, Norway, was made through the Tjuvholmen project. To increase marine life, various artificial reefs modules were used. The pictures show (A) the modules before setting out in 2008 and (B) the modules in place. The successful initiative transformed a previous muddy sea bottom desert to a lively marine area with many types of fish and shellfish. Part of the project was also a natural cleaning of the water through the use of nearly 2000 mussel ropes and more than 300 cleaning modules in concrete.

(A)

(B)

Figure 1.31 (A) The Pablo Picasso–Carl Nesjar relief, *The Fishermen*, government building, Oslo, Norway. (B) Rock sculpture in Baimai Spring Park in Shandong
Province, China.

Figure 1.32 Snow shelters, Japan.

(A)

Figure 1.33 (A) Concrete stairway in condominium building, Asker, Norway.

(B)

Figure 1.33 (Continued) (B) Stairway components for air traffic control tower, United
States.

(C)

Figure 1.33 (Continued) (C) Elegant polished concrete spiral stairway, building of Engineering Society, Oslo, Norway.

(D)

Figure 1.33 (Continued) (D) Precast concrete stairs for a summer cabin, Sponvika, Norway.

(A)

(B)

Figure 1.34 The water tower in Newton-le-Willows in England was built in 1904 with a capacity of 300,000 gallons (1.36 million litres). The tower was demolished in 1979. (B) Tank numbers 14 and 15 at a winery in Duero Valley in Portugal are concrete and store wine for fermenting, which will later be port wine.

(A)

Figure 1.35 (A) Several-hundred-year-old trees at Confucius Temple, Qufu, Shandong Province, China.

(B)

Figure 1.35 (Continued) (B) Tree at the Ivar Aasen Museum, Fjærland, Norway. (Photo courtesy Jan Eldegard.)

(A)

(B)

Figure 1.36 (A) The Great Wall, China. (B) Berlin Wall, Germany.

(C)

(D)

Figure 1.36 (Continued) (C) Sound barrier wall, Oslo, Norway. (D) Basement wall in split concrete blocks, Asker, Norway.

(A)

(B)

Figure 1.37 Water channels, Japan.

(A)

Figure 1.38 A 100 m high 3 MW turbine windmill in Finland, where the bottom 45 m is precast concrete from Consolis/Homifuste.[14]

(B)

Figure 1.38 (Continued)

Chapter 2

Environmental issues

2.1 GLOBAL/REGIONAL/LOCAL ASPECTS

Looking at the various environmental challenges that the cement and concrete industry is facing, we see that some of these challenges are local in nature, and that they can only be solved locally—some of them have regional solutions, and some, for example, energy and emission challenges, are global in nature, challenges, and solutions. However, even with global challenges, the possible effects and resource availability of the tools to improve the situation vary considerably from one place to another. Also, the economical realities and local traditions and technology levels add to the variability and complexity of the solution equation. It is therefore important that we have both a wide number of alternatives in the toolbox and the necessary modesty in striving for more sustainable solutions when we attack the problems. One of the most important intentions of this book has been to try to contribute to such understanding and versatility.

Even if the most sustainable solution in one country does not necessarily have to be the best solution in another, we can most often find interesting alternatives, ideas, and inspirations in solutions and technologies from other parts of the world. Interchange of information about solutions, research, case examples, and progress is therefore one of the most important bases for intensified efforts for an even more sustainable future for the concrete industry.

2.2 RATING SYSTEMS

In a presentation at the ACI, Concrete Sustainability Forum IV, in Cincinnati, Terry Neimeyer[15] claimed that there were more than 900 rating systems for sustainability in the world. Actually, we have the definite impression that the many systems and abbreviations for them are causing a problem for the technical community, and that the differences between them are probably only really understood by a limited group of specialists.

The rating systems for more sustainable or "green" buildings have become increasingly popular as tools for tracking and measuring the environmental achievements in design and buildings and other structures. In 1990 the Building Research Establishment (BRE) in the UK developed and presented the Building Research Establishment Environmental Assessment Method (BREEAM) for office buildings. That was a pioneering work being a model for many other similar systems. It also laid the foundation for the Green Building Initiative (GBI) and Green Building Councils. Later came the BREEAM versions for shopping centers in 1991 and industrial buildings in 1993.

Possibly the most well-known or referred to rating system today, besides BREEAM, is the American Leadership in Energy and Environmental Design (LEED) system (which originates from the BREEAM concept), which was developed through the establishment of the U.S. Green Building Council (USGBC) in 1993. The first LEED standard was formed in 1999.[16]

In some countries we also find modifications of these systems. In Norway, for example, the major partakers in the building industry, like owners, developers, contractors, architects, consultants, and material suppliers, decided to use a modified BREEAM system, called BREEAM-NOR, especially suited to Norwegian conditions and standards. After long discussion and negotiations, this system was introduced in Norway in October 2011.

Another example is the Evaluation Standard for Green Building (GB 50378-2006) enforced in China since June 1, 2006, which includes six indices as follows: land saving and outdoor environment, energy saving and energy utilisation, water saving and water resource utilisation, material saving and material resource utilisation, indoor environmental quality, and operation management (for residential buildings) and life cycle comprehensive performance (for public buildings). Under each index are the adopted solutions or measures specified in three levels (basic, updated, and optimum) for certain buildings, which can be used for quantitative scoring evaluation.

Interest in the use of these rating systems has had a formidable spread to all parts of the world for the last decade (Figure 2.1).

A very important property or significance with the rating systems is that the rating from these systems starts where the regulatory minimum ends (Figure 2.2).

One of the best things with the most popular rating systems is that even if they are very versatile and include a great number of environmental challenges, they are rather nonacademically, simple mathematically, and press the user to think sustainable and holistic. Their use is also bringing in a market argument that inspires increased use of the systems.

The systems encourage designers and owners to minimise disruption of local ecosystems by ensuring efficient use of water, energy, and other natural resources and try to improve health factors such as indoor air quality.

The green building rating systems help designers and decision makers design and build buildings that minimise negative effects on the environment.

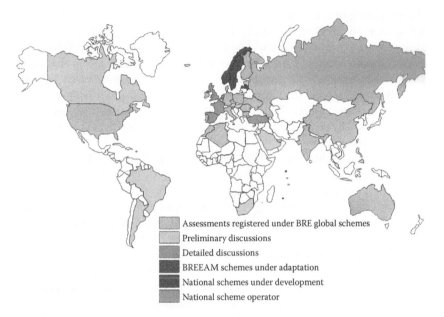

Figure 2.1 Use of BREEAM-related rating systems.[17] (From Bramslev K., BREEAM–Nor kort innføring. Presented at Breeam-Nor seminar, Lysaker, Norway, December 12, 2011.)

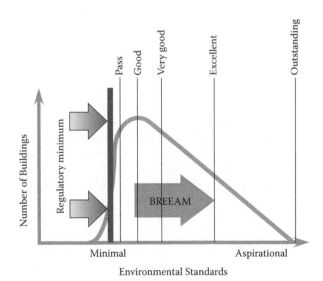

Figure 2.2 The environmental rating system starts where the regulatory minimum ends.[18] (From Eldegard J., *BREEAM*, Report 3, Norwegian Concrete Society, Environmental Committee, October 2011.)

The weighting of the various environmental factors differs somewhat from one system to another, naturally because conditions might differ somewhat from one part of the world to another. In a sustainability aspect, the weighting values can of course be questioned, but we must accept that there is no exact formula for the importance of various environmental aspects, and a holistic attitude to improved sustainability will often be a trade-off between various aspects. We just have to trust the evaluations, qualifications, and common sense of the institutions that at any period in time are issuing the weighting values.

An interesting thing with the systems is that they are clearly market driven. Property developers see that the value of the property increases and that renting income and demand increase with a higher system rating. Also, financing institutions seem to give advantages to plans with a high rating.

An important thing learned from these systems is that the building materials themselves are of far less importance than how the materials are utilised toward a more sustainable design.

However, the importance of the rating systems must not be underestimated, even for the practical utilisation of a more sustainable practice regarding the production and use of the building materials, for example, concrete. In a thoughtful article at a conference in Coventry, UK, in 2007, Christian Meyer from Columbia University, New York, gave a somewhat philosophical analysis on the degree of recycling in the U.S. construction industry.[5] His primary message is that in a free-market society, business decisions are driven by economic incentives, i.e., mostly profit, subject to government regulations and interventions.

He claims that in the United States it is estimated that about 150 million tonnes of construction waste are generated annually, and that concrete debris constitutes slightly over 50% of this. Almost 60% of this amount or 45 million tonnes is landfilled, with tipping fees as high as 50 USD per cubic yard. (The U.S. annual production of concrete is estimated to be well over 500 million cubic yards.)

He is of the clear opinion that the prime reason for the United States lagging behind most developed countries in recycling of building material is that virgin building materials are cheaper. However, he states: "At present we are experiencing a fundamental change in public attitudes, spearheaded by a vocal and growing environmental community that has taken a strong foothold in the building and construction industry. This development is best illustrated by the exponential growth of membership and the significant influence exerted by the U.S. Green Building Council, to name just one, maybe the largest pertinent organization."[5] He further claims:

> Other economic factors are even less quantifiable, such as the novelty aspects and the label "environmental friendliness." For example, the LEED rating system of the U.S. Green Building Council is gaining

increasing popularity among builders and developers, who recognize the value, both tangible and intangible, of a LEED rating. It is said that apartments in the solaire, the first "green" high-rise residential building in the United States, fetch higher rents than comparable units in conventional high-rise buildings, simply because of the cachet associated with living in a "green building." The LEED rating is based on credit points for a variety of environmentally friendly attributes such as energy-saving measures and use of recycled materials.

The rating systems evaluate how a structure or building is functioning in a sustainable manner from a holistic perspective. The different systems weight the various environment aspects somewhat differently, but principally at about the same magnitude. The weighting list for BREEAM-NOR is as follows:[17]

- Management: 12%
- Health and well-being: 15%
- Energy: 19%
- Transport: 10%
- Water: 4.5%
- Materials: 14%
- Waste: 7.5%
- Land use and ecology: 10%
- Pollution/CO_2:   8%

After having passed a certain minimum standard you will, depending on how good the design is in the various environmental categories, be able to earn credits. In the BREEAM system these credits will be weighted versus the weighting list to calculate the final score. Finally, the result is evaluated as (BREEAM-NOR):

Acceptable:		(≥10%)
Pass:	≥30	(25%)
Good:	≥45	(40%)
Very good:	≥55	(55%)
Excellent:	≥70	(70%)
Outstanding:	≥85	(85%)

In a paper in 2007, Mahrer and Kramer give some insight into the point crediting possibilities in the LEED-NC, Version 2.2 rating system.[16] The system is today in use not only for office buildings, but also for schools, multiunit residential buildings, manufacturing plants, laboratories, and many other building types. LEED-NC is divided into 5 main categories concerned with the quality of sustainability accounting for 64 of the possible

Table 2.1 LEED-NC certification levels

Level	Points required
Certified	26–32
Silver	33–38
Gold	39–51
Platinum	52–69

Table 2.2 Available points in LEED-NC rating system

Category	Points
Sustainable sites	14
Water efficiency	5
Energy and atmosphere	17
Materials and resources	13
Indoor environmental quality	15
Innovation and design process	5

69 points. The last 5 points are earned through innovation and the use of LEED-accredited professionals on the project (Table 2.1).

The six categories for possible points earning are: sustainable sites; water efficiency; energy and atmosphere; materials and resources; indoor environment quality; and innovation and design process. In principle, sustainable use of the construction materials can receive points from all these categories, with the exception of water efficiency (Table 2.2).

The rating systems claim market advantages in the use of the systems:[18]

- Operation cost decrease
- Building value increase
- Return on investment improvement
- Occupancy ratio increase
- Rent ratio increase

2.3 EVALUATION SYSTEMS/TOOLS

Even if we experienced the first serious public attention to environmental issues more than half a century ago, probably first with a demand to clean emission and smoke from industrial chimneys, it was not until the 1980s that we got the first general environmental evaluation systems.

In a lecture in Cincinnati, Ohio, in 2011, one of the speakers claimed that we today can find more than 900 evaluation and rating systems around the world.

A central principle in modern evaluation systems is to evaluate the effect on the environment of a product from start to finish: from the production of raw materials to product production, to actual use, to destruction/recycling/deposits. This is the lifetime effect, which also is called the birth-to-grave principle.

The *environmental product declaration* (EPD) systems might be found in three levels:

Type 1: Specifies if a product meets certain defined criteria, e.g., ISO 14021. An example of this is swan labels. The specifications do not take into consideration raw material processes.

Type 2: A self-declaration system. This is a system where a producer declares its product, without a neutral third-party evaluation. The system is specified in ISO 14024.

Type 3: A system based on birth-to-grave evaluation by a neutral third party, according to ISO 14040-43. In this declaration the environmental footprints of a product are specified with respect to a number of issues. There are also rules, for example, for presentation and qualifications of the third-party evaluator.

In a paper from 2000, Vold[19] comments on why a type 3 EPD system is needed. In society, private and public buyers will have a guarantee or exact information about the products they are buying. The EPD system gives:

- The buyers, designers, and advisors useful information in choice of products and services
- Data for use in further environmental evaluations
- Increased attention around the EPD of a product; challenges the producer, and inspires development of more environmentally friendly products
- Increased amount of data and information about raw materials, secondary products, and final products

She claims that a good EPD system must:

- Be based on international standardised rules and systems for the content of the declaration, and have a high degree of transparency in data and predictability, and use data of good quality
- Have availability of data from the industry, based on a common data platform
- Be easy to understand, even for nonexperts
- Have branch agreements about a common platform regarding comparison of data
- Be economically efficient in combination with other environmental evaluation systems

As an example of the use of EPD systems, we refer to a report from a Fib task group.[20] They refer to a French EPD carried out by four organisations:

- Fib: Fedration de l'Industrie de Beton
- UNPG : Union National des Producteurs de Granulates
- BPE: Syndicat National du BetonPret a l'Emploi, SYNAD
- CIMbéton: Centre d'Informationsur le Cimentetses Applicationes

Three examples are given:

1. Single-family house for four people with walls of concrete blocks and a service life of 50 years.

   CO_2 emission: Production of concrete blocks 1.7 tonnes
   Standard service life   350 tonnes (200 times more)

2. Sewage network with 3 km of 400 mm concrete pipes, capable of serving 5000 inhabitants. This is measured against daily activity of 5000 people in 50 years.

   CO_2 emission: Production of concrete pipes 102 tonnes
   Standard service life   420,000 tonnes (4000 times more)

3. Five-story residential building with 20 units housing 50 people, using 800 m^3 of ready-mixed concrete. This is measured against the daily activities of the people in 50 years.

   CO_2 emission: Production of concrete   180 tonnes
   Standard service life   5000 tonnes (30 times more)

The examples above might be used to illustrate that the emission from the materials used has far less effect on greenhouse gas (GHG) emission than the structures they are used for, but the referred use of this particular evaluation tool gives little help to evaluate other sustainability issues.

An important factor, and sometimes the most important factor, when comparing various design solutions or choice of building materials, is the time perspective. Typically, heavier or more durable materials will always profit on longer-life evaluation time. An important factor in this respect is changes in society and the type of building or structure to be evaluated. A typical commercial building might more often change in ownership and variation in needs and user specification. A hospital might have the possibility to adapt to new technology. Flexibility in design and adaptability to changes will be more important to some structures than others. Such important facts are very seldom taken into consideration.

2.3.1 LCC, LCI, and LCA

Life cycle cost (LCC) analysis might be a powerful tool to compare different design alternatives and materials on an economical basis. The analysis makes an account of all expenditures incurred over the lifetime of the structure analysed. The costs at any given time are discounted back to a fixed date, based on assumed rates of inflation and the value of the money. The LCC is equal to the construction detail investment cost plus the present value of future repair, maintenance, and replacement costs over the lifetime. The typical European evaluation period is 60 years, but for special structures, deviating time expectancy has been used. For the investment in offshore windmills we have seen 25 years being used as an evaluation period, mainly due to investment safety reasons. For municipal harbour structures and typical public structures that are constructed and meant to be landmarks for centuries, even 300 years has been used.

Solutions that have low investment/initial cost often require higher costs during the lifetime of the structure. High-performance concrete has the advantage in LCC to have lower maintenance cost than many other material alternatives. Typically, durable materials have advantages in an LCC analysis versus only looking at the investment cost, and more so the longer the evaluation time that is chosen.

Life cycle inventory (LCI) analysis is often the first step of an LCA, and accounts for single-material or individual-component environmentally important data flow to and from the material or component over the lifetime. Typical data are energy consumption, and emissions to air, land, and water associated with producing the material or product. It is important to define the system boundaries of the LCI, where the inventory starts and ends, to be able to give the correct input data for a later LCA.

Life cycle assessment (LCA) is an environmental assessment of the life cycle of a product, component, or a group of products, or a design component. The LCA looks at all aspects of the product's life cycle, from the first collection of raw materials from nature, through the nature and processing of the raw materials into a product, to the actual use of the product and its effect on the environment, and finally to the final demolition, recycling, or disposing it back to nature. Several individual LCIs might be background data for the LCA.

The environmental aspects that are normally looked at in LCA are:

- Ecology
- Acidification
- Biological diversity
- Ecotoxicological impact
- Eutrophication (the deterioration of the aesthetic and life-supporting qualities of lakes and estuaries, caused by excessive fertilisation from effluent high phosphorous, nitrogen, and organic growth substances)[21]

- Global warming
- Habitat alteration
- Ozone layer depletion
- Photo-oxidant formation
- Human health
- Toxicological impact
- Nontoxicological impact
- Occupational health
- Resources
- Energy use
- Land utilisation
- Material depletion
- Water

LCA has also been used on, for example, whole buildings, sewer pipelines, and roads. Typical results are that the environmental impact of the production of the various building materials has limited effect on the final result compared to other factors, like energy use in a building during its lifetime, excavation, transport and handling of materials in the pipe trenches, the life expectancy of pipes, the traffic on a road, etc. Most LCA studies of buildings seem to show that 10 to 30% of the energy use and GHG emissions come from the production phase, while 70 to 90% come from the use of the building.

In a brochure/report from 2009, the European Concrete Paving Association (EUPAVE) gave sustainability data for concrete roads in a very instructive way. Through CIMbéton in France, Centre d'Energetique de l'Ecole des Mines de Paris, France, it carried out an LCA study to look into differences between four types of concrete pavements, one type of composite pavement, and one type of asphalt pavement for a twin-lane dual-carriage motorway of 1 km length and service life of 30 years. The interesting report claims that without including the impact of the traffic, the bituminous structure is more favourable than concrete for wastes, greenhouse gases, eutrophication, and toxicity to humans. The concrete structure, on the other hand, is more favourable for the remaining indicators: energy, water, natural resources, radioactive wastes, acidification, ecotoxicity, smog, and odour.

LCA can also be used in analysing parts of structures to optimise design. In Figure 2.3 we give a Danish example showing CO_2 emissions from production of hollow-core precast components.[22]

Another example from the same source (Centre for Green Concrete, Denmark)[23] is where they compare the CO_2 emission by construction of a bridge pillar over a 100-year perspective with different types of reinforcement (Figure 2.4).

However, when the usage phase, i.e., the traffic, is taken into consideration, an entirely different picture emerges. With the sole exception of the

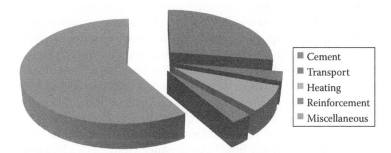

Figure 2.3 CO_2 emissions from production of precast hollow-core components.[23] (From Hasholt M.T., et al., *Anvisninger i grøn betong*, Center for grøn betong, Taastrup, Denmark, December 2002.)

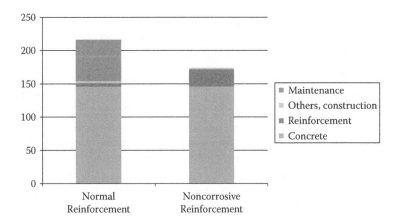

Figure 2.4 CO_2 emission from a bridge pillar in a 100-year perspective, with different types of reinforcement.[23] (From Hasholt M.T., et al., *Anvisninger i grøn betong*, Center for grøn betong, Taastrup, Denmark, December 2002.)

"solid waste" indicator, the impact of the traffic is *at least* 10 times greater than all the other phases of the lifetime of the road. Some more examples are shown in Section 2.4.

Of the many evaluation systems—as mentioned earlier, claimed to be more than 900—we mention some examples from various parts of the world:

EIO-LCA (Economic Input-Output Analysis-Based LCA): An evaluation method used by the U.S. Department of Commerce, based on publicly available environmental data, mostly from the U.S. Environmental Protection Agency, such as the Toxics Release Inventory (TRI) data, hazardous waste data from Resource Conservation and Recovery Act (RCRA) reporting, etc.[24]

In a paper in 2001, Horvath[24] gave examples from bridge analysis of various designs where the indirect impacts, including upstream suppliers, i.e., suppliers of suppliers, are also taken into account. He reminds us of the importance of evaluating not only the service life but also the flexibility of structures to be restructured when their design might be obsolete before the end of the actual service life, as well as the reuse rate of the materials at the end of the service life.

ecoMA: A Japanese material flow simulation system, developed to evaluate the effect of technical and social aspects on the flow and then propose the optimum flow.[25] In a paper from 2007, Noguchi and Fujimoto[26] described the main characteristics of ecoMA as a modelling of social, time, and geographic factors of resource flow to concrete structures. In the system, a set of decision makers called agents is categorised into three types:

- PL agent as a model for material plants
- TR agent as a model for transportation companies
- GV agent as a model for governmental organisations

Lime (Lifecycle Impact Assessment Method Based on Endpoint Modelling): Another Japanese system. In a paper from 2007,[27] Kawai and Fujiki described the use of the system for evaluating the environmental impact of a precast concrete vegetation retaining wall, compared to an *in situ* concrete retaining wall. The case they evaluated is an embankment protection wall of 8 m height and 120 m length. The solution they favoured is based on 7 m high and 1.4 m deep hollow precast blocks piled on top of each other, with a top and bottom adjustment, filled with soil emitted at the construction site and planted. Their LCA includes the complete construction works as well as the net primary productivity (NPP) of the land use. The authors claim that the planted hollow precast concrete retaining wall showed favourable numbers on all environmental aspects evaluated.

JSCE—Recommendations: Sakai[28] mentions Recommendations of Environment Performance Verification for Concrete Structures by JSCE (Japan Society of Civil Engineers) as a methodic evaluation system where LCA as well as testing and other evaluations are included. As an example of its use, he mentions the evaluation of a highway reinforced concrete underpass versus a road intersection. Evaluating the effect of emissions from the construction of the underpass versus pollution from the traffic, he explains that the traffic pollution counterbalances the construction emissions; after 9.3 years for CO_2, 7.6 years for NOx, and 12.2 years for SOx.

ATHENA®: According to the U.S. Green Council's *Sustainable Concrete Guide*,[29] the LCA programs in North America are often those of the ATHENA Sustainable Material Institute and National Institute of

Science and Technology. The ATHENA Impact Estimator for buildings uses LCA methodology to estimate the environment of a building or building system. These tools have a limited number of options relating to concrete mixture designs, and they do not have the possible advantages of utilising the thermal mass.

BEES® (Building for Environmental and Economic Sustainability): Another North American evaluating system, using LCA methodology.[29] Concrete variables are limited to three compressive strengths that do not include cement substitution options.

EcoQuantum: The Interfaculty Environmental Science Department (IVAM) of the University of Amsterdam, Netherlands, has developed an LCA calculator called EcoQuantum. Various tools have also been developed to calculate the many variables needed for a good LCA evaluation.[30]

ECO-it: MiSA is a Norwegian Company founded by a group of enthusiastic associates from the Program for Industrial Ecology at the Norwegian University of Science and Technology (NTNU) in Trondheim, Norway. It offers LCA analysis for products and structures based on a software tool it calls ECO-it. The company also offers its software for sale.[31]

LISA (LCA in Sustainable Architecture): An Australian system claiming to be a streamlined LCA decision support tool for construction (Figure 2.5).[32] The system has been used to optimise designs from structures of all sizes to fences.

Umberto: The first version of UMBERTO was developed in cooperation between universities in Hamburg and Heidelberg, Germany, in 1994. Later, other German universities joined as partners. Umberto is a software tool for material and energy flow calculation and analysis based on graphical modelling of process systems.[33]

Equer: Equer is a French LCA tool for buildings. The output of the software is an ecoprofile including the following indicators:[34]

- Exhaust of abiotic resources
- Primary energy consumption
- Water consumption
- Acidification
- Eutrophication
- Global warming
- Nonradioactive waste
- Radioactive waste
- Odours
- Aquatic ecotoxity
- Human toxity
- Photochemical ozone (smog)

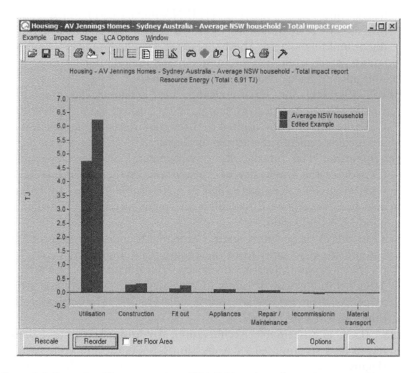

Figure 2.5 Example of illustrations from LISA.[32] (From http://www.lisa.au.com/ (accessed April 6, 2012).)

Eco-Bat 3.0: Eco-Bat is a Swiss LCA software tool. The database claims to contain information about more than 100 building materials.[35]

LEGEP: LEGEP is a German LCA software tool with descriptions in English but printouts in German. The system is based on DIN 276, and LCC on DIN 18960. The environmental assessment comprises the material flows (input and waste) as well as an effect-oriented evaluation based on ISO 14040-43.[36]

Some systems have been developed to optimise special environmental items more thoroughly. A few examples are:

- The Danish Building Research Institute promotes its PC tool for analysing buildings and installations, including tools for simulating and calculating thermal indoor climate, energy consumption, daylight conditions, synchronous simulation of moisture and energy transport in construction and spaces, calculation of natural ventilation, and electrical yield.[37]

- SOS is a Canadian computer-based system to reduce GHG emission by optimised use of supplementary cementing materials. The development involved three nongovernmental organisations (led by Ecosmart), three universities, two cement companies, four governmental organisations, and nine engineering and construction companies.[38]
- Giacomo Lorenzoni advertises on the Internet a program for the dynamic thermal properties of a multilayer wall, called ThermalWall.[39]

On the Internet are found a number of LCA software systems for general use, not necessarily aiming at building structures.

2.4 ISO METHODOLOGY/STANDARDS

Following the UN Conference on Environment and Development (held in Rio de Janeiro in 1992), on June 1, 1993, ISO's Technical Committee 207 (TC 207) held its first plenary meeting. TC 207, with five subcommittees (SCs) for each category of standard, was directed to establish environmental standards in five areas of environmental management: environmental management systems, environmental auditing and related environmental investigation, environmental labelling, environmental performance evaluation, and life cycle assessment.

ISO first launched the Environmental Management System (EMS) (ISO 14001) in 1996.[25,40] So far the entire family of ISO 14000 standards has evolved fast and become a worldwide recognised comprehensive aid in finding solutions to the range of environmental challenges facing business, government, and society today, and not only provides environmental benefits, but also significant tangible economic benefits, such as:

- Reduced raw material/resource use
- Reduced energy consumption
- Improved process efficiency
- Reduced waste generation and disposal costs
- Utilisation of recoverable resources

At present the ISO environment management-related published standards, as shown below, mainly address the issues of environment management system, environmental performance, environmental labels and declarations, life cycle assessment, greenhouse gas (GHG) accounting, verification and accreditation, environmental communication, and environmental aspects in product standards.

ISO14001:2004: *Environmental Management Systems—Requirements with Guidance for Use*

ISO14004:2004: *Environmental Management Systems—General Guidelines on Principles, Systems and Support Techniques*

ISO14005:2010: *Environmental Management Systems—Guidelines for the Phased Implementation of an Environmental Management System, including the Use of Environmental Performance Evaluation*

ISO14006:2011: *Environmental Management Systems—Guidelines for Incorporating Ecodesign*

ISO 19011:2002: *Guidelines for Quality and/or Environmental Management Systems Auditing*

ISO14015:2001: *Environmental Management—Environmental Assessment of Sites and Organizations (EASO)*

ISO 14020:2000: *Environmental Labels and Declarations—General Principles*

ISO 14021:1999: *Environmental Labels and Declarations—Self-Declared Environmental Claims (Type II Environmental Labelling)*

ISO 14024:1999: *Environmental Labels and Declarations—Type I Environmental Labelling—Principles and Procedures*

ISO 14025:2006: *Environmental Labels and Declarations—Type III Environmental Declarations—Principles and Procedures*

ISO/DIS14031:(1999): *Environmental Management—Environmental Performance Evaluation—Guidelines*

ISO/TS14033:2012: *Environmental Management—Quantitative Environmental Information—Guidelines and Examples*

ISO14040:2006: *Environmental Management—Life Cycle Assessment—Principles and Framework*

ISO14044:2006: *Environmental Management—Life cycle Assessment—Requirements and Guidelines*

ISO14045: *Environmental Management—Eco-Efficiency Assessment of Product Systems—Principles, Requirements and Guidelines*

ISO/CD14046: *Life Cycle Assessment—Water Footprint—Requirements and Guidelines*

ISO/TR14047:2003: *Environmental Management—Life Cycle Impact Assessment—Examples of Application of ISO 14042*

ISO/PRFTR14047: *Environmental Management—Life Cycle Assessment—Illustrative Examples on How to Apply ISO 14044 to Impact Assessment Situations*

ISO/TS 14048:2002: *Environmental Management—Life Cycle Assessment—Data Documentation Format*

ISO/TR14049:2000: *Environmental Management—Life Cycle Assessment—Examples of Application of ISO 14041 to Goal and Scope Definition and Inventory Analysis*

ISO/PRFTR14049: *Environmental Management—Life Cycle Assessment—Illustrative Examples on How to Apply ISO 14044 to Goal and Scope Definition and Inventory Analysis*
ISO14050:2009: *Environmental Management—Vocabulary*
ISO14051:2011: *Environmental Management—Material Flow Cost Accounting—General Framework*
ISO/TR14062:2002: *Environmental Management—Integrating Environmental Aspects into Product Design and Development*
ISO14063:2006: *Environmental Management—Environmental Communication—Guidelines and Examples*
ISO 14064:2006: *Greenhouse Gases—Parts 1: Specification with Guidance at the Organization Level for Quantification and Reporting of Greenhouse Gas Emissions and Removals; Part 2: Specification with Guidance at the Project Level for Quantification, Monitoring and Reporting of Greenhouse Gas Emission Reductions or Removal Enhancements*
ISO 14065:2007: *Greenhouse Gases—Requirements for Greenhouse Gas Validation and Verification Bodies for Use in Accreditation or Other Forms of Recognition*
ISO 14066: *Greenhouse Gases—Competency Requirements for Greenhouse Gases Validators and Verifiers*
ISO/WD 14067: *Carbon Footprint of Products—Part 1: Quantification; Part 2: Communication*
ISO/AWI 14069: *GHG—Quantification and Reporting of GHG Emissions for Organizations (Carbon Footprint of Organization)—Guidance for the Application of ISO 14064-1*

Note: DIS = Draft International Standard, TR = Technical Report, TS = Technical Specification, AWI = Approved Work Item, WD = Working Draft, CD = Committee Draft.

Take life cycle assessment (LCA), one of the most commonly used tools and probably the best tool measuring sustainability, as an example. LCA, as defined by ISO 14040:2006(E), is a life cycle approach and methodologies for the compilation and evaluation of the inputs, outputs, and potential environmental impacts of a product system throughout its life cycle. It addresses the environmental aspects and potential environmental impacts (e.g., use of resources and the environmental consequences of releases) throughout a product's life cycle from raw material acquisition through production, use, end-of-life treatment, recycling, and final disposal (i.e., cradle to grave).

LCA assists in:

• Identifying opportunities to improve the environmental performance of products at various points in their life cycle

- Informing decision makers in industry, government, or nongovernment organisations (e.g., for the purpose of strategic planning, priority setting, product or process design or redesign)
- The selection of relevant indicators of environmental performance, including measurement techniques
- Marketing (e.g., implementing an ecolabelling scheme, making an environmental claim, or producing an environmental product declaration)

There are four phases in conducting an LCA study, as detailed in ISO 14044:2006(E):

1. Goal and scope definition phase
2. Life cycle inventory (LCI) analysis phase
3. Life cycle impact assessment (LCIA) phase
4. Life cycle interpretation phase

Relationships based on ISO 14044:2006(E) between elements within the interpretation phase and the other phases of LCA are shown in Figure 2.6.

It should be noted that LCA, as one of several environmental management techniques (e.g., risk assessment, environmental performance evaluation,

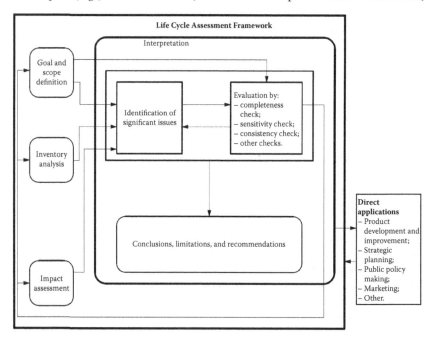

Figure 2.6 Relationships between elements within the interpretation phase and the other phases of LCA.

environmental auditing, and environmental impact assessment), might not be the most appropriate technique to use in all situations. It provides the life cycle approach and methodologies for a product system, yet without addressing typically the economic or social aspects of the target, for which we have to rely on the life cycle cost assessment (LCCA), a commonly used tool in civil engineering applications to make better investment decisions by finding the lowest long-term cost that can meet the desired need.[41]

Within the construction industry, LCA is mainly concerned with the environmental impact of materials and covers the process from winning the raw materials, through processing, construction, and planned lifetime maintenance, to demolition and disposal/recycling. Although from the life cycle impacts point of view 9 to 98% impact results from the building operations, while 2 to 10% impact is from material extraction, manufacturing, and construction, it is worthwhile to measure and compare how concrete works with sustainability compared with other materials specifically in terms of the energy consumption carbon footprint.

LCA of concrete must be carried out in the context of a particular structure or part of a structure, i.e., a functional unit; this could be the whole structure or just a floor slab, a foundation pile, a concrete bridge abutment, or a precast beam. It is essential that the functional unit for the options to be compared meet exactly the same overall specification; otherwise, the LCA is not valid.

The process tree for LCA of concrete is outlined in Figure 2.7, detailing the cradle-to-grave life cycle of the concrete in the functional unit and what specification items are included.

The LCA can be quantified as 1 m^3 of concrete in that structure or for all the reinforced concrete in the structure. Alternatively, it could be for 1 m^2 of a precast block wall.

An example of life cycle assessment of concrete is given by Sjunnesson.[42] The study is done for two types of concrete, ordinary and frost-resistant concrete, and has an extra focus on the superplasticisers. The utilisation phase is not included since the type of construction for which the concrete is used is not defined and the concrete is assumed to be inert during this phase.

The results show that it is the production of the raw material and the transports involved in the life cycle of concrete that are the main contributors to the total environmental load. The one single step in the raw material production that has the highest impact is the production of cement. Within the transportation operations the transportation of concrete is the largest contributor, followed by the transportation of the cement. The environmental impact of frost-resistant concrete is between 24 and 41% higher than that of ordinary concrete due to its higher content of cement. Superplasticisers contribute with approximately 0.4 to 10.4% of the total environmental impact of concrete, the least to the global warming potential (GWP) and the most to the photochemical ozone creation potential (POCP).

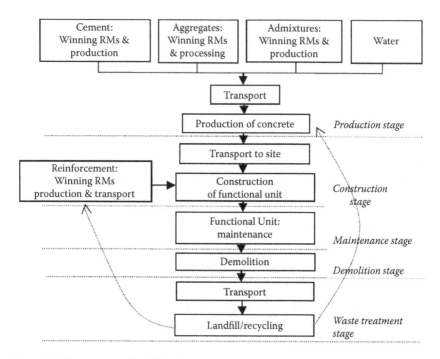

Figure 2.7 Process tree for LCA of concrete.

Craig D. Weiland[43] attempted to quantify the environmental impacts of several pavement replacement and rehabilitation strategies using life cycle assessment by conducting a comparative LCA of three different replacement options for an aging Portland cement concrete (PCC) highway: replacement with a new PCC pavement, replacement with hot mix asphalt (HMA) pavement, and cracking, seating, and overlaying (CSOL) the existing pavement with HMA.

The environmental impacts have been assessed according to common environmental concerns, such as global warming potential and acidification. For all assessments, only air emissions (common criteria pollutants) and energy usage were quantified due to data source limitations and study scope.

The results seem to indicate that CSOL uses the least amount of energy and contributes the least to the environmental impacts studied in this LCA: CSOL used 44% less energy than the HMA option and 15% less energy than the PCC option, and contributed 62% less than PCC to global warming and 44% less than HMA to global warming. The largest contributor to all of the environmental impacts is the production of the paving materials (bitumen, HMA, PCC).

The PCC replacement option was a greater contributor to global warming potential (CO_2) and human health criteria air pollutants (particulate emissions), while the HMA replacement option was the largest contributor to acidification, eutrophication, and photochemical smog (SOx and NOx emission), as well as the largest consumer of energy. Recommendations are given, such as use of fly ash or slag in PCC and use of recycled asphalt pavement (RAP) in HMA, to lower the energy consumption and environment impact.

Athena Institute[44] presented a slightly different result from the above regarding the CO_2 emission when conducting an LCA of concrete versus asphalt. The CO_2 intensities for concrete and asphalt are 674 and 738 t/km, respectively.

Concrete sewer pipes have also been proved, though controversial, to have markedly better environmental performance than PVC pipes across a range of impacts, according to a life cycle assessment published by the Dutch association of concrete pipe manufacturers, VPB.[45]

Gajda, of the Portland Cement Association,[46] demonstrated that concrete systems (integrated concrete frame (ICF)) reduce energy by 17% compared with wood frames, using LCA to evaluate the energy use of single-family houses with various exterior walls as residential framing systems.

Guggemos and Horvath[47] conducted a partial LCA (including material extraction, manufacturing, construction) on a concrete frame versus steel frame structural system. The result showed CO_2 emission intensities for concrete and steel of 550 and 620 kg/m².

2.5 VARIATION IN FOCUS

Evaluating an issue will always have a subjective side depending on the view platforms you might have. In this book we have tried to narrow in on the various topics from different sides, but still we know that treating such a complicated matter as sustainability will never be a completely objective analysis.

After a few years of discussion, the Norwegian Concrete Society decided in 2001 to start work on a database of the various publications related to environmental topics regarding cement and concrete. Early in 2002, a working group drew up a program, drafted the framework for the base, and established contact with potential owners. From the start, it was decided that the base should be an independent project, owned by interested companies and institutions in the cement and concrete industry, and financed and controlled by the owners. The Norwegian Concrete Society administrated the base.

In the spring and summer of 2002, the work started by collecting contributions to the base from the libraries of the owners. The owners at this stage were 19 companies and institutions in Norway, including the Ready Mixed Concrete Industry Association, Precast Concrete Industry Association, Building Research Institute, other research institutes such as SINTEF

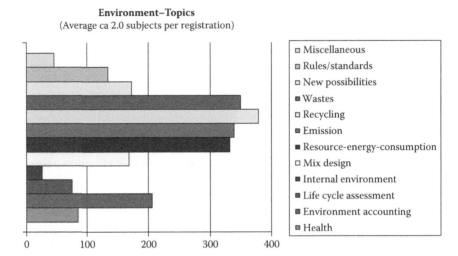

Figure 2.8 In June 2011 the Norwegian database, miljobasen.no, contained 1165 references. The references are characterised by 12 main topics. More than one topic can be registered by each reference (average of 2.0).

and NORUT, the largest contractors, precast and ready-mixed concrete companies, the largest consulting companies and material suppliers of, for example, cement and silica fume, and some government institutions.

In February 2003, the Norwegian Concrete Society and the control committee for the base arranged the first information meeting about the base. At this stage, the base had been on the net for a trial period of 2 months, and had collected nearly 500 references.

Different segments or parts of the cement and concrete industry have quite different focuses in the context of environmental challenges.

An example of this is that when we collected references from different areas of the concrete environment issues in Norway, asking the different contributors to forward documents, or information about documents, that they had in their libraries and felt should be in the base, hardly any (less than 10%) of the 300 to 400 first documents were the same (Figure 2.8).

2.5.1 Different sectors of the concrete industry tend to focus on different aspects

We learn from this exercise that even within the concrete industry, the focus, and what is regarded as most important on environmental issues, might change considerably from one part of the industry to another.

Several years ago, we introduced the term *triple focus* when analysing the CO_2 emission from concrete.

To improve evaluation of status, needs, and possibilities regarding CO_2 emission, a triple-focus system for evaluations might be helpful. An important reason and consequence is that the various tools we have to possibly improve the present situation have considerably different effects, depending on which focus we have:

- Lifetime expectancy perspective
- Social/political accepted reduction strategy
- Recommendations from IPCC

2.5.2 Focus: Lifetime Expectancy Perspectives

Concrete is one of the most important pillars for a sustainable development. Development of infrastructure and modern living quarters in the former, nonindustrialised world is the most important reason for the increase in cement consumption during the last 10 years or so.

Life cycle analysis based on a birth-to-grave evaluation shows that concrete structures and solutions might be a sustainable answer to the challenges. However, analysis shows that there are interesting possibilities for improvements.

2.5.3 Focus: 2020

In January 2007, the EU Commission presented its climate and energy package solution. In short, this document states that the EU will reduce its greenhouse gas emissions by 20% by 2020, compared to 1990.

Many of the various actions considered are still under discussion, but there is a high degree of acceptance of the final goal. The final decisions will set the framework for all the cement (and concrete) industry in Europe for the future years regarding CO_2 emission reduction.

The minimum 20% emission reduction by 2020 is about to become not only the target, but a minimum rule as working conditions for society in general, and consequently also for the cement and concrete industry far outside Europe.

2.5.4 Focus: 2050

In his speech to the UN general assembly on September 24, 2007, the chairman of the IPCC said: "We have the time to 2015, if we want to stabilize the emissions, then they have to be reduced considerably." He ended his speech referring to Mahatma Gandhi: "'A technological society has two choices; first we can wait until catastrophic failures expose systematic deficiencies, distortion and self-deception. Secondly, a culture can provide social checks and balances to correct the distortion prior to catastrophic failures.' May

I submit, it is time for us to move away from self-deception and go to the second of the two Choices."

The IPCC published four climate reports in 2007. One of the most important conclusions from the reports, apart from us needing to achieve control over the emissions by 2015, which seems to be taken care of by the political actions taken (Focus 2020), is that we have to reduce the human-created greenhouse gas emissions by 50 to 80% before 2050.

As we will treat later in the book, it is necessary to evaluate all three focuses simultaneously when we look at emissions, if we want good results.

2.6 TRADITIONS/TESTING

Over a decade ago, P. Kumar Mehta started advocating for a more holistic approach to concrete technology and education as an important pillar in a more sustainable development. In an article in *Concrete International* in 1999,[48] he gave examples of his philosophy, saying:

> It may be noted that a holistic approach should not be confused with a systems approach, which is commonly practiced in the resolution of complex problems. An integrated approach considering both structural design and structural durability aspects is a good example of a systems approach. However, it is not comprehensive enough to be called holistic. The holistic approach has its roots in the idea that the whole exists before the parts. For instance, the holistic approach would consider society as a whole, and the concrete industry as a part of the whole. Therefore, in addition to providing a low-cost building material, the concrete industry must assume responsibility for other social needs, for example, conservation of the earth's natural resources and safe disposal of polluting wastes produced by other industries, as discussed previously. In short, if sustainable development is a wheel, and conservation of concrete-making materials and durability are spokes, then the holistic approach to concrete technology is the kingpin of the wheel.

Holistic thinking in concrete design and technology has increased considerably over the last decade or so. An example of this is the so-called rating systems, where the quality of a building is judged from a number of tools that might make the structure more sustainable. Sustainable actions and holistic design might often be a trade-off between various arguments. For example, in designing the thickness of a partition wall, reducing the thickness reduces the resource use and carbon footprint, while increasing the thickness reduces noise pollution and might also lead to increased energy efficiency. Finding the optimum solution is the sustainable and holistic challenge. However, this solution cannot be standardised, as the conditions might vary considerably from one type of building to another, from one type of climate to another, and from one part of the world to another. Another example is the durability of concrete. There is little doubt that increased durability of concrete in general will in the long term reduce the carbon footprints of concrete structures, and save resources. However, it is also a fact that more than two-thirds of all the concrete that is demolished is not due to lack of durability, but changes in society. It would be a waste of resources to spend resources to make this concrete more durable than necessary. Holistic thinking would be both to differentiate more in the durability specifications and to request analysis about the needed life expectance of the structure, and to design the structures more tolerant, robust, and flexible to allow for reuse without demolishing.

We have in the national and international codes and standards a well-documented and accepted regime for testing and specifying cement and concrete. There is probably time for looking at this regime to analyse if it in some areas might be sustainability-unfriendly.

An example is a possible adjustment of the ages for testing and specifying concrete strength, and the temperature under which one tests, which might be both more realistic to real life and not so unfavourable to new, slower, and more environmentally friendly binders.

With a rather conservative standardisation regime, it is not likely that such changes can take place in a short time frame. But hopefully it might be possible to see some modifications with alternative methods in a reasonable time.

More than half a century ago, some wise professors managed to make an international agreement for how and when we test cement and concrete, so that we can compare results across national borders. China, for example, has for many years been using "standard cement" for testing the properties of concrete admixtures, and there is also "reference cement" specifically developed for comparison of mortar strength tests in labs. This is very useful and has a great advantage in understanding, development, and spreading concrete technology. But, we think it is now time to critically have a closer look at these rules again—to see if they are up to date with the technology we use today, and have the right perspective with respect to an efficient sustainable development.

2.6.1 Example I

The most important parameters and comparison and approval of cement and concrete are testing of, for example, compressive strength after 28 days, although the massive concrete for dams normally uses the parameters of 90 days even 180 days as criteria for acceptance. In addition, often testing also takes place after 1, 3, 7, or 14 days. Concrete normally has the following critical ages for strength demand:

1. **Deforming time:** Normal deforming time is often 15 to 18 hours after casting in industrial concrete production, and possibly a bit later for *in situ* concrete, depending on the strength development and designated strength grade. Request for strength from 15 to 25% of characteristic strength for precast industrial products is normal to avoid damage during handling and storage, and 40 to 60% of characteristic strength for prestressed concrete, before releasing the stressing wires.
2. **Time of first loading:** This might be when precast concrete components are erected, when the moulds have to be taken away on a bridge, etc. Often this first load will be something close to the dead load that the structure is designed for, or something in excess of 50% of the critical load. The time aspect differs considerably from one type of concrete usage to another, but often in the order of 1 week after casting.
3. **First time with maximum load:** This seldom happens until 3 to 6 months after casting.

More realistic time cycles for testing, more adjusted to real life and better adjusted to useful user information, than the regime we have today, might, for example, be to test cement and concrete for a characteristic strength after 90 days, and possibly have additional testing after 18 hours or 7 days. This will also reduce the possibility for stopping development and allow for the use of more sustainable-friendly binders that might develop their strength more slowly and with a higher increase in strength over a much longer time period than 28 days. We have seen examples of good sustainable cement alternatives that might have the same strength as normal Portland cement after 28 days, but have 20 to 25% higher strength after 3 months. The typical example of such a new binder is high-belite cement, introduced in Chapter 3. We also know from numerous examples that slower strength development might be favourable regarding the durability of concrete, for example, by the use of retarding admixtures.

For practical reasons, and in particular to be able to compare with existing technical data, research, and general knowledge, a new test regime must not come instead of the regime we have today. A more modern and more sustainable test regime would be to have a subsidiary system to ensure that testing does not give competitive disadvantages to the most sustainable alternatives.

2.6.2 Example 2

Standard testing is normally done at 20°C. Hardly any concrete is cast at this temperature, nor does most of the concrete experience the same 20°C after placement due to the hydration heat liberation and accumulation. The standardised temperature is thus of very little pedagogic and even practical value. A typical common characteristic of the more sustainable alternatives is that they are bit "slower" than previous binder practice. Lower heat development also normally results in durability advantages. Subsidiary testing temperatures, more adjusted to the climate and conditions in which concrete is produced and experienced, should be discussed regarding their effect on concrete performance.

2.6.3 Example 3

In Chapter 4 we mention various recycling alternatives. In Section 5.1 we give some comments regarding a question that seems to pop up in most countries in the world—possible aggregate shortage. There is globally probably not any real shortage of aggregate, in the widest sense of the word, but we can in more and more countries, at least near urban areas, in the future wave farewell to collecting the best possible aggregate just outside our doorstep. Future sustainable aggregate consumption will bring to concrete, not only as rare tests in the laboratory, but also at an industrial scale, a wide range of alternatives: aggregates that will be good enough to make high-quality, durable, and sustainable concrete, but they will have a number of variations in properties. We will have to accept "other types" of aggregate, as full or part or partly substitutes for "normal" aggregates, and even as ternary and quaternary aggregate blends. Crushed stone aggregates from various and "new" mineral resources, wastes, or secondary products from mineral activity, industrial wastes, household wastes, recycled concrete based on different recycling technologies, or other, partly reactive industrial recycling products must be used more in the future, with variation in volume in different parts of the world.

Several questions about present testing routines could be raised regarding the future developments in aggregate use. One of them, considering the fact that there is unified standard sand for mortar strength testing, is: Would it be easier for international communications, when testing, if all tests included a reference base mix with an international standard concrete aggregate?

Chapter 3

Emissions and absorptions

3.1 GENERAL

Planet earth has a rare and unique atmosphere. The experts tells us that in the early years, some millions of years ago, the atmosphere was more like that on Venus and Mars, mainly CO_2. So no doubt, the CO_2 level has been much higher than today, but the difference is that there were no living creatures on the planet, and far from any human beings. Today we are enjoying an atmosphere with approximately 21% oxygen and 78% nitrogen + some other gases, among them about 0.03% carbon dioxide. At the same time, we have a rare balance in temperature. While the cosmos might experience some temperatures from several thousand degrees plus, down to a few hundred minus, we seldom find temperatures on earth outside ±50°C, and most people feel uncomfortable with temperatures below –20°C and above 40°C. Probably most people would set the range much narrower (Figure 3.1).

Now climate experts tell us that climate is about to change, and in the worst case will destroy the delicate balance of the environment.

In the many discussions, meetings, workshops, seminars, and conferences arranged over the last decade or two, to discuss cement and concrete technology and sustainable development, the CO_2 challenge has often been the most central topic, and sometimes the only topic to be discussed. The importance of the climate/emission topic is indisputable. At the opening session of the United Nations General Assembly on September 24, 2007, the chairman of the Intergovernmental Panel on Climate Change (IPCC), Mr. Rajendra Pachauri, gave an alarming speech.[49] He said:

> To start with, let me say that we, the human race, have substantially altered the Earth's atmosphere. In 2005 the concentration of carbon dioxide exceeded the natural range that has existed over 650 000 years. 11 of the warmest years since instrumental records have been kept occurred during the last 12 years and therefore climate change is

Figure 3.1 TELLUS (from the Roman goddess Tellus).

accelerating. In the 20th century the increase in average temperature was 0.74 degrees centigrade; sea level increased by 17 cm and a large part of the Northern Hemisphere snow vanished. Particularly worrisome is the reduction in the mass balance of the glaciers and this has serious implications for the availability of water; something like 500 million people in South Asia and 250 million people in China are likely to be affected as a result.

He also argued that "the Arctic region is warming twice as fast as the rest of the globe."

Among the warnings he described was

that 20 to 30 percent of plant and animal species are in danger of extinction if temperatures exceed 1.5 to 2.5 degrees centigrade.

Projections for this century tell us that at the lower end of feasible trajectories, we have a best estimate of 1.8 degrees centigrade as the increase in temperature by the end of the century and at the upper end of feasible scenarios we get 4 degrees centigrade. The inertia of the system that we have is such that climate change would continue for decades and centuries even if we were to stabilize the concentration of gases that are causes of this problem today, which means that the adaptation is inevitable.

There might be a difference in opinion on whether the emission challenge is the most important in the list of environment challenges for the concrete

industry. However, it can hardly be doubted that it is of uttermost importance, and that both the problems and solutions are of global interest.

Nearly daily we read or hear about phenomena that are claimed to be
caused by the climate change. Many of the cases have long-term effects
on the lives of millions of people. For example, the *Norwegian Journal
for Climate Research*[50] wrote in 2008, with reference to the International
Centre for Integrated Mountain Development (ICIMOD), Nepal, and
Chinese Academy of Science, about the melting of the glaciers in the
Himalayas. They say that 5.5% of the Chinese glaciers in the Himalayas
have melted during the last 24 years, and that prolongation of the action
will result in two-thirds of the Chinese glaciers being gone by 2050, and
nearly all will be gone by 2100. The gigantic rivers running into China,
Bangladesh, and India that are fed from the Himalayan water battery will
inflict 300,000 million people in Nepal, Bhutan, India, and Bangladesh,
and probably some 650,000 million people in total.

There are also secondary effects of the melting of this "Asian water tower,"
among them the glacier lake outburst flood (GLOF). When the glacier melts,
a lake is often created at the end of it. Sometimes, when the lake gets too
full, the natural dam at the end of the lake breaks and creates a GLOF.
Researchers report about 200 such lakes in the Himalayas. The effect of the
GLOF is detrimental and is sometimes called a mountain tsunami.

In a cover story on December 16, 2011,[51] *China Daily* gave some exam-
ples of the economic aspects of a climate change. Referring to an Asian
Development Bank (ADB) seminar on mitigation and global warming at
Manila, Philippines, the delegates were told that the "worst is yet to come."
Natural disasters in 2010 caused 109 billion USD in economic damages,
threefold more than 2009. This year's (2011) figure will be much higher.
A study shows that global warming is likely to cause rice yield potential to
decline by up to 50%, on average, by 2100, compared to 1990 in Vietnam,
Thailand, the Philippines, and Indonesia, and a large part of the dominant
forest and woodland could be replaced by tropical savanna and shrubs with
low or no carbon sequestration potential. If the world continues "business
as usual" emissions trends, the cost to these countries each year could equal
a loss of 6.7% of their combined gross domestic product by 2100, more
than twice the world average.

Stanford Business had an editorial in November 2005[52] that opened
like this:

> Climate change may prove to be the most important business issue of
> the 21st century. Every sector—from finance to energy to consumer
> products—will be impacted by changes in regulations, weather pattern,
> and commodity prices. On balance, climate change will probably have
> a devastating impact on ecosystems and economies—especially in the

poorest parts of the world. However, important solution exists to miti-
gate the impact as the global economy moves to a new energy future. The
question is no longer "Is there a human-caused climate change?" Now
the questions are: 1) "How intense will the impact be?" And 2) "what
are feasible and profitable solutions to the intensity of those impact?"

Stanford Business continues to remind us of IPCC predictions that, with
90% confidence, an average increase in global average temperature from
2.3°C to 4.5°C is expected by the end of this century. They remind us
that the 10% uncertainty is possibly both below and above these figures.
They also remind us that the effect of the climate change also includes that
some scientists have predicted as many as 100 million refugees fleeing the
impacts of the climate change

Emission of greenhouse gases (GHGs) in general, and CO_2 in particular,
from production of cement and concrete is a very wide technical, practical,
political, and social challenge, with very wide variation in effect, and the
presence of possible tools and actions, with variation in regional possibilities, cultures, and standpoints in the industry. Several excellent books have
explained the chemical reactions in Portland cement and blends with various supplementary materials. We have spent little time on this aspect, but
instead tried to illustrate the versatility in possibilities and the wide global
interest from the cement and concrete industry to find more sustainable
alternatives (Figure 3.2).

We have chosen to handle the emission challenge in the concrete industry
in the following sections.

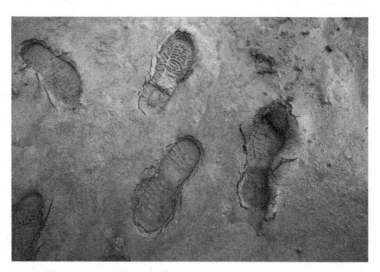

Figure 3.2 What will be the footprints in the future? (Photo courtesy of Jan Eldegard.)

3.2 CO₂ EMISSION FROM CEMENT AND CONCRETE PRODUCTION

The main greenhouse gas emissions from production of concrete come from the production of cement. However, it is very important, particularly in a life cycle evaluation, to also take into consideration how we are using concrete, as most analysis shows that the greenhouse emission from the use of a concrete structure is far more significant than from the production of the materials.

In 2009, the world produced 3 billon tonnes of cement.[53] China is by far the biggest producer, and India, as of 2006, is number two. The United States is third.

Cement production is the most important factor when evaluating emission from the production of concrete. Several authors mention that about 90% of emissions from the production of concrete come from cement production.[54–56]

Sakai,[57] in 2008, referred to an investigation involving 19 ready-mix concrete plants in Japan, where the emissions from the raw materials ranged from 71 to 97%, with an average of 85%, of the total. Other publications have given lower percentages.

A Fib Commission report,[21] referring to an investigation in Finland, gives the ranges of energy consumption in various concrete plants (Table 3.1).

Punkki et al.[58] have, with reference to Parma Oy, Finland, shown example calculations where the cement represents 62.7% of the total emission in precast concrete production (Table 3.2).

Table 3.1 Examples of energy consumption in concrete production plants[20]

Type of plant	Average (MJ/m³)	Range (MJ/m³)
Concrete element plant	790	400–1700
Ready-mix plants	520	160–700
Concrete product plants	350	200–700 (e.g., casting)
Multiproduct plant	580	300–1500

Table 3.2 Example of average CO_2 emission of precast concrete production, Finland[58]

	CO₂ kg/kg	% of total
Cements	0.095	62.7%
Admixtures	0.002	1.2%
Reinforcement and metal fixings	0.012	7.9%
Insulation materials	0.002	1.4%
Aggregates	0.003	1.7%
Production process	0.025	16.2%
Transport	0.013	8.7%

In future evaluations that we have done, it will be assumed as an average that emissions from cement production represent 85% of the total emissions from the production of concrete.

The situation regarding emissions varies considerably from one part of the world to the other, from plant to plant and between countries. Consequently, reports about how much CO_2 is emitted from each tonne of cement deviate somewhat: figures from slightly below 0.6 to 1.2 tonne CO_2 per tonne of cement have been reported.[59]

Examples of evaluations of global averages are:

0.84: Jahren, 2003 and 2007[56,59]
0.87: World Business Council for Sustainable Development (WBCSD), 2002[60]
0.87: Sakai, 2007[61]
0.8–0.9: Damtoft, 2007[62]
0.8 (2006), 0.75 (2012): WBCSD, 2009[63]
0.81–0.90: Gelli, 2003[64]

Gartner and Quillin[65] in 2007 mention 0.94 tonne of CO_2 per tonne of clinker in the United States (with reference to the U.S. Geological Survey), global overall average data of 0.88 tonne per tonne of cement, and European average values of 0.63.

Ge and Wang in 2010 claimed that the figures for the United States were 0.90 to 0.98.[66]

Figure 3.3 shows the CO_2 intensity based on cement and clinker in China since 1978.[67] A big decrease in clinker CO_2 emission intensity started in 1998, mainly due to the replacement of the vertical shaft kiln process with new dry technology, which leads to the remarkable improvement of both capacity and energy efficiency. The further decrease in the cement CO_2 emission intensity over the last three decades can be attributed to the enhanced use of blended materials, mainly industrial wastes in Chinese cement, which has also resulted in the steady decrease in the clinker factor of Chinese cement.

Figure 3.4 gives the average clinker factor of Chinese cement in recent years. Taking the data of 0.62 in 2011 as an example, estimation, excluding some 5% of gypsum, shows that as high as 680 million tonnes of blending materials, mainly industrial wastes, is interground in the 2.06 billion tonnes of Chinese cement yearly output, considerably reducing CO_2 emission.

In a report dated November 2008, the WBCSD[68] claimed that based on a database including 18 member companies of the Cement Sustainability Initiative (CSI), the CO_2 emission for each tonne of cementitious products has been reduced from 752 kg in 1990 to 715 kg in 2000 and 661 kg in 2006 for the member companies. CSI in early 2012 reported about 22 member

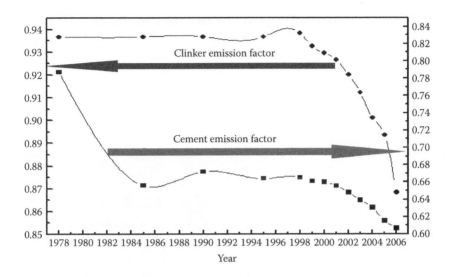

Figure 3.3 The CO₂ emission intensity based on cement and clinker.

companies. The group has in recent years produced 20 to 27% of the world consumption (Figure 3.5).

A comparison between the WBCSD CSI Get the Number Right (GNR) database and Chinese cement data gives the numbers shown in Table 3.3.

Most single reports refer to values from 0.75 to 0.9, with minimum single number of 0.71 for Portland cement clinker and 0.5 for blended cements for general use.[59]

The number for the Norwegian cement production in 2007 was 0.69 in total and 0.63 for the largest and best plant.[70]

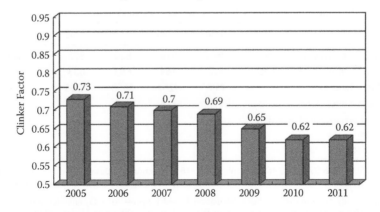

Figure 3.4 Clinker factor of Chinese cement. Calculated by author with original data from Chinese Cement Association.

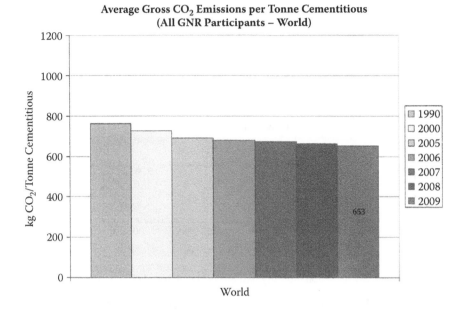

Figure 3.5 Development among CSI members.[69] (From WBCSD, http://wbcsdcement.org/
GNR-2009/world/Graphs/GNR%20-%20indic_ (accessed February 2, 2012).)

Table 3.3 Comparison between CSI and Chinese cement data

Item	Unit	1990	2008	2009	2009 China
Gross CO_2 emission/t clinker	kg/t	916	860	853	853/865/893[a]
Gross CO_2 emission/t cement	kg/t	764	665	653	566[a]
Kiln fuel efficiency	MJ/t	4267	3658	3586	3567[b]
Electricity consumption	kWh/t	114	110	107	91.1[b]
Use of alternative fuel including biomass	%	2	11	12	—
Clinker factor	%	83.0	76.1	75.6	65.4[b]

Note: CSI GNR 2009 covers 46 companies with 918 production facilities worldwide and 608 million
tonnes of clinker and 789 million tonnes of cement, representing 26% of global cement production.

[a] Project study outcome and author's calculation, in which figures for CO_2 emission intensity were
derived from varying kiln capacities of over 5000 tpd, 3000–5000 tpd, and below 3000 tpd, respectively. The figure for power consumption excludes the electricity from waste heat recovery for
power generation.
[b] Industry average data, Chinese Cement Association.

These figures show a considerable improvement over the last decade or so. At a conference in 1998 the following figures were given:

Canada: Emission figure = 0.88 per tonne of clinker in 1996,[71] and 1.1 from wet process and 0.89 from dry process.[72]

United States: The figure dropped from an average 1.16 in 1974 to 1.09 in 1990, with a prediction of 0.96 in 2010.[73]

Australia: The figure for 1990[91] is equal to 0.92 for clinker and 0.74 for cement. The estimate for 1996[97] was 0.89 for clinker and 0.68 for cement, respectively, and estimates for 2000–2001 were 0.85 and 0.60, respectively.[74]

A reasonable and simple assumption of the CO_2 emission from global concrete production in 2009 is:

$3 \times 0.85/85\% = 3$ billon tonnes

(85% of the emission from cement clinker, and 0.85 tonnes of CO_2 per tonne of cement).

At the same time it is reasonable to believe that about 10% (or possibly more—see Chapter 3, Section 3.4) of this is absorbed in concrete again in the carbonation process,[75] giving a net amount in the order of 2.7 billion tonnes of CO_2 emitted. This represents 6 to 7% of the total global man-made emissions of CO_2.

Many countries report lower figures. As an example, similar amounts for Norway are that the total emissions amount from Norwegian cement production in 2007 was 1.19 million tonnes. The emission from the production of concrete from the total national emissions was 2.9%.

Another example of the positive ongoing development is that the national consumption of blended cement (CEM II–V) increased from just over 50% in 1990 to 70% in 2005, while the consumption of pure Portland cement (CEM I) was reduced from nearly 50% to 30% in the same period.

3.3 EMISSION OF OTHER GREENHOUSE GASES

Even though the quantity of the other greenhouse gases is small compared to CO_2, some of them are regarded as more damaging. The relative damage index of greenhouse gases is as follows:

CO_2	1
Methane	20
Nitrous oxide (NOx)	200
Fluorine	15,000

As a comparison of the importance between the different greenhouse gases, Delong Xu gave the following figures for emissions from cement production for the world's largest cement producer, China, in 2007:[76]

CO_2:   800 million tonnes
SO_2:   0.4 million tonnes
NOx: 1.57 million tonnes

Roughly, these figures represent approximately half the emissions from cement production in the world.

In a paper at a CANMET/ACI symposium in Ottawa, Canada, in 1998, Klein and Rose[77] gave good insight into the topic in general and the Canadian situation in particular. They stated:

> NOx emissions from cement kilns are primarily generated from the high combustion temperature at the main burner. They have a wide range of values, depending upon type of kiln and fuel used. The average emission rate is about 3.7 kg NOx per tonne of clinker. Total Canadian NOx emissions are in the range of 30–35 kilotonnes, depending on the production split between gas-fired plants (higher NOx) in western Canada, and mostly coal fired plants in eastern Canada. A table below [Table 3.4] summarizes finding in the Radian study, based on averages from various research activities and from Canadian and international industry data. In similar kilns, coal firing produces less NOx than natural gas, since thermal NOx is the dominant mechanism. Note that the difference is less for precalciner kilns. Pyroprocessing does require high temperature and resulting NOx emissions in the kiln firing zone, and emission monitoring has been used to optimize this clinker burning zone. Emission preventing potential from combustion can be realized by minimizing fuel use, and by transferring the heat input to the feed end of a shorter kiln. There is limited experience with backend NOx emission control methods. There is evidence that waste derived fuels such as tires and solvents tend to decrease these emissions.
>
> While the raw material and fuel input of SO_2 into a kiln is in the range of 5–12 kg/tonne, the limestone acts as a natural scrubber trapping 90–99% of the SO_2 in the clinker produced. Emission rates are

Table 3.4 Average NOx emissions from cement kilns (kg/tonne)

Fuel	Wet kiln	Long dry kiln	Preheater	Precalciner
Gas	9	7–9.5	5.6	1.7–3
Coal	1.5–4	2.4–4.6	1.5–2.8	1.4–2

very site-specific, having even larger variability than NOx emissions; 1–10 kg SO_2/tonne in wet and long dry kilns, 0.5–2 kg SO_2/tonne in preheater and precalciner units. NOx and SO_2 emissions are often inversely related in the clinker process.

They further reported that new national emission guidelines state that emission from large new natural gas or coal-fired cement kilns (capacity larger than 1500 tonne/day) should not exceed 2.3 kg of NOx per tonne of clinker production, based on a monthly average time period.

As seen from Figures 3.6 and 3.7, emission from the other GHGs is relatively small compared to the emission of CO_2. However, their importance in a GHG context must not be underestimated.

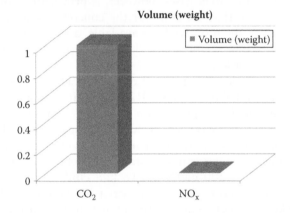

Figure 3.6 NOx emission compared to CO_2 emission in weight in cement production.

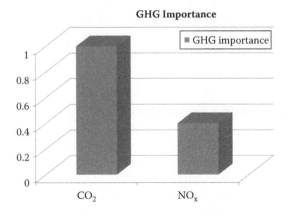

Figure 3.7 NOx emission compared to CO_2 emission as GHG importance in cement clinker production.

Table 3.5 Case study of NOx reduction in a new dry process line using varying De-NOx technology

Main De-NOx measures	Emission level mg/Nm³(NO₂,10% O₂ as basis)	Reduction ratio
Basic emission	880	0
Low NOx burner	~830	~5%
Staged combustion/low NOx calciner	~600	~30%
Selective noncatalytic reduction (SNCR)	<350	>60%
Staged combustion + SNCR	<350	>60%
Selective catalytic reduction (SCR)	Under development	

The emission of the other gases, however, is principally more a cement production challenge than an issue for the concrete industry. Important also is that most of the tools we have available, and the efforts that have been done to reduce CO_2 emission, will also reduce emission of the other GHGs, in the same order.

To reduce NOx is less complicated than CO_2. The Norwegian cement company Norcem stated in January 2012 that it is investing in a urea-based cleaning system at its Dalen plant, with 80 to 90% cleaning efficiency.

In Norway, a NOx tax has been introduced to the industry. However, the tax goes to a fund that supports financing the best cleaning projects.

Current Chinese cement emission standards define the SO_2 and NOx emissions as less than 200 mg/Nm³ (0.6 kg/t clinker) and 800 mg/Nm³ (2.4 kg/t clinker), respectively. A more strict limitation of 500 mg/Nm³ (1.5 kg/t clinker) for NOx emission from the cement industry has already been implemented in Beijing and will be applied in China soon. As shown Table 3.5, Sui[78] gave the results of a case study of NOx reduction in a new dry process line using the de-NOx technology developed by China Sinoma International Engineering Co. Ltd.

We discuss the possibility for concrete to reduce NOx emission in Section 5.4.

3.4 ABSORPTION/CARBONATION

The absorption of CO_2 in concrete is an important factor in evaluation of the total emission footprint of concrete.

There is little doubt that concrete absorbs CO_2, but there are some various views about which chemical processes are in action, how much carbonation we can expect (how much CO_2 is absorbed), the effects of the conditions that speed up or slow down the reactions, and how the use of cement substituting materials might reduce the uptake, as they already have consumed some of the potential for the reaction. Sometimes researchers tend to forget the difference between depth of carbonation measured, percentage degree of carbonation, and amount of CO_2 absorbed. As an indication of how pozzolanic materials "compete" in using calcium hydroxide we refer to Sellevold and Nilsen,[79] showing examples of how increasing the content of silica fume by weight of cement is reducing the calcium hydroxide content, and eliminating the calcium hydroxide at roughly 24% of silica fume.

There are also reasons to distinguish between what we might call normal carbonation and early/accelerated carbonation.

The reaction mostly referred to is, in many ways, that the carbonation's reaction is the reverse of the calcination's reaction from the cement kiln, where carbon dioxide from the air reacts with calcium hydroxide in the concrete, to form stable carbonates. It is as if nature is restoring its own balance:

$$Ca(OH)_2 + CO_2 \rightarrow CaCO_3 + H_2O$$

However, carbonation is complex, and it is a surface phenomenon. To carbonate inside the surface, the carbon dioxide or ions must travel through carbonated layers. This slows down the process. Carbonation is slower in both dry and wet conditions. The speed of carbonation seems to be highest at a relative humidity of 60 to 80% in the concrete.

Early accelerated carbonation has been explained by Swedish and Russian scientists as

$$3CaO.SiO_2 + (3 - x)CO_2 + yH_2O \rightarrow xCaO.SiO_2.yH_2O + (3 - x)CaCO_3$$

$$\downarrow$$

$$+ zCO_2 \rightarrow (x - z)CaO.SiO_2.yH2O + zCaCO_3$$

Maries[80] argues that this is oversimplified. He explains his understanding of the complicated reaction with the following table:

No.	Process step	Mechanism
1.	Diffusion	CO_2 in air
2.	Permeation	CO_2 in concrete
3.	Solvation	$CO_2/g \rightarrow CO_2(aq)$
4.	Hydration	$CO_2(aq) \rightarrow H_2CO_3$
5.	Ionisation	$H_2CO_3 \leftrightarrow H^+/HCO_3^-/CO_3^-$
6.	Dissolution	$CaSiO_5, CaSiO_4 \rightarrow CA^{++}, SiO_4^+$
7.	Nucleation	$(CaCO_3), (C\text{-}S\text{-}H)$
8.	Precipitation	$CaCO_3 \downarrow, C\text{-}S\text{-}H \downarrow$
9.	Secondary carbonation	C-S-H

The action of CO_2 takes place even at small concentrations, such as are present in the air, where the CO_2 content is about 0.03 to 0.035% by volume. In an unventilated laboratory, the content might rise above 0.1%; in large cities it is on average 0.03%, and exceptionally up to 0.1%.[81] However, carbonation also needs moisture to take place, as the actual reactive agent is carbonic acid because the CO_2 gas itself is not reactive.

We also know that in the pozzolanic reaction, the silicon dioxide in substituting materials like granulated blast furnace slag (GBFS), fly ash, or silica fume, reacts with the calcium hydroxide to form stable silicates. Therefore, the concretes with pozzolanic additions might have somewhat less reaction potential for CO_2^- uptake, and as a result, a smaller amount of CO_2 is required to remove all the $Ca(OH)_2$.

However, this relationship is somewhat complicated. Researchers have found that due to the lower calcium hydroxide level, the carbonation reaction might increase in speed in the presence of, for example, fly ash, slag, and silica fume. On the other hand, the use of pozzolans might reduce the porosity of the concrete and consequently slow down the carbonation. As a result, the measured relative carbonation speed might differ between different mix designs. Another complication is that researchers have found that at high concentrations of CO_2, the carbonation of $Ca(OH)_2$ is followed by the carbonation of C-S-H.[81]

To many concrete technologists it sounds strange to look for increased carbonation to achieve increased sustainability. Increased carbonation in concrete could in general result in more corrosion of the reinforcement, shorter life expectancy, and the opposite of a sustainable development.

However, a fully carbonated concrete is a stronger concrete, with some 20 to 30% higher strength, often a desired effect, when corrosive reinforcement is not present.

Use of carbonation to produce stronger nonreinforced products is an old invention, aiming to save cement. As early as 1855 the first patent for purposeful carbonation was presented. Since then researchers and inventors have renewed the idea many times. Based on modern mixture design and machinery technology, this might get a renaissance. A high degree of carbonation is achieved by letting concrete producers cure in a moist and CO_2-rich atmosphere. The practical utilisation would particularly be effective for typical nonreinforced machinery-produced products, like roof tiles, paving stones, building blocks, retaining wall blocks, traffic barriers, and nonreinforced pipes in the smaller dimensions. For these products, we would receive a double effect with respect to CO_2 reduction by saving of cement and increased CO_2 absorption.

Research has shown that fully carbonated concrete could bind above 0.3 kg of CO_2 per kg of cement and get a strength increase of 30%. (Djabarov[82] has reported about using carbonation to improve the strength of lightweight aerated concrete with a strength increase from 50 to 100%.)

Theoretically, if utilised to its optimum, this could lead to 0.3/0.7 = 0.4 kg reduction in emission per kg of cement compared with production of these products with ordinary Portland cement, without any carbonation.

Unfortunately, there are several limitations to this:

- In practical concrete production, it would be difficult to achieve a fully carbonated structure in the time frame available for modern production. The effect would also greatly depend on the products in question. Our evaluation is that a total effect of 30% (0.3 kg/kg) would be a realistic maximum limit.
- Concrete products will always carbonate somewhat over time under any circumstances. Consequently, the existing carbonation has to be deducted from this when we try to calculate the environmental improvement potential of such a tool.
- Analysis of the end use of cement indicates that cement consumption for nonreinforced "machine produced" products range from 10 to 15% of the cement consumption in a typical industrial country to over 50% in some less developed countries.

Some examples:

- Cornelissen[71] reported that 13% of the cement production goes for nonreinforced concrete products in Canada.
- Paving stone accounts for 5%, roof tiles 3%, and blocks 2% of the cement consumption in Sweden.[83]
- Figure 3.8 shows a typical Norwegian split of the cement consumption of various sectors.

The products to be industrially carbonated have to be produced in a CO_2-rich environment. We doubt that transport of CO_2 to and investment

Percentage

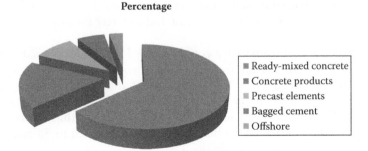

- Ready-mixed concrete
- Concrete products
- Precast elements
- Bagged cement
- Offshore

Figure 3.8 Norwegian cement consumption split on segments, 2001.[84] (From Norcem, *Sementforbruk på segmenter*, Norcem, Heidelberg Cement Group, Oslo, Norway, February 2003.)

in the necessary equipment at the currently very fragmented location of the concrete product production industry around the world will get much acceptance in a shorter time perspective. The practical utilisation of this absorption potential and technology in an environmental context would be in the future to build larger production units for concrete products downstream from the major emitters, like the cement industry or coal-fired power plants.

In 2001, Jacobsen and Jahren[83] made an evaluation of the degree of natural carbonation in Norwegian ordinary Portland cement (OPC) concrete. The evaluation assumed that OPC contains 65% of CaO and that 32.3% of the CaO in OPC is carbon able. It was further assumed that the total carbonation depth reaches a maximum of 10 to 40 mm, depending on the concrete quality. The average was calculated to 27 mm for historical concrete and 15.5 mm for present-day concrete. Ninety percent of all concretes were assumed to carbonate unidirectionally as an infinite slab with a thickness of 10 cm (or from both sides of a slab with a thickness of 20 cm). This means that 90% of 8 m^2 of exposed surface per m^3 of concrete was consumed.

Based on the above assumption, the estimations resulted in conclusions that 11% of the CO_2 from the clinker production in Norway is absorbed again by carbonation of the historic concrete, and about 8% of the present concrete, concluding with an estimate of 8 to 11% binding of the CO_2 emission from cement production by carbonation.

As a comparison, Maries[80] argues from a chemical evaluation of the case

> that up to 15% by weight of the CO_2 released during the manufacture of cement can be "fixed" by applying carbonation to many types of concrete and related materials containing cement or lime, including those used in processes for solidifying waste streams from other industries. Moreover, the use of CO_2 to neutralize the effluent from any type and size of concrete plant could consume an additional few percent of CO_2 by weight of all cement produced. Thus carbonation processes have the potential to make a definite contribution to the overall reduction of CO_2 emissions in the cement and concrete industry.

In a paper at a sustainability conference at Lillehammer in 2007, Pade et al.,[85] referring to a study from the Nordic countries, argue for a much higher carbonation uptake. The paper argues that 28% of the concrete in Norway and 37% of the concrete in Denmark are carbonated after 70 years. Then in a life cycle evaluation, they assume that the concrete is demolished and crushed, so that the figure is increased to 58 and 86%, respectively, after a life cycle (Figure 3.9).

From these figures they calculate on a life cycle basis the percentage of CO_2 uptake from carbonation compared to what is emitted from calcinations in Denmark, Norway, Sweden, and Island: 57, 33, 33, and 42%,

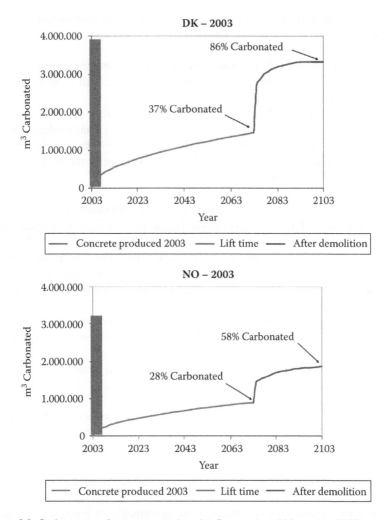

Figure 3.9 Carbonation of concrete produced in Denmark and Norway in 2003, according to Pade et al.[85] (From Pade C., et al., The CO_2 Uptake of Concrete in the Perspective of Life Cycle Inventory. Presented at International Conference on Sustainability in the Cement and Concrete Industry, Lillehammer, Norway, September 16–19, 2007.)

respectively. It is important to remember that this percentage is not from the total emission, but only from the calcinations emission, and consequently must nearly be divided by 2 to be comparable to the figures given previously, which are referring to total CO_2 emission from cement production.

The referenced study is very interesting, and brings up some very interesting aspects and throws interesting and important lights on the carbonation question. However, we are afraid that some people have drawn drastic conclusions from it. Among others, the following three considerations must be taken into account when comparing to other studies:

1. The life cycle evaluation does not take into consideration that most concrete has a much longer life expectancy, and is not crushed after demolishing.
2. The production of concrete is far from constant over time, so comparison with the emission figures might be somewhat misleading.
3. Carbonation depth is measured by methods measuring the change in pH value in concrete. This does not necessarily mean that all the calcium hydroxide has reacted with CO_2.

There have been many evaluations of the carbonation question. In 1998, Jahren[86] referred to various publications with figures ranging from 5 to 25% CO_2 uptake compared to the emission from cement production. In a report in 2008, Baetzner and Pierkes[87] from the European Cement Research Academy evaluated figures from a number of different estimates after a very thorough literature search, and gave the following conclusions:

1. Carbonation of concrete occurs unavoidably, but is no risk for concrete buildings, which were built under consideration of the appropriate rule.
2. CO_2 uptake over a usage time of 50 to 70 years can be up to 60 to 100 kg per tonne of cement. This is comparable to about 10% of the total emission of CO_2 in production of cement.
3. For the time after demolishing of the concrete structure, the binding potential is somewhat larger. This is due to the fact that the area of free surface increases considerably through demolishing and crushing. For the crushed concrete, an additional 10 to 15% of the total emission of CO_2 from cement production can be absorbed through carbonation.

The CO_2 uptake in concrete is, environmentally at least, a two-sided issue. While increased carbonation decreases the CO_2 in the air, the carbonation might make reinforced concrete more vulnerable to corrosion attack, and thereby reduce the durability of concrete, which might be negative to sustainability.

However, Maries[80] argues that accelerated carbonation, despite an extensive reaction during carbonation, lowers the pH three or more units. He claims that there is no need for concern over the durability of embedded reinforcing steel. However, several researchers contradict this.

In addition to strength increase and CO_2 uptake, carbonation also has other advantages. Maries,[80] referring to Berger et al., explains that shrinkage control by carbonation was adopted commercially in the U.S. block industry in the 1950s by recirculated flue gas curing. This resulted in reduction in initial and in-service shrinkage and improvement in the volume stability to wetting and drying in service. Because of the low partial pressure of CO_2 in the process, rapid strength development was not observed. The drying shrinkage was claimed to be reduced by up to 25%.

Collepardi et al.[88] presented an interesting paper at a sustainability conference in Milano in 2003 that throws some interesting light on this question. They reported a test with Portland cement concrete, as well as with various substitution levels of slag, fly ash, and limestone powder. Their conclusions are interesting in many respects. Probably the most important finding and conclusion was that they did not find any significant difference in carbonation rate with different raw materials for concretes with the same strength level with replacement by mineral additions up to 50%. However, they measured the carbonation rate, i.e., the level where concrete gets such a low pH it is vulnerable to corrosion attack, and not the CO_2 uptake. We know that concrete with mineral additions is considered more durable and more resistant to chemical aggressions, including alkali-aggregate reaction, sulphate attack, etc., partly due to reduced porosity. The tests indicated higher carbonation with increased mineral replacements, but as mentioned, no significant difference was recorded on the same strength level. From the tests it is not unreasonable to believe that the pozzolanic reaction has reduced the amount of calcium hydroxide, and thereby the ability of the concrete to absorb CO_2, but the sum of the two reactions will increase the reduction in pH.

The industrial application of accelerated carbonation can, according to Maries,[80] be traced back to patents granted over a century ago: Chaudot and UK Patent No. 96 (1855), "An Improved Stucco," and Rowland and U.S. Patent No. 109,669 (1870), "An Improvement in the Manufacture of Artificial Stone." He further claimed that very little mention is found in the literature until the 1950s, when proposals were made for recycling flue gases in curing chambers to partially carbonate concrete blocks, and thus control their shrinkage in service.

There are many reports that show that the carbonation of concrete increases the strength of concrete through the transformation of $Ca(OH)_2$ to stable calcium silicates. A possible strength gain or a potential for cement reduction up to theoretically 25 to 30% has been argued. Many have therefore tried to utilize this potential industrially in nonreinforced products. However, this is not that easy. In a paper in *Concrete International* in 2009, Shi and Wu[89] reported on tests with industrially produced lightweight concrete blocks with lightweight aggregate. They compared results from switching the curing process of the blocks from a steam chamber to a CO_2 chamber. The freshly produced blocks were typically allowed to rest for 2 to 6 hours after direct demoulding. When they were steam cured they were placed in a steam chamber where the temperature rise was approximately 55°C. The temperature was slowly reduced to about

45°C after 20 hours, before the blocks were taken out after 25 hours. For CO_2 curing the blocks were first transported to the laboratory, where they were left to air-dry for about 4 hours. Then they were brought into a CO_2 curing chamber. The chamber was closed with a vacuum pump and then pressurised with CO_2 to 70 or 140 kPa, and the blocks were left there for 2 hours. Tests were also performed on blocks from both types of curing after mist curing afterwards. Of the results recorded, the CO_2-cured blocks had slightly lower compressive strength than the steam-cured blocks, both without and with later steam curing; however, the shrinkage and water absorption in the CO_2-cured blocks were significantly lower in both cases. Even if the energy consumption in the CO_2 process was lower, the economical analysis was not in favour of the process.

Even though the results were disappointing from an absorption point of view, they were interesting. However, the results might not be surprising, as the carbonation curing went on for a rather short period of time.

A paper by Shao and Lin from Montreal, Canada, in *Concrete International* in 2011 gave some interesting results from curing of three different types of concrete in a CO_2-rich curing chamber.[90] The three types of concrete they tested were:

- A dry cast concrete with a water/cement ratio of 0.26, typically for machine-produced products like pavers, interlocking stones, blocks, etc.
- A lightweight aggregate product with a w/c ratio 0.48
- A normal weight product with a w/c ratio 0.48

All products were produced with 426 kg/m³ of Portland cement. In their very special process setup they measured a carbon uptake of the cement content of about 12% after 7200 seconds for the dry mix product and the normal concrete, where two-thirds of the uptake in the dry mix product was recorded after less than 2 minutes. The uptake in the lightweight aggregate concrete was about 7% after 7200 seconds. In their conclusion, the authors emphasized the environmental benefits, but due to the cost of CO_2, they stressed the importance of having a probable production downstream from a carbon source as, for example, a cement plant. They are also of the opinion that an economical production will depend on governmental carbon policy and the market value of CO_2. They further stressed that large-scale implementation still faces technical challenges. Initial curing in open air evaporates water and allows diffusion of gas into concrete, but it also leads to drying shrinkage. Early carbonation itself also leads to shrinkage due to exothermic water loss. For nonreinforced precast products, early-age dimensional change may not be harmful.

In 2007 Shao and Zhou published a paper[91] where they tested early carbonation in a flue gas with 25% carbon dioxide concentration, which they claim represents an as-received flue gas concentration. The process was carried out at a constant pressure of 5 bar and in a period of 2 hours. They concluded that the flue gas CO_2 uptake was in the range of 7 to 9% based on dry cement as a reference. To investigate the maximum possible CO_2 uptake by cement, loose powder samples of the same mass were carbonated under the same condition. The uptake could reach approximately 17%. A microstructure of the carbonated cement and concrete indicated that the material was mainly composed of poorly crystalline calcium carbonate incorporated in the calcium silicate hydration products.

Owens et al. from Belfast, Northern Ireland, in 2007 presented a report from accelerated carbonation by two different methods.[92] They claimed, with reference to Ngala and Page (1997), that flue gases have approximately 10 to 15% carbon dioxide content, and with reference to Parrot (1987), that the most effective carbon dioxide percentage range was between atmospheric content (0.03%) and 5%. One of the methods they tried was saturating the mixing water with CO_2 with dry ice. The results did not show any significant differences from the control concrete. In the other production method tested, they used a controlled curing chamber with 0.03, 1, and 5% carbon dioxide concentrations. They reported that they found increased compressive strength for all samples at all ages; 1% CO_2 concentration was shown to have a notable impact on compressive strength, especially after 3 days, compared to samples tested at atmospheric condition. The strength increase at 5% concentration was only slightly higher than those at 1% concentration.

Ye et al. from Jinan, China, in 2011 reported on a study on how the different cement clinker minerals reacted to early carbonation curing.[93] The mineral composition of the clinker tested was:

C_3S 56%
C_2S 21%
C_4AF 12%
C_3A 7%
Others   4%

After mixing the clinker minerals with water, the objects were subjected to CO_2 (99.9%) and water in an autoclave with a temperature of 25°C and a partial pressure of 0.2 MPa. They recorded an uptake of 0.168 g CO_2 gas per g of C_3S and 0.125 g for C_2S. Uptake in the other minerals was smaller. The compressive strength of the clinker with carbonation curing was considerably higher than the control after 2 hours and 1 day, and slightly higher after 3 days.

In a paper from 1998, Djabarov[82] reported a very efficient use of the carbonation possibility in lightweight cellular concrete. In the introduction he mentioned that production of carbonated cellular concrete is based on a 40-year-old (in 1998) Bulgarian patent. The foamed lightweight concrete precast blocks were cured utilising smoke gases from lime and cement kilns with CO_2 contents of 18 to 24% and 12 to 18%, respectively. The treatment took place for 20 to 40 hours, depending on CO_2 concentration, bulk density, and product dimension. In a comparison with blocks of bulk density of 400 kg/m³ cured in air for 28 days, the strength increase with CO_2 curing was from 0.7 MPa to 1.6 MPa.

The same comparison for blocks with a bulk density of 800 kg/m³ was an increase from 2.0 kg/m³ to 3.1 kg/m³, or 55%.

In a paper from 2008, Cheng, from the University of Jinan, China, mentioned the possibility to sequester CO_2 through carbonation in various types of slags for the production of building materials.[94] For use of this type of steel slag in concrete blocks, pavers, boards, and precast components with propylene fibre reinforcement, in addition to reducing CO_2 (and calcium hydroxide), the early strengths might increase and shrinkage might be reduced.

Zhao et al., from Jinan, China, in a paper from 2010[95] reported on early carbonation of concrete, with 20 to 80% replacement of Portland cement with steel slag. The carbonation took place in an autoclave filled and continuously refilled with CO_2 at a temperature of 74°C and a pressure of 0.15 MPa. After 2 hours they observed a weight increase of 10.44% in a mortar with slag:cement ratio of 3:2, and a corresponding compressive strength of 20.06 MPa.

As a conclusion we will remark:

- When looking at the total CO_2 emission balance from the concrete industry, it is important to reduce the emission figures by 10% from the natural carbonation/absorption of CO_2 in concrete. A 10% absorption figure might be a conservative estimation. However, with the various uncertainties mentioned earlier, scepticism should be shown for calculations with much higher general figures unless they are documented and are referring to specific projects.
- The possibility to use carbonation as a potential tool to reduce global emission further is real, but we doubt that even strong efforts can result in total global reductions in excess of 1 to 3%.

In our evaluation, we have tried our best to evaluate the six to eight variables that have to be taken into consideration:

1. Carbonation speed: Mainly depending on concrete quality and climatic conditions.
2. Carbonation depth: Mainly depending on age of concrete/time.
3. Carbonation volume: Mainly depending on available reactive calcium hydroxide present.

4. Structural shape/thickness: As carbonation is a surface phenomenon, the percentage absorption will depend on structural shape.
5. End use of cement/concrete: The cement or concrete quantity for each type of use is an important factor in the equation.
6. Surface protection: Some concrete has been given surface protection that might slow down carbonation.

In addition:

- The possible effect of early carbonation must be evaluated.
- In a life cycle assessment (LCA), the amount of concrete that is crushed and allowed to carbonate more fully must also be evaluated after use, and demolition must be taken into account.

3.5 THE TOOLS AND POSSIBLE ACTIONS

In the following, we mention 17 tools that we have in the toolbox to reduce CO_2 emission from concrete production. The first 10 are supplementary cementing materials:

1. Fly ash
2. Blast furnace slag
3. Silica fume
4. Metakaolin
5. Rice husk ash
6. Natural pozzolans
7. Other ashes and slags
8. Limestone powder
9. Other supplementary cementitious materials

3.5.1 Increased utilisation of supplementary cementing materials

Robert E. Philleo, a reputed American researcher, gave the following introduction to the supplementary materials at a conference in Trondheim, Norway, in 1998:[96]

The available supplementary materials have similar effects on concrete, but differ in many respects including composition, geographical distribution, amount of processing required, properties, economics, method of use, and specification requirements. Blast-furnace slag, when properly granulated or palletized, is a cementitious material which reacts directly with water in the presence of an activator. In the absence of a special chemical activator, calcium hydroxide produced in the hydration of Portland cement serves as the activator. The other commonly used supplementary materials, fly ash and condensed silica fume, are pozzolans, which require the calcium hydroxide as a necessary reactant in the pozzolanic reaction. High-calcium fly ash is both cementitious and pozzolanic. Natural pozzolans and rice husk ash are acceptable materials not in a widespread use. Condensed silica fume is a premium product for special uses such as high strength, low permeability, and high electrical resistivity. Fly ash and blast-furnace slag have the potential to reduce the cost of concrete and to provide such benefits as reduced heat of hydration, resistance to sulfate attack and inhabitation of the alkali-aggregate reaction. Blast-furnace slag provides the greatest energy saving in that it may be used with little or no Portland cement. Alkali-activated slag can also be made to hydrate at low temperatures.

Philleo's general technical introduction can be used even today, a quarter of a century later.

Increased use of substitute/supplementary cementing materials, blended into the cement, or used as a separate component blended in the concrete mixtures, is by far the most important tool to reduce CO_2 in a global evaluation, in particular when looking at a rather short time effect.

As can be seen in Figure 3.10, the use of blended cements has had a considerable increase over the last 10 years in Europe. The same tendency can be found in many places in the world.

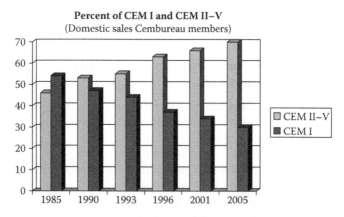

Figure 3.10 Development in use of blended cements in percent in Cembureau countries.

Table 3.6 Potential in reduction in million tonnes of CO_2 emission from substituting materials[59]

Supplementary materials	Immediate (2004)	2010	2020
Fly ash	400	700	1200
Blast furnace slag	25	55	75
Silica fume	0.1	0.3	0.3
Rice husk ash	3	6.5	6.5
Lime stone powder	150	170	200
Metakaolin	—	—	—
Natural pozzolans	25	40	50
Other ashes	—	20	2
Other industrial wastes	—	40?	200?
Total	603	1032	1734

Source: How to Improve the CO_2-Challenge: A World Wide Review. In *International Conference on Sustainability in the Cement and Concrete Industry,* Lillehammer, Norway, 2007.

Should supplementary materials be used in blended cements, or directly by the concrete producer? There is no easy answer to this question, as conditions vary considerably from one part of the world to the other, not the least regarding what alternatives are available in a logistically reasonable and sustainable manner. Also, the local culture in raw material supply has a major influence on the strategies now and in the future.

The blending percentages of supplementary cementing materials have also increased.

In 2003, with an update in 2006, Jahren[56,59] made an evaluation of what potential CO_2 emission reduction potential the various supplementary materials represent (in addition to what is already utilised for cement/concrete production or other purposes) (Table 3.6).

However, it is important to remember that even if fly ash is by far the most important supplementary material in a global evaluation, there are considerable differences of importance between the various materials from different countries, and even within regions of some of the countries.

In Europe, Portland–fly ash cement represents about 6% of the total cement production.[3] The cement that has had the highest growth in the last 10 years is Portland limestone cement, where up to 20% of the clinker can be substituted with limestone powder. This product type represents 24% of the cement production in the Cembureau area. Which supplementary material will be used most in Europe in the future will be decided depending on availability, price, logistic factors, industry structure, and traditions.

From the above it can be seen that the use of cementing supplementary materials alone can solve the challenge to reduce emission by 20% by 2020 (from a consumption of 2.9 billion tonnes, and an emission of 0.85 tonne per tonne of cement):

Scenario 1: Substitute potential (1.73 billion tonnes in 2020) utilised 50%.
  A.   Cement consumption 2.9 billion tonnes:

$$\text{Anticipated emission} = 0.85 \times (2.9 - (1.73 \times 0.5))/2.9 = 0.60 \text{ tonne}$$
$$CO_2 \text{ per tonne of binder}$$

  B.   Cement consumption 3.5 billion tonnes:

$$\text{Anticipated emission} = 0.85 \times (3.5 - (1.73 \times 0.5))/3.5 = 0.64 \text{ tonnes}$$
$$CO_2 \text{ per tonnes of binder}$$

Scenario 2: Substitute potential utilised 70% in 2020.
  A.   Cement consumption 2.9 billion tonnes:

$$\text{Anticipated emission} = 0.85 \times (2.9 - (1.73 \times 0.7))/2.9 = 0.50 \text{ tonne}$$
$$CO_2 \text{ per tonne of binder}$$

  B.   Cement consumption 3.5 billion tonnes:

$$\text{Anticipated emission} = 0.85 \times (3.5 - (1.73 \times 0.7))/3.5 = 0.56 \text{ tonne}$$
$$CO_2 \text{ per tonne of binder}$$

Scenario 3: Substitute potential utilised 90% in 2020:
  A.   Cement consumption 2.9 billion tonnes:

$$\text{Anticipated emission} = 0.85 \times (2.9 - (1.73 \times 0.9))/2.9 = 0.39 \text{ tonnes}$$
$$CO_2 \text{ per tonne of binder}$$

  B.   Cement consumption 3.5 billion tonnes:

$$\text{Anticipated emission} = 0.85 \times (3.5 - (1.73 \times 0.9))/3.5 = 0.47 \text{ tonnes}$$
$$CO_2 \text{ per tonne of binder}$$

Portland cement is generally composed of 63 to 70% CaO, 19 to 24% SiO_2, 3 to 7% Al_2O_3, and 1 to 5% Fe_2O_3, by feeding the cement kiln by limestone, clay, or slate, as well as some iron ore.

All materials with a content of CaO, SiO_2 (amorphous), and Al_2O_3 are in principle an interesting possible substitute or supplement material for Portland clinker. The interest for their utilisation grows with:

- The quantity of the mentioned minerals per unit weight
- The reduction in other constituents (impurities)
- The reduction in possible harmful substances
- The smallest possible size of the particles or the grindability of the source

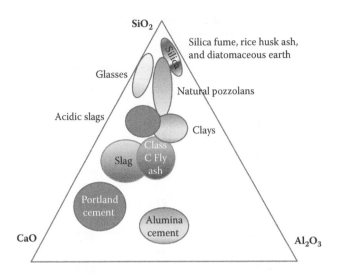

Figure 3.11 Some of the cementitious materials in a CaO-SiO$_2$-Al$_2$O$_3$-system.

- The stability in constituency of the source
- The available quantity of the source
- The logistic availability of the source
- The price

As seen from Figure 3.11, most of the possible supplementary cementing tools are found in the CaO-SiO$_2$-Al$_2$O$_3$ diagram. However, not all of the materials are cementitious.

3.5.2 Fly ash

Table 3.6 shown earlier clearly shows that the increased use of fly ash in concrete is the potentially most important tool we have to reduce CO$_2$ emission with proven technology, in a global context.

Coal is still, and for many years to come, the most important energy source for increasing in the electricity production in the world. Every week, one new coal-fired electricity plant is opened somewhere. This also gives a formidable source for fly ash and reducing CO$_2$ emission from the increased concrete production in the world. Malhotra,[97] referring to *Weather Makers* by Tim Flannery, gives the following data for expected construction of new coal-fired electricity plants in the mentioned period:

1999–2009: 249
2010–2019: 483
2020–2030: 710

Table 3.7 World's largest producers and
          consumers of coal, in million tonnes
          of oil equivalents[98]

Country	Production	Consumption
China	989.8	956.9
United States	567.2	564.3
Australia	199.4	54.4
India	188.8	204.8
South Africa	136.9	94.5
Russia	127.6	105.9
Japan		120.8
Indonesia	81.4	
Poland	69.8	57.7
Germany	54.7	85.7
Kazakhstan	44.4	
South Korea		53.2

Source: Pocket World in Figures, The Economist, 2007.

It is also interesting to observe that we find most of the countries with the strongest growth in cement production among the countries with the largest coal consumption (Table 3.7).

The world reserves of coal do not seem to be any serious limitation to the development. In a paper in 1998, Yeginbali, referring to the Turkish Industry and Development Bank,[99] claimed that the total global coal reserves was estimated to be 1089 billion tonnes, equal to 230 years' consumption at that time. Later the consumption increased considerably, but still there should hardly be a serious shortage of reserves in this century (Table 3.8 and Figure 3.12).

Fly ash is a by-product of the combustion of pulverised coal in thermal power plants. The dust collection system removes the fly ash, as a fine residue from the combustion gases. Fly ash particles are typically spherical, ranging in diameter from less than 1 micron up to 150 microns (0.015 mm).

Table 3.8 World reserves of fossil energy according to Yeginbali[99]

Source	Energy (10^{21} J)
Coal	180
Oil shale	22
Oil	14
Natural gas	11

Source: Yeginbali A., Potential Uses for Oil Shale and Oil Shale Ash in Manufacturing of Portland Cement. Presented at CANMET/ACI Symposium on Sustainable Development of the Cement and Concrete Industry, Ottawa, Canada, October 21–23, 1998.

Figure 3.12 Realistically, coal-fired electricity plants will still be the most important electricity source in the foreseeable future.

The type of dust collection equipment used largely determines the range of particle sizes in a given fly ash. Particles from plants using electrostatic filters are normally finer than those from plants using old mechanical collectors. The quality of the coal will have an influence on the chemical composition of the fly ash. Typically, fly ash from the combustion of subbituminous coal contains more calcium and less iron than fly ash from bituminous coal.

Most importantly, fly ash exhibits pozzolanic activity. The ASTM defines a pozzolan as "a siliceous or siliceous and aluminous material which itself possesses little or no cementitious value but which will, in finely divided form and in presence of moisture, chemically react with calcium hydroxide at ordinary temperature to form compounds possessing cementitious properties."[100]

Malhotra and Ramezanianpour explain that the term *fly ash* was first used in the power industry ca. 1930, and the first comprehensive data on its use in concrete in North America were reported by Davies in 1937. Real noticeable interest for use in concrete developed in the 1970s.[100]

Table 3.9 Development in world production and utilisation of fly ash[102]

	Production (million tonnes)	Utilisation (%)
1959	100	2%
1969	200	15%
1992	459	33.3%

Source: Manz O.E., et al., Worldwide Production of Fly Ash and Utilization in Concrete. Presented at Third CANMET/ACI International Conference on the Use of Fly Ash, Silica Fume, Slag and Natural Pozzolans in Concrete, Trondheim, Norway, June 1989.

According to Owens,[101] the first practical use of fly ash in the UK of some significance was in 1955, where 20% replacement with fly ash was used in the last part of the Lednock Dam.

The use of fly ash as a supplementary material grew in importance, in particular in the 1970s to 1980s.

In a paper in 1994, Manz et al.[102] gave the data shown in Table 3.9 on the development. Manz et al.'s review included reports from 32 different countries.

It is important to remember that not all coal combustion products (CCPs) from different boilers can be utilised directly in concrete, and that CCPs and fly ash are utilised for other purposes.

In a paper in 2002, Aglave and Feuerborn[103] claimed that the production of CCPs in Europe in 2000 was 59 million tonnes:

- 66.0% fly ash
- 18.0% flue gas desulphurization (FGD) gypsum
- 9.5% bottom ash
- 4.0% boiler slag
- 1.7% fluidized bed combustion (FBC) ash
- 0.8% spray dry absorption (SDA) product

This amount was utilised as follows:

- 53.1% construction industry and underground mining
- 34.3% restoration of open-cast mine quarries and pits
- 6.2% temporary stockpile
- 6.4% stockpile

For more technical information about the technical utilisation of fly ash in concrete, we refer to the thousands of reports and papers over the last few decades. As a basic knowledge base we recommend the book *Fly Ash in Concrete* by V.M. Malhotra and A.A. Ramezanianpour.[100]

In addition to the fly ash produced each year, there are still millions of tonnes of fly ash that have been stockpiled over the years. According to Malhotra,[104] in the UK alone, 130 million tonnes could be recovered from these stockpiles. We know from various places in the world that fly ash is now recovered from some of these stockpiles.

There obviously exist some practical challenges in the utilisation of fly ash, and some of them produce some obstacles that might affect the environmental advantages, for example:

- In typical cold countries, the highest energy consumption, and consequently the highest energy production and highest production rates, of fly ash is in the winter season. The concrete industry has its top season in the summer. Maximum utilisation creates a logistic challenge that has to be met by increased silo capacity, etc.
- Some countries or regions do not have any local fly ash. If fly ash can be transported on ship, the environmental transport impact might be limited. Otherwise, transport must be taken into consideration in a total sustainability evaluation.
- Not all available fly ash is directly suitable for use in concrete. However, some of this might find use for other purposes. In addition, several technologies have been developed to beneficiate fly ash that fails to meet specifications for fineness or carbon content requirements. Technologies have also been developed to upgrade fly ash to a higher fineness and a more sophisticated product.

According to Jain,[105] one of the most important obstacles for even higher utilisation of fly ash in cement in India is the distance from the fly ash source to the cement production. However, as can be seen from Figure 3.13, the development in use of fly ash in cement is impressive.

It is important to remember that in the same period, the cement consumption in India grew considerably. Only from 2003–2004 to 2009–2010 did the production increase from 115 to 202 million tonnes per year, to place India as number two in the world in cement consumption, and with a growth rate higher than that of China. Jain claimed in 2011 that 80 million tonnes of the total Indian fly ash production of 160 million tonnes is now utilised, where half of the utilisation goes for blended cements. Jain also claimed that all available sources of blast furnace slag already are utilised.

A special type of fly ash use that might be very interesting for special purposes is what has been called high-volume fly ash concrete (HVFAC). V.M. Malhotra and P.K. Mehta have been particularly energetic in promoting this special type of concrete and its superb sustainable properties. Mehta and Manmohan claimed that HVFAC is superior to conventional concrete with regard to the following:[106]

- Constructability
- Durability
- Sustainability
- Cost

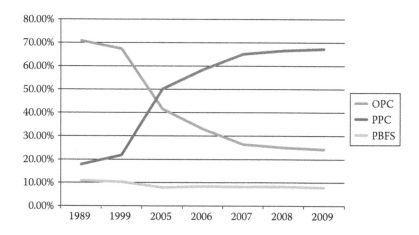

Figure 3.13 Development in types of cement in India, according to Jain.[105] OPC = ordinary Portland cement, PPC = Portland pozzolanic cement (fly ash), PBFS = Portland blast furnace slag cement. (From Jain A.K., Fly Ash Utilization in Indian Cement Industry: Current Status and Future Prospect, in *Concrete Sustainability through Innovative Materials and Techniques*, Roving National Seminars, Bangalore, January 10–14, 2011, pp. 46–51.)

Compared to conventional concrete, HVFAC must obtain at least 50% fly ash as cement replacement by mass. It has a water/cementitious ratio generally below 0.4, and may or may not contain any chemical and air-entraining admixtures. Superplasticizers are needed for low w/cm ratio concrete mixtures (generally less than 0.35 w/cm) designed for high strength, or high workability in the case of heavily reinforced structures.

In the mentioned paper, the authors refer to three practical examples (CS = compressive strength):

- Kauai Temple, Hawai: CS = 3000 psi (21 MPa)
- BAPS Temple, Houston, Texas: 4000 psi (28MPa)—28 days
- C. Barker Hall, Berkeley, California: 4470 and 4730 psi (31 and 33 Mpa)—28 days; 5330 and 5610 psi (37 and 39 MPa)—56 days

In particular, the reported durability of the HVFAC is impressive.

The work by Malhotra and Mehta, and CANMET, Canada, in the 1980s has had an impact in many parts of the world. As an example, Desai reported the use of HVFAC in a road in Ropar, Punjab, India, in 2007 (Figure 3.14).[107] The reported road case was a two-lane, 7 m wide road with a length of 750 m and a slab thickness of 300 mm.

Malhotra claims that high-performance HVFAC is superior to normal Portland cement concrete in almost all aspects,[108] and gives the

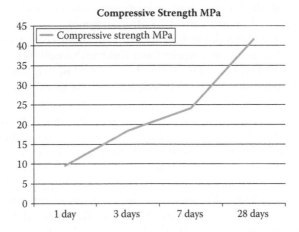

Figure 3.14 Average compressive strength of road concrete according to Desai.[107] (From Malhotra V.M., High-Performance High-Volume Fly Ash Concrete, *Concrete International*, July 2002.)

following typical characteristics for high-performance HVFAC used for highway pavements:

- A minimum of 50 to 60% ASTM Class F fly ash by mass of cementitious materials
- Low water content, generally less than 130 kg/m³ of concrete
- Cement content not more than 200 kg/m³ of concrete, but generally about 150 kg/m³
- Very low dosage of a superplasticiser, if needed
- Air-entraining admixtures, only if freezing and thawing conditions are anticipated
- Low w/cm ratio, generally less than 0.35

We also like to mention a few reports from a conference in Beijing, China, in 2011: Researchers from Dalian, China,[109] tested fly ash replacements of 50 to 70% in various concrete blends with w/cm ratios from 0.4 to 0.55 and reported various efficiency factors for fly ash with age of testing and other blending factors. They reported compressive strengths after 28 days, ranging from 23.5 to 56.7 MPa, and after 360 days, ranging from 61.6 to 78.5 MPa.

Researchers from Shandong and Beijing, China,[110] reported using very high dosages of fly ash combined with other wastes as blast furnace slag and fluorogypsum, and only 5 to 10% of Portland clinker. Their compressive strengths after 28 days ranged from 47 to 51 MPa for various blends and for the reference mix with only Portland cement. They reported rather low 3-day strengths for their high-substitution concrete, and with a rather low pH of 12.26, decreasing to 11.75 after 30 hours.

A special use of high dosages of fly ash is in roller-compacted concrete dam projects. We mention a report from Kokkamhaeng[111] reporting about a dam project in Thailand. The 300 m long and 24 m Pak Mum Dam was constructed with roller-compacted concrete with 58 kg of Portland cement and 124 kg of fly ash (substitution = 68%) in the concrete.

As fly ash has changed from being a problematic waste material to an important cost and emission reduction tool in the cement and concrete industry, increased emphasis is directed toward producing a better product. As one of the problems with some fly ashes is too high carbon content, another driving force has also been to utilise the energy potential better. Bruce W. Ramme from Wisconsin has in a number of reports told about progress in beneficiation technology. A paper by Ramme et al. from 2005[112] reported on various processes by the Wisconsin Electric Power Company (WE Energies). These processes allow the use of not only better-quality fly ash instead, consequently reducing the need for landfill, but also possibly ash from existing landfills. In addition to product improvements for the cement and concrete industry, the technology also has other environmental benefits: reduction of NOx emission, reduction of ammonia and mercury contents, etc.

Similar developments also take place in other countries. In 1998, van den Berg and Moret from the Netherlands[113] reported about a national processing plant in Rotterdam-Maasvlakte that was put in operation in 1995 with a capacity of 250,000 tonnes of fly ash per year. In addition to serving a local power plant, the fly ash is transported to the processing plant by bulk truck and barge. In the upgrading processing plant the fly ash carbon content is reduced to less than 5% and the fineness is improved to over 70% less than 45 micron. The author claimed that this was the first plant of its kind in the world.

Various ideas to process fly ash to a more advanced mineral admixture have been tried. An example is a report from Lu at the Southwest University of Science and Technology, Shenyang, China,[114] about processing highly reactive superfine fly ash by using a "superheated steam jet mill." In the process Lu et al. could produce a product down to an average particle size of 1.597 microns, but used a product with a particle size of 10 microns in testing. Lu claimed that in 2007 China generated 330 million tonnes of fly ash from its power plants.

Somewhat similar is the process described by Justnes[115] called EMC technology. The technology was developed by Vladimir Ronin at Luleå University, Sweden, in the early 1980s; it is a high-intensity activation process to increase the reactivity of OPC with high filler or pozzolan replacement. The process involves multiple stages of vibratory or stirred ball mills with custom-designed grinding circuits to obtain target properties. Justnes reports that when the EMC technology was used on a binder with 30, 50, and 70% replacement with fly ash, even if the setting time was not changed, the 3-day compressive strength increased considerably—for 30% replacement as much as three times the base mix.

3.5.3 Blast furnace slag

The first modern blended cements were blast furnace slag cement, and were produced in Germany in 1901, but work on the use of slag as a binder or binder supplement started in 1886 by Prussing. According to Aitcin,[116] in 1900 the German cement industry did not very much like the arrival of a competitive product on the market, and started to refuse to sell clinker to the metallurgical industry. The steel industry had no other choice than to build its own Portland cement clinker kilns. The confrontation went on until 1909, when the two first blended cements containing 30 and 70% of slag were introduced on the market. After that, the two industries decided to cooperate to produce the first standard on slag-blended cement.

Slag cement was thereafter produced rapidly in Belgium and the Netherlands. In France, it was not until 1918 that the first slag cement started to be utilised.

Prior to 1975 there was no significant use of slag as a cementing material in North America. Since then, large cement companies using separately ground slag began in Canada in 1976, and in the United States in 1982.[117]

Blast furnace slag has also been for many decades recirculated in high dosages and large volumes as low-strength concrete (0.3 to 0.4 MPa) in securing and filling of old mine shafts.

Ground blast furnace slag (GBFS) is probably the second most important cement substitution tool worldwide, in volume.

As can be seen from Table 3.10, the biggest growth in steel production is in the same countries that have the biggest growth in cement consumption: China and India.

In 1998, Mehta[119,120] stated that the yearly production of BFS in the world was 100 million tonnes, where more than 50 million tonnes came from China, India, and Europe.

Fidjestøl[121] in 2008 estimated the total available global volume to be about 130 million tonnes, slightly growing.

A reliable market search in 2002[57] stated that 100 million tonnes of granulated blast furnace slag (GBFS) was produced in Asia per year, representing 55% of the world production. The granulation ratios at the time were:

- 75% in Asia
- 60% in Europe
- 48% in North America

Table 3.10 World's 10 largest crude steel
producers in 2009, according
to World Steel Association[118]

Country	2000	2009
China	129	568
Japan	106	63
India	27	63
Russia	59	60
United States	102	58
South Korea	43	48
Germany	46	33
Ukraine	32	30
Brazil	28	27
Turkey	14	25

Source: World Steel Association, Steel Statistical
Yearbook 2010, accessed December 19, 2011.
(http://www.worldsteel.org/statistics/crude-steel-
production.htm)

In a study commissioned by the World Business Council for Sustainable Development in March 2002,[60] the figures for BFS in 2020 were estimated as shown in Table 3.11. It is possible that the development in steel production in the world has been underestimated.

Looking at the slag numbers, it is also important to have in mind the difference in practice in different parts of the world regarding granulation rate and the rate of use in cement/concrete. In 2001 Hardtl gave the global comparison shown in Table 3.12.[122]

Hardtl explained the rather large differences in granulation rate by differences in the traditions in the steel industry, technical development, general economic conditions, and ecological awareness. He pointed out that the highest granulation rates, more than 80%, are found in China, Germany, Belgium, and the Netherlands. He is of the opinion that a realistic vision in the future is that the granulation rate might change to nearly 100%.

Looking at slag as a tool for emission reduction, it is important to take into consideration the energy used in grinding, when comparing with other alternatives.

Comparing GBFS from different parts of the world, there might be considerable differences in quality due to the differences in raw materials, production processes, and technological levels of process equipment. Typically, GBFS is normally of higher quality in countries like Japan, South Korea, Australia, and Taiwan, than in China and India. Asian and Australian GBFS normally has a higher content of Al_2O_3 than similar products from Europe and America.

Table 3.11 Estimated available BSF in 2020 in million tonnes estimated by World Business Council for Sustainable Development in 2002[60]

Country	Million tonnes GBFS in 2020
United States	16
Canada	3
Western Europe	27
Japan	15
Australia and New Zealand	1
China	20
Southeast Asia	3
Korea	3
India	4
Former Soviet Union	13
Eastern Europe	4
Latin America	7
Africa	2
Middle East	1

Source: Sakai K., Standardization and Systems for Sustainability in Concrete and Concrete Structures. Presented at International Conference on Sustainability in the Cement and Concrete Industry, Lillehammer, Norway, September 16–19, 2007.

Table 3.12 Comparison between total production of blast furnace slag (BFS), granulated blast furnace slag (GBFS), granulation rate, cement consumption, and percentage use in cement in 1999–2000 according to Hardtl[122] (volume numbers in million tonnes)

Region	BFS	GBFS	Granulation rate	Cement consumption	%
Asia	99.6	74.8	75%	962	7.8%
America	25.2	12.0	48%	241	5.0%
Europe	56.4	33.8	60%	301	11.2%
Africa	2.3	2.3	100%	79	2.9%
Oceania	1.7	1.0	52%	8	11.1%
World	**185.2**	**123.9**	**67%**	**1591**	**8.2%**

Source: Hardtl R., Utilization of Granulated Blast Furnace Slag. A Contribution to Sustainable Development in the Construction Industry. Presented at FIB Symposium, Concrete and the Environment, Berlin, Germany, October 2001.

GBFS is traded by specialist companies, traditionally mostly based in Europe and the United States.

In an evaluation in 2003, Jahren[56] estimated the total quantity of GBFS available for use at that time to be approximately 125 million tonnes per annum, where 90 million tonnes were used already. The potential additional CO_2 emission reduction from increased use of GBFS as supplementary material was estimated to be 25 million tonnes in 2002, 55 million tonnes in 2010, and 75 million tonnes in 2020.

3.5.4 Silica fume

The first paper on silica fume in cement and concrete was by Berhardt[123] in a Norwegian concrete publication in 1952. It took a long time before the next publication, by Markestad in 1969,[124] from a Nordic concrete congress. Tests reported in both these papers, however, are rather far from the silica fume technology we use today with respect to both dosage rates and the use of admixtures (Figure 3.15).

In the 1970s, however, testing and practical utilisation increased rapidly, first in Norway and later in neighbouring countries and countries like Japan, Canada, and the United States (Figure 3.16).

The use of silica fume in concrete production became a rather normal technical procedure in Norway in the early 1980s (Figure 3.17).

One of the first important studies of silica fume in concrete outside Norway was reported in Japan by Nagataki in 1979.[126]

Figure 3.15 The Lake Mjøsa Bridge from Biri to Moelv, Norway, built 1983–1985, was the first major bridge project in the world that used silica fume in concrete.[125] (From Jahren P., Long-Term Experience with Silica Fume in Concrete. Presented at Ninth CANMET/ACI International Conference on Fly Ash, Silica Fume Slag and Natural Pozzolans in Concrete, Warsaw, Poland, May 20–25, 2007.)

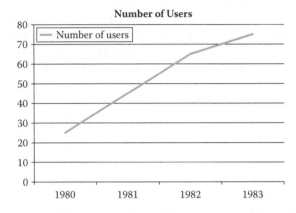

Number of Users

Figure 3.16 Number of users of silica fume in Norway.[125] (From Jahren P., Long-Term Experience with Silica Fume in Concrete. Presented at Ninth CANMET/ACI International Conference on Fly Ash, Silica Fume Slag and Natural Pozzolans in Concrete, Warsaw, Poland, May 20–25, 2007.)

Figure 3.17 Grain silo at Buvika in middle Norway, built 1979. It was the first slip forming project with silica fume in the concrete.[125] (From Jahren P., Long-Term Experience with Silica Fume in Concrete. Presented at Ninth CANMET/ACI International Conference on Fly Ash, Silica Fume Slag and Natural Pozzolans in Concrete, Warsaw, Poland, May 20–25, 2007.)

At the first CANMET/ACI International conference on fly ash, silica fume, slag, and mineral by-products in Montebello, Canada, in 1983, Jahren[127] reported from an investigation done in July 1982, where he found 132 articles, reports, and other publications about silica fume in concrete, where 107 originated in Norway, and less than 20% of the reports were in English. It might be typical to the spread of knowledge that Khayat and Aitcin, in an overview paper in 1991,[128] mention a literature search by Aitcin early in 1983, where he only found 47 papers dealing with silica fume.

Iceland was the first country that started regular production of blended cement with 7.5% silica fume, more than 30 years ago.

Some of the earlier test and promotions for the use of silica fume, over 20 years ago, were clearly directed toward cement replacements. Today we believe that most of the silica fume in concrete is added for other purposes, like higher strength, better durability, etc. Silica fume as a direct cement substitute to reduce considerable amounts of CO_2 emission has little merit; however, silica fume is still a very important tool in the toolbox, because of its versatility in use to solve various technical challenges, and also as a possible important and effective full ingredient in ternary and quaternary blends.

Fidjestøl[121] in 2008 estimated the total volume of silica fume in the world to be less than 1.5 million tonnes.

3.5.5 Metakaolin

In quite a few places in the world, metakaolin formed through calcinations of kaolinite has been used as a pozzolan. Kaolinite is a clay (kaolin or China clay) that is important to the ceramic industry, but also extensively used as a filler in rubber products. After calcinations at a relatively low temperature, less than 800°C, kaolinitic clays are capable of being converted into pozzolans. The process is complicated by the fact that the heat treatment is variable depending on the nature and fineness of the clays. The substitution possibilities are reported variable from 20 to 50%, depending on the kaolinitic content of the clays.

Kaolinite $(Al_2Si_2O_5(OH)_4)$ is transformed into metakaolin $(Al_2Si_2O_7+H_2O)$ according to a hydration reaction at 600°C[129] (650 to 850°C according to Sabir[130]). Metakaolin can also be formed from other sources containing kaolinite, e.g., paper sludge and red mud, a residue from the aluminium industry. Marvan et al.[131] reported good results with 30% substitution of calcined laterite with kaolin content. Lateritic soils are traditional materials for road and airfield construction in many countries in Africa and Asia. A 1968 paper by Rossouw and Kruger[132] reported that the Indian "Standard Specification for Burnt Clay Pozzolana" is IS 1344 from 1968.

Sabir[130] claims that the optimum replacement dosage for metakaolin is 10% for a water/binder ratio of 0.35 at 20°C. At higher curing temperature and higher water/binder ratios, an optimum level of cement replacement by metakaolin is 5%. Of other results, we mention tests in India[133] claiming that 10% addition of metakaolin can increase early compressive strength by 25 to 50%, while tensile strength does not increase by the same ratio.

Some authors claim that the use of metakaolin can give special environmental advantages in concrete production. Researchers from Iran[134] claimed, in a paper in 2007, that 5% cement replacement with highly reactive metakaolin did not improve the gas permeability of concrete, while 15% of cement replacement decreased the gas permeability of concrete by 70%.

A considerable amount of CO_2 is normally emitted in the production of metakaolin. Although it might be an interesting and effectual alternative as a pozzolan to achieve special advantages, we do not consider metakaolin an important net tool in global reduction of CO_2 emission, and have not included it in further comparisons.

3.5.6 Rice husk ash (RHA)

By weight, rice is the single most important food item in the world, next to water (Figures 3.18 and 3.19).

According to Zhang and Malhotra,[135] prior to 1970 RHA was usually produced by uncontrolled combustion, and the ash was generally crystalline

Figure 3.18 Rice field before harvesting.

Figure 3.19 In the harvest season, the asphalt on the local road is a popular work platform—India.

and had poor pozzolanic properties. In 1973, Mehta published the first of several papers describing the effect of pyro-processing parameters on the pozzolanic reactivity of RHA. By burning the rice husk under controlled temperature and atmosphere, a highly reactive RHA is obtained. According to Mehta,[136] rice husk makes up about one-fifth of the harvest of dried rice. Burning the husk, 20% of the husk becomes ash, which corresponds to about 40 kg of RHA per tonne of rice harvested. The husk surrounds the paddy grain, and during milling, 78% becomes rice, while 22% is received as rice husk. This husk is often used as fuel in the rice mill to generate steam for the parboiling process. The husk contains about 75% organic volatile matter, and the balance of the weight of the husk (25%) is converted into ash during the firing process, and is known as RHA.[137] RHA contains some 85 to 90% of amorphous silica.

According to Azevedo,[138] in 1999 Brazil produced 5.06 million tonnes of rice, giving about 0.2 million tonne of RHA that in most cases did not have a specific application, and its "disorderly disposal generates serious problems related to environment contamination." According to Bui et al.,[139] Vietnam produces about 30 million tonnes of rice. It was further recorded that 1 tonne of rice gave 0.3 tonne of rice husk. Burning this, 20% was RHA.

Reports state that there is the possibility of achieving a volume of 4% of RHA for each tonne of rice produced.

The following six countries each represent more than 5% of the world rice production in 2003: China, 31%; India, 23%; Indonesia, 8%; Bangladesh, 6%; Vietnam, 5%; and Thailand, 4%. Asian countries represent well over 90% of all the rice production in the world (the consumption figures vary): China, 36%; India, 22%; Indonesia, 9%; Bangladesh, 6%; and Vietnam, 4%.

However, there is a long way to go for utilisation of RHA in concrete. In a status report at a conference in Hyderabad, India, in 2005, Jain[140] claimed that India produces 22 million tonnes of rice, which can give 4.5 million tonnes of rice husk ash. He claims that very little progress has been achieved in utilising this potential. Two small power plants (below 5 MW) in Andhra Pradesh are based on rice husk and produce ash. In addition, a few cement plants have started using rice husk as fuel in their kilns. However, he is optimistic, and predicts that rice husk utilisation as fuel and rice husk ash as pozzolan are likely to accelerate in the coming years.

A major challenge in utilising RHA to its full extent is to use the right type of combustion equipment and procedure. Another obstacle for the full utilisation of this very interesting tool is the fragmented production system, causing a serious logistic challenge regarding utilisation. A report by Joseph et al.[141] has concluded that there are some 80,000 rice mills in four Southeast Asian countries alone.

We have estimated a total potential for some 16 to 20 million tonnes of RHA in the world, most of it located in countries where the major growth in cement production is predicted.

Several evaluations indicate that up to about 35% of the potential RHA can be utilised as a cement replacement material. From the reports studied, only small amounts of the potential have been utilised so far. We have anticipated a possible future utilisation of 30% of the available potential sometime in the future. The conclusion from our evaluations is that the RHA tool, in practical terms, amounts to a possible substitution of approximately 6 million tonnes of Portland cement clinker in Asia, and about 0.5 million tonne in the rest of the world.

Two of the main challenges in effective utilisation RHA are logistics and adequate burners to receive the optimum burner temperature and time. A paper from researchers at Amirkabir University of Technology, Teheran, Iran,[142] claims that an optimum burner temperature is 650°C with 60 minutes burning temperature. With higher temperatures, the SiO_2 turns crystalline and the amorphous reactive silica gets less reactivity. With reference to Mehta, they also claim that the first patent of RHA in concrete was taken out in 1924.

In a paper at a conference in 1998, Horiguchi et al.[143] reported about test with RHA to improve abrasion resistance of concrete. The tests were performed with local RHA from northern Japan, one of the largest rice-producing regions in Japan. The RHA was produced by controlled combustion in the laboratory and added in dosages of 5, 10, and 15% of cement weight. They reported about a reduction of wear depths in their tests of 49% with 15% addition of RHA.

In a paper at a conference in Jinan, China, researchers from Delft University of Technology in the Netherlands also address the importance of the particle size of RHA.[144] In tests with cement replacements of 10 and 20%, they tested various particle sizes of RHA ground down to 22.5, 15.5, 9.0, 5.6, and 3.6 micron. The authors claim:

- The addition of RHA results in less heat release and decelerates the hydration of Portland cement.
- The finer the RHA, the higher is the total heat evolution, and the lower is the porosity, in particular with higher dosages of RHA.
- The pore size distribution of cement blended with RHA particle size of 15.5 microns is similar to that of the control mixture.
- The porous structure of RHA observed in the study collapses gradually when its mean particle size is reduced from 22.5 to 9.0 microns.
- A reasonable mean size of RHA particles was proposed to be from 5.6 to 9.0 microns.

Nigeria produces about 2 million tonnes of rice each year. A report by Alhassan and Mustapha, from Federal University of Technology, Minna, Nigeria,[145] tells about the use of RHA as a supplementary material in cement stabilisation of soil in roads with the aim to save cement, and thereby improve sustainability performance of the work. Tests were done with 2 to 8% of cement content and 2 to 8% of replacement with RHA. The optimum replacement level seems to be in the order 2 to 4%.

3.5.7 Natural pozzolans

Natural pozzolans are found in many places in the world, and have been used as binders for a very long time.

The background for our modern word, *pozzolan*, comes from the town of Pozzuoli, northeast of Naples in Italy, where red "pozzolanic" deposits from the volcano Vesuvius, which erupted in 79 AD, were found, excavated, and used from the third century BC. Greeks had up to two centuries earlier used high siliceous volcanic Santorin earth, from the island of Santorini, as a pozzolan, in limited amounts. The powerful volcano that erupted on the island in 1500 BC produced lava and ash with corresponding waves of 80 to 100 m.[146]

The famous Roman architect, engineer, and writer Vitruvius (i.e., Marcus Vitruvius Pollio) wrote his famous publication *De architectura* about 25 BC (Figure 3.20). In Book II, Chapter VI, he writes about pozzolan:[147]

> The reason for the strength of this material can be found in the fact that the ground in the hills is warm and it has warm springs. This could not have happened if not the mountain underneath had big fires with burning sulfur or asphalt. The fire and the heat from the flames come from the inside through cracks and make the soil light, and the pumice that can be found there is like a swamp free from moisture. In this way, the three substances will be mixed together in fire.
>
> That it is burning hot in these areas can be proved by the fact that in the mountains near Baiae, that belongs to Cumaeans, there are excavations which have been made into baths. What we call "swamp-stone" or "Pompeii pumice" seem to have been reduced from another type of stone to the type of stone we can find. The type of stone found in this region cannot be produced elsewhere, but can be found in Aetna and in the mountains in Mysia, called "burnt area" by the Greeks.

With their pozzolan concrete, the Romans created a giant step forward in the history of concrete by (Figures 3.21 to 3.23):

• Creating more durable structures
• Allowing new architecture as arches and bridges

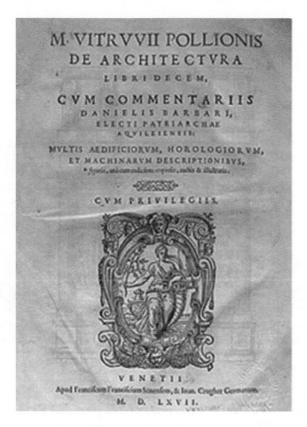

Figure 3.20 The front page of Daniele Barbaro's famous edition of Vitruvius's book from 1567.

- New versatile use of the material in roads, bridges, water channels and baths, buildings, dams, etc.
- Spreading the technology to the whole Roman Empire
- Documenting their technology.

Mehta[146] has compiled a very interesting introduction to the history of concrete through the use of natural pozzolans. We have been tempted to refer to this history, but instead we refer the reader to the reference.

Among the natural pozzolans, those of volcanic origin consist of glassy and amorphous materials arising from the deposition and alteration of volcanic dust and ash. Other high-silica minerals also show pozzolanic activity. Best known are possibly the diatomaceous earths, in which the main active component is opal, an amorphous form of hydrous silica.[146]

Trass has been utilised in Germany for 2000 years, and is standardised in DIN 51043. Trass is also found in Romania and Russia.

Figure 3.21 The aqueduct in Segovia, Spain.

Uzal and Turanli,[148] in 2001, reported on the use of high-volume natural pozzolan cements from Turkey with 55% pozzolan addition. Already in 1984, 28 of 33 cement plants in Turkey produced pozzolan-modified Portland cement, amounting to 50% of the national cement production. A wide variety of natural pozzolans were in use. In a report from 1992, comparative tests were made with 15 different natural pozzlans used by the cement industry in Turkey.[149] In 1992, Mazlum and Aköz[150] also reported on tests in Turkey with diatomite enriched through leaching with nitric acid.

In a report in 2003, Jahren[56] reported about activity with natural pozzolans as a supplementary cementing material in countries like Algeria, Brazil, Canada, China, Colombia, Denmark, Egypt, England, France, Germany, Chile, Greece, India, Iran, Italy, Japan, Mexico, New Zealand, Romania, Russia, Turkey, and the United States. Most of the substitution rates reported were in the range of 10 to 30%. The report from 2002 states that from the available data it is neither possible nor reasonable to give accurate estimates of the volume of the use of natural pozzolans, or the potential for increased use in the future. However, as a rough estimate from the available data, the present use at the time was estimated to be between 22 and 30 million tonnes. Evaluating the enormous resources reported, the

Figure 3.22 The Pantheon temple in Rome, Italy, built in 127 A.D..

world potential for annual use of natural pozzolans was estimated to be in excess of 80 million tonnes in the future.

WBCSD[151] in 2010 claimed that 300 million tonnes was available in 2003, but only 50% was used.

However, there are some practical limitations when we look at natural pozzolans as a tool to reduce CO_2 emission. The energy involved in bringing the raw material to a stadium of possible comparison with Portland clinker must be taken into account. Several reports tell about variation in reactivity and chemical composition of the natural deposits. This results in need for homogenisation, silos, and possible modification of reactivity versus Portland clinker. In addition, the emission from the energy needed for grinding has to be taken into account.

As an example we mention that Salazar reported about tests with natural pozzolan in Colombia with an efficiency factor compared to Portland clinker between 0.3 and 0.4.[152]

Volcanic ash is natural pozzolan, with considerable variation in particle size and, to some extent, chemical composition. As an example we mention that Canadian researchers studied the properties of volcanic ash from the

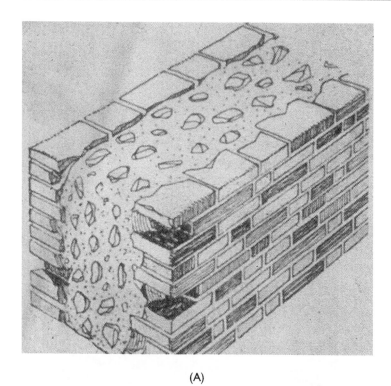

(A)

(B)

Figure 3.23 The Romans preferred to use bricks as moulding material for their concrete.

Rabul area of the East New Britain Province of Papua New Guinea to see if it was suitable as a cement replacement. The ash had a SiO_2 content of 59.32%, CaO content of 6.10%, and Al_2O_3 content of 17.54%. The fineness of the material was 285 m²/kg, and 12% was retained on a 0.045 mm sieve. They tested variable substitution ratios from 2 to 75%. The test results were interesting, but probably mostly of local interest. However, of general interest we note the comment that the compressive strength was sharply reduced at substitutions higher than 40%.[153]

In a paper in 2010, Long et al., from Central South University, Hunan, China,[154] claimed that due to intensified construction activity there is shortage of fly ash and slag in some parts of China. In many provinces natural pozzolans are found. In particular, some areas of Yunan Province, such as the Baoshan region, are rich in natural pozzolanic rock. The purpose of the reported study was testing of pozzolanic powders ground by tuff and pumice taken from the Baoshan region as a potential natural admixture to concrete used in high-speed railway construction. Tests were done with 30% cement replacement and compressive strength as well as sulphate attack. The testing of resistance to sulphate attack showed considerable improvements versus the control sample with Portland cement, even if the compressive strength after 28 days was considerably lower.

Several researchers have tried to produce an independent binder or geopolymer from natural pozzolan. As an example, we mention a paper from a conference in Coventry, UK, in 2007, where researchers from Teheran, Iran, and Sheffield, UK,[155] reported about a natural pozzolan from Iran called Taftan, activated with KOH. In their conclusion they claim:

- The curing temperature for achieving the highest strength of the pozzolan was 60°C. However, 40°C was found to be adequate for achieving strength in the range suited for structural concrete.
- Comparable strengths can be achieved with 60°C curing and autoclaving at 2.5 MPa for 3 hours.

The strengths reported in the paper were in the 30 to 50 MPa range.

A rather special type of natural pozzolan is perlite. In the concrete world or the building industry, perlite is probably mostly known as an aggregate for lightweight concrete, or as a raw material for fire protection, in particular for steel structures. Perlite has the ability, when heated to about 900°C, to expand up to 18 times its original volume. Grinding of expanded perlite gives a light powder of density 60 to 120 kg/m³. In a paper at a conference in Coventry, UK, in 2007, Erdem et al.,[156] from Middle East Technical University, Ankara, Turkey, reported on a test with expanded perlite as a possible use in blended cement. They claimed that the total reserve of perlite in the world is around 6700 million tonnes, with two-thirds of this in Turkey. There are also large deposits of perlite in the Hubei Province in China, with the largest no-metal mine in China.

There is no doubt that perlite is an interesting material, but partly due to the energy needed to expand the raw material, and partly because of the price competition for other and more advanced utilisation, we do not believe that perlite will play any significant role in reducing CO_2 emission from cement and concrete.

3.5.8 Other ashes and slags

Testing of the use in concrete as supplementary material of the many possibilities of ashes and slags available has been going on in many countries with increased interest in the last two decades. Below we have tried to give some examples of the versatility, and tried to evaluate the emission reduction from this possible tool.

In the book *Supplementary Cementing Materials for Concrete* from 1987, Douglas and Malhotra have written a state-of-the-art paper about nonferrous slag-Portland binders.[157] They reviewed 34 different reports from various parts of the world from the period 1962 to 1982. The review gives the clear impression that there is an interesting potential in nonferrous slags, but the chemistry and practical complications in the use vary considerably with type of slag and various production methods. Judging the reported results versus the normal general requirements for cement and concrete, it looks like the potential is much greater in typical mine fill concrete and concrete for special purposes, than as a major general tool for reduction of GHG emission in concrete. However, these "specials" might locally be very important.

In a paper in 1995, Madej et al.[158] reported on tests with mortars with substitution with two types of nonferrous slags and two ferroalloy slags and compared them with mortar with Portland cement and blast furnace slag substitution. They reported that the early compressive strength of mortars with slags is influenced negatively, when compared with plain mortar of the same w/cm ratio. From 10 to 30% substitution by mass of the cementitious material with nonferrous and ferroalloy slags, however, either comparable or slightly higher strength to the mortar with BFS of the same content is achieved. In this range, the strength gains up to 15% over the control mortar that was obtained for the blended mortars. They also tested special curing conditions as autoclaving, in an autoclave with a maximum temperature of 180°C and a pressure of 1 MPa for 3 hours, and hot water curing at 70°C for 24 hours, and achieved compressive strengths up to 11 MPa, depending on type of slag. They claimed that the mortars with nickel slag and copper slag, respectively, provide high sensitivity and extremely diffusivity to the carbon dioxide and chloride ion penetration.

3.5.8.1 Sewage sludge incineration ash (SSIA)

The Danish Centre for Green Concrete has for years been a "leading lady" in analysing and looking for new and more sustainable solutions. At a conference in 2003,[159] Damtoft et al. presented a paper where they reported their efforts to find alternatives to fly ash, as the Danish government has decided to gradually phase out coal electricity production. They explain that SSIA differs from normal fly ash in having a higher CaO content and lower SiO_2 and Al_2O_3 content. From their testing they report that they succeeded in getting a 56-day strength, higher than 85% of their reference concrete. They also got acceptable results in their carbonation and chloride testing.

3.5.8.2 Ferroalloy slag

Jean Pera and researchers from the National Institute of Applied Science of Lyon, France, have been very active in investigating alternatives in recycling and binders from waste and by-products. In 1998 they presented a paper on the use of manganese slag as a binder.[160] Manganese is an important alloy used to increase hardness and resistance to the wear of steel. They refer to a forecast to increase the consumption of manganese ferroalloy of 1% per year and a world consumption of 7.0 MT in 2000. Typically this slag has a much lower portion of CaO than blast furnace slag, and a high portion of MnO. The paper states that the use of manganese-rich ground granulated slag in concrete is feasible, and its presence, up to 80 kg/m^3, does not affect the mechanical strength of concrete. In another paper,[161] Pera and Ambroise report on silicomanganese slags containing about 11% MnO ground to a Blaine fineness of 360 to 600 m^2/kg. They report about reactivity of 50/50 blends with normal Portland cement that is lower than that of the equivalent GBFS. In lower concrete qualities (C25) the slag reduces the water permeability. They find the substitution particularly interesting in self-levelling concrete.

Duricic et al.[162] in 1998 gave a report about an impressive research program with manganese slags in former Yugoslavia. The tests were performed from 25 laboratories with 12 different cements. The conclusion from the tests was that the usefulness of the slag was confirmed. The total amount of slag potential in the country was recorded to be 40,000 tonnes per year.

We also mention a report from the French researchers Cabrillac et al. in 1995[163] on the use of magnesium slag, not really as a substitute, but as a binder for production of building blocks. In their paper they mention that there is one producer in France, with an annual volume of 90,000 tonnes of by-products. They also mention that 6 tonnes of waste is obtained from each tonne of magnesium produced. In their report, they stress the importance of vibration in the block production process. By vibration the compressive strength after 28 days increased from 6 MPa to 7.5 MPa.

3.5.8.3 Barium and strontium slag

In 2008, Cheng[94] reported on a successful test with barium and strontium waste residues as the main raw materials to produce cement.

3.5.8.4 Other types of slag

As mentioned in Chapter 4, Section 2.10 on recycling, Collepardi et al.[164] conducted successful tests with replacement of 15% of cement with grinded slag from production of metallic lead. Roper et al.[165] used copper slag as a substitute for low-strength mine fill purposes. The slag was a waste product derived from reverberatory furnaces used for melting of copper concentrates. Hwang and Laiw[166] reported that in Taiwan, 75% of the copper slag is utilised, while the rest is discarded on the seashore. Tests were made on the use of the slag as an aggregate, not as a cement replacement.

Qi and Peng in 2012[167] reported on a test with 0 to 60% cement replacement with phosphorous slag. The slag comes from the production of yellow phosphorous. They claimed that 5 to 6 million tonnes of phosphorous slag is discharged annually in China, with a negative environment effect both for the disposal of land and regarding pollution. The slag has contents of SiO_2 and CaO. In the sample tested, the SiO_2 content was 40.33% and the CaO content 45.42%. The report contains an evaluation of the pozzolanity index of the slag, varying with age, replacement level, and type of strength measured. At 90 days' compressive and flexural strength testing with a substitution rate of 20%, the strength was higher than the reference without replacement. Otherwise, the strengths reported were lower.

Douglas et al.[168] reported studies at CANMET with various types of nonferrous slags. They reported that the production of nickel, copper, and lead in Canada generates over 4 million tonnes of slag. They concluded that most of the slags tested had pozzolanic activity indices higher than 75% of the control OPC mortar, and that they might be regarded as interesting potentials in the future. Douglas and Malhotra[169] reported that Canada produces some 4.1 million tonnes of nonferrous slags per annum. Some of this is used for railroad ballast and engineering fill. The current accumulation at that time amounted to about 17 million tonnes. Based on literature studies and tests, the report concludes that nonferrous slags appear to have considerable potential for partial replacement of Portland cement in cemented mine backfill and in concrete. However, the more interesting results seem to be achieved when the slags are milled to a rather high fineness. Looking at nonferrous slags as a potential to reduce CO_2 emission, the energy involved in the preparation must be taken into consideration.

Yeginbali[99] reported interesting experiences in the use of ash from oil shale in Jordan and Turkey. Oil shale is one of the most important energy sources in the world, ranked second to coal, and with larger reserves than oil and natural gas. Blended cements with such ash were found to comply with relevant Turkish standards.

With the purpose to improve concrete durability Warid Hussin and Abdul Awal, at the universities in Malaysia and Bangladesh,[170] tested 30% substitution of cement with palm oil fuel ash (POFA). POFA is a waste obtained from ash when burning oil husk and palm kernel shell as boiler fuel to produce steam for electricity generation and the palm oil extraction process. They mentioned that around 200 such oil mills are under operation in Malaysia alone, where thousands of tonnes of such ash are produced.

3.5.8.5 Ashes from co-combustion

Co-combustion of waste to replace coal in coal-fired power plants has been tried out in a number of countries, with the benefits to the operator to achieve:

- A cheaper, high-calorific material replacing part of the fuel
- Reduction in the CO_2 emission from the combustion of coal

Of the different wastes that have been tried out, we mention biomass pellets, bone meal, cacao shells, citrus pellets, coffee grounds, demolition wood, hydrocarbon gas, paper sludge, pet coke, poultry dung, sewage sludge, straw, and soil contaminated by tar. The fly ash product coming out of these processes might have slightly different effects on the concrete when used. Typically there might be differences in several of the chemical substances, where changes in the SiO_2 and CaO contents seem to be of most importance. Tests in different countries seem to confirm that the use of fly ash in concrete from co-combustion processes is manageable, but will increase the importance of quality control.

A paper by Van den Berg et al. in 2001[171] gave a state-of-the-art report on the situation in Europe regarding co-combustion of coal and other materials. They divided the potential materials into seven main groups:

- Vegetable biomass (wood chips, straw, olive shell, and other vegetable fibres)
- Green wood and cultivated vegetable biomass
- Municipal sewage sludge
- Bone meal
- Paper sludge
- Petroleum coke
- Virtually ash-free liquid fuels and gaseous fuels

The report mentions more than 20 co-combustion examples in the years 1995–2001, with the percentage of coal substitution ranging from 0.9 to 20% (70% of the cases from 5 to 10%). They claimed that concrete tests with these kinds of ashes do not show significant different concrete properties from the control test with the use of coal ash.

We also mention that the EU produces in the order of 200 million tonnes of municipal solid waste (MSW) per year.[172] By incineration this might be transformed into municipal solid waste incineration fly ash (MSWIFA).

Of the ash types tested, be refer to a paper by Zhou at a conference in Jinan, China, in 2010[173] on tests with ox dung ash. In many ways, the paper represents true environmental holistic thinking that has received more and more acceptance in the concrete industry over the last decade or so. We quote a few sentences:

> In the north-west cold area of China, climate is dry with low rainfall, but the grass resources are enough to nurse cows so that stockbreeding develops well. In some areas, people use oxes dried dung to make fires to eat and warm themselves. Because the surface area of burned ox dung ash is very large and average volume particle radius is only about 45 microns, when the wind blows on the grasslands, these particles fly in the air easily, polluting the environment. We have made use of the burned ox dung ash to develop an autoclaved aerated concrete block in order to prevent the environment from pollution and to economize the resource.

Zhou further reports about where the concrete is tested, with a cement replacement with ox dung ash of about 10%. The test results show better properties of the final material with ox dung ash than without.

3.5.8.6 Wood ash

Naik[174] claimed that 70% of the wood ash generated in the United States is landfilled. An additional 20% is applied on land as a soil supplement. The remaining 10% has been used for miscellaneous applications. He tested wood ash in 15, 25, and 35% replacement of cement, and found the pozzolanic contributions of wood ashes to be significant, and that blends to achieve a compressive strength up to 50 MPa at the age of 28 days can be made. However, he reports that typical wood ashes in Canada and the United States have higher than normal desirable loss on ignition.

3.5.8.7 Fluidised bed ash

In a paper from 1995, Brandstetr and Drottner gave an introduction to the use of ash from fluidised bed coal combustion.[175] They explained that the ash from this process with combustion at 850°C is quite different from "normal" fly ash with combustion at a temperature above 1250°C. They claimed that there are only a few of this type of plant in the world (only four), and consequently, there is a lack of standards from this type of combustion product. However, they claimed that the ash has a chemical composition close to that of natural pozzolans. They proposed using the ash either as backfill in mines, in combination with lime in ternary blends, or as a 10% replacement in Portland cements.

Quantification of the potential emission reduction from this tool (other ashes and slags) is difficult, as the available data are limited. However, based on collected data and relating these to fly ash and cement consumption data in the countries where data could be found, Jahren in 2003 made an estimation[56] of the potential. He concluded that the potential in 2020 would be in the order of 10% of the potential emission potential from fly ash. In this figure he did not include the effect from materials used as alternative fuel directly in cement production.

3.5.9 Limestone powder

Limestone filler or powder has been used as cement replacement for many years, both in blended cements and directly in concrete. Apparently, the first modern pioneering work regarding use in regular blended cement was done in France in the 1970s.[176]

In addition to cement replacement properties, limestone powder can contribute to an increase in the machineability of machine-produced concrete products as concrete blocks, lightweight aggregate blocks, paving stones, etc. It can also contribute somewhat to an increase in early strength, particularly when the filler has finer particles. Limestone powder has also successfully been used as a replacement in masonry cements, and as an effective filler in self-compacted concrete. As an example of the above, Lieberum[177] reported on German limestone cement, complying with DIN EN 197-1 and CEM II/A-LL 42.5 R Portland limestone cement, which gives better properties regarding:

- Higher flow (bordering on self-compacting concrete)
- High early strength
- Low compaction energy

Most technical reports, e.g., those by Moir and Kelharn[176] and Irassar et al.,[178] refer to cement replacement properties in dosages in the range of 10 to 15% (a few times higher), with similar or better behaviour than ordinary Portland cement concrete. This coincides with our own practice from use in Norway for 30 to 40 years.

During the mid-1980s, proposals were made to incorporate Portland limestone cements in the new European cement standards. This led to investigation also in the UK. By 1997, the following European countries had national standards with reference to preliminary European Standard

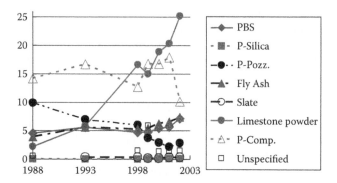

Figure 3.24 Percentage development in European blended cements. The development in limestone cements is remarkable.

ENV 197-1, allowing 6 to 20% (* = also 21 to 35%) limestone filler in the cement:[176] Austria, Belgium,* Denmark, France,* Germany, Greece,* Italy,* Luxemburg, Netherlands, Portugal, Sweden, Switzerland,* and the UK (Figure 3.24).

As in other examples, Argentina[179] implemented its Portland limestone cement standard in 1991, allowing up to 20% of limestone powder, and the Australian standard allows up to 7.5% of limestone (Australian Standard AS 3972-2010).[180]

Even if limestone fillers have been utilised as a close to standard ingredient in the mixture design in many parts of world for 30 to 40 years, the U.S. cement standard practice has been rather conservative in this respect. It was only in 2004 that the ASTM International C150 standard specification for Portland cement was modified to allow the incorporation of 5% mass fraction of limestone in ordinary Portland cements. Bentz et al.[181] presented in 2009 an interesting paper where they analysed limestone cement based on Power's model. Their findings were interesting, as they concluded that larger substitutions than 20% were acceptable, in particular for lower w/cm ratios. However, they concluded that there was a decrease in compressive strength of 7% for a replacement level of 10%, and on the order of 12% for a 20% replacement level. They added that these strength losses could be compensated by slight reductions in w/cm.

Kong et al.[182] reported in 2010 about 40% replacement with limestone powder in the building of a roller-compacted dam at Jinghong Hydropower Station. In addition, substantial quantities of slag and fly ash were used. They claim that large amounts of limestone powder will accelerate cement hydration and participate in the hydration, producing calcium carboaluminate crystals, whether single used or mixed with other mineral materials.

Several researchers have stressed the importance of the amount and type of grinding of the limestone powder on the various properties in concrete. In a report in 2010, Sakai et al.[183] described tests with low-heat Portland cement and limestone powder with six different degrees of grinding. Among their findings was that the packing fraction of limestone powder increased up to a maximum at replacement values of 10 to 30%, and that the highest packing fraction was obtained with the smallest particle size.

At the same conference in Jinan, China, Wang et al.[184] reported on the effects of various dosages of limestone powder. Among their findings was that incorporation of limestone powder significantly increased the cement hydration speed, and that limestone particles filled in the pores of the cement paste to form a dense microstructure of hardened cement paste. Partial replacement with limestone powder gave positive effects, in particular on the early strength, up to 20% replacement. Then the compressive strength decreased with increasing limestone powder content.

Dong et al. from Wuhan, China,[185] tested replacement of limestone of various fineness and combinations with fly ash. They claimed that adding limestone powder increases early-age strength, and limestone powder in combination with fly ash can improve fluidity and the strength at later ages. The speed-up of strength at an early age is more distinct, with a larger specific surface area of the limestone powder. This will also improve strength characteristics at later ages in combinations with fly ash.

Limestone fillers are available from a wide range of sources, i.e., from cement production, from limestone quarries, and as secondary products from various mineral operations. They are simple to use, normally easily available, and will in most cases give little negative environmental effects regarding transport, energy consumption, etc. The question of environmental implications regarding resource utilisation will vary considerably from one part of the world to the other.

As a global average, based on the above factors, we evaluate this tool to have a further potential for emission reduction of 5 to 8%, somewhat varying with the present practice. With a global cement consumption of 3 billion tonnes of cement, this represents some 120 to 200 tonnes of potential CO_2 emission reduction.

Also from personal experience from several interesting projects, we strongly feel that this tool represents an interesting potential for the future that is not fully developed. However, we cannot expect rapid important global volume effects regarding emission reduction, as it will take time to establish technological acceptance comparable to that of other alternatives. Based on the indications we have seen so far, we are convinced that the tool has the potential sometime in the future (some decades from now) to represent something in the order of up to 10% of clinker replacement.

3.5.10 Other supplementary cementitious materials

In some areas of the world, additional alternatives might be rather interesting for the future, and include a wide range of possibilities.

Toledo Filho et al.[186] investigated the potential use of crushed brick. In this respect they refer to what they call "the first man-made cement," 3600 years ago, where pottery fragments were mixed with lime. Tests with substitution of 20 to 45% of grinded crushed bricks were done with acceptable results. They report that the availability of such a substitute was 18 million tonnes in 2000 in Brazil, compared with a cement production of 38 million tonnes in 1997, expected to grow to 72 million tonnes in 2010. The authors claim that a 40% replacement of cement by crushed brick is possible.

We have from various parts in the world recorded tests with crushed glass as a possible substitute. Shao et al.[187] tested glass with varying fineness with interesting results. According to ASTM C593, glass with a fineness of 0.038 mm can be regarded as a pozzolan.

Quite a few researchers have been working with this alternative, in particular over the last decade. We mention a paper from researchers from the University of Sherbrook, Quebec, Canada,[188] who reported from testing a clean glass ground to a fineness of 350 and 400 kg/m³. In their conclusion they say:

> Incorporating glass powder as a partial replacement has shown interesting pozzolanic potential as an alternative cementitous material. The results reported in the study show that:
>
> - Workability of concrete is not affected by the presence of glass powder despite its high fineness.
> - The presence of glass powder does not influence the stability of the entrained air.
> - The duration of setting of binary concretes is very close to that of the control concrete.
> - The incorporation of glass powder improves the long-term mechanical properties of binary and ternary mixtures.
> - The pozzolanicity of glass powder is more influenced by the water dosage.
> - Glass powder alone or combined with other cementitious materials decreases concrete permeability. The permeability of binary concrete is affected by the water dosage.
> - Binary and ternary concretes with both w/cm = 0.55 and 0.40 ensure good freeze-thaw resistance.
> - The addition of glass powder as a cement substitution does not increase the expansion due to alkali-silica reaction.

Norway has had a very workable return system for glass for many years. About 20,000 tonnes of glass are available for the building industry per annum. Other uses absorb the rest.[189] Based on a per capita evaluation, and considering that most countries would have a less efficient recirculation system than the Norwegians, the potential global CO_2 emission from crushed glass as a substitute to cement will be less than 10 million tonnes, not taking into account the energy consumption used for processing. However, this does not mean this crushed glass cannot be an interesting option in some parts of the world.

Harald Justnes, from SINTEF in Norway,[190] has in several papers enthusiastically reported on the possibility of using marl as a cement substitute. Marl is clay contaminated with small particles of calcium carbonate. However, marl has to be calcined to be an effective pozzolan. In a paper in 2010, Justnes refers to tests showing that 800°C might be an optimum calcining temperature. Even if this also means CO_2 emission, this is a great step forward compared to Portland clinker. In the same paper, Justnes shows possible substitution ratios up to 50%, where 35% seems to be an approximate optimum value.[191]

> Marl = An argillaceous, nonindurated calcium carbonate deposit that is commonly grey or blue-grey. It is somewhat friable, and in some respects resembles chalk, with which it is interbedded in some localities. It is formed in some fresh-water lakes, partially by the action of some aquatic plants. The clay content of marls varies, and all gradations between small amounts of clay (marly limestone) and large amounts (marly clay) are found. Marlstone, or marlite, is an indurayed rock of the same composition as marl. The marlstones are not fissile but blocky and massive with sub-conchoidal fracture.[191]

Jahren and Lindbak, in 1998, at a conference in Bangkok, Thailand, reported on a pozzolan developed as a by-product during development of a process for extracting a flocculant for wastewater treatment from minerals.[192] The flocculant product, with a pozzolan as a by-product, was developed in the laboratory in 1990–1992. From the chemical process came a by-product consisting of a pure unstable amorphous SiO_2 with microspores in the 0.001 to 0.02 mm range. Concrete tests with the product showed results comparable to those of silica fume. The name Sican was proposed for the product. The raw material for the chemical process is anorthosite rock, which has to be processed with a strong acid at a temperature below 100°C to get the ion exchange that forms the main product.

At the same conference, Chandra from Chalmers University of Technology, Sweden, reported on colloidal silica, made from an aqueous silicic acid suspension.[193] The products were compared with silica fume in mortar tests.

3.5.10.1 Ternary and quaternary blends

We cannot leave the possible tools under the label "supplementary cementing materials" without mentioning the interesting and sometimes very effective use of ternary and quaternary blends with alternative cementitious materials.

Many researchers have for the last few decades been actively trying to utilise this opportunity to optimise the best properties from the various materials and find combinations that might get synergies that lead to fascinating improvements. Professor Arezeki Tagnit-Hamou at the University of Sherbrook, Canada, received an award for his work in this field at the Fall 2011 ACI Convention in Cincinnati, Ohio. In a paper coauthored with Laldji,[194] he reported results with the use of ternary and quaternary mix designs using glass frit. Despite the lower early strength of the mixes, the strength developments after 91 days of hydration varied from 1.05 to 1.25 times those of the control. Permeability was also reduced and varied from 35 to 17.7% of the control. In the reported tests, various mixes with water/cementitious materials of 0.45 were investigated. The other cementitious materials that were combined with the glass frit in the various mixes were fly ash, blast furnace slag, and silica fume.

Thomas et al.[195] claim that the first ternary blended cement in North America was introduced in 1999. They report a test with blended cement in Canada utilising blast furnace slag and silica fume. They claim that an optimum blend will be with silica fume in the order 4 to 5%, and slag contents in the range of 20 to 25%, in order to produce cement suitable for the widest variety in application. In certain cases it may be necessary to add additional supplementary cementing materials at the concrete mixer, such as fly ash or additional slag, when low heat of hydration is required.

Another well-known concrete technologist who for years has argued for increased utilisation of ternary blends is Per Fidjestøl, Norway. Optimisation of the best properties of different supplementary cementitious materials (SCMs) gives fascinating synergy effects. In a paper in 2008,[121] he shows examples of combinations of 75% Portland cement, 25% fly ash, and 5 to 8% silica fume, with considerable improvements in material properties, like better durability, lower permeability, and improved strength development.

Theodor A. Burge from Switzerland, in a paper in 1995, reported a very special use of the advantages of ternary blends.[196] He was looking for a quick hardening concrete repair material for use in roads, bridges, and runways, where it is important that traffic will be stopped for as little time as possible. He claimed to have found an acceptable solution in blending Portland cement, oil shale ash, and silica fume, also adding high dosages of superplasticisers and an effective accelerator. With this special combination he claims to be able to get strength in 2 hours that is comparable to what is achieved with oil shale ash alone after 7 days, and 24-hour strength comparable to 28-day strength.

At a conference in Hyderabad, India, in 2005, Tiwari and Jha[197] reported tests with ternary blends with fly ash and silica fume, with the purpose to speed up the strength development of the fly ash concrete. They report concrete with a compressive strength of 75 MPa after 28 days with a cement replacement of 50%.

3.5.11 Improvements and more efficient cement production

In a global evaluation, about half the emission of CO_2 from cement production comes from the thermal and mechanical energy consumption in production.

The potential for improvements in reduced CO_2 emission and reduced energy consumption in cement production is considerable, but differs a great deal from one part of the world to the other, depending on the present state of the cement production technology (Figure 3.25).

Since China produces about half of the cement in the world, the possible potential improvements in China are of special importance. Cheng from Jinan University, in a paper in 2008, presented some data about the present status (Figure 3.26).[94]

A conversion of the more old-fashioned kilns to modern large kilns might lead to a reduction in energy consumption of about one-third. Cheng[94] reports a considerable change in the production structure, where only 12% of the annual cement output came from modern dry process kilns in 2000, and that the share in 2005 had grown to 47%. This resulted in a 3.1% annual decline rate in energy consumption per tonne of cement produced. However, there is still a long way to go.

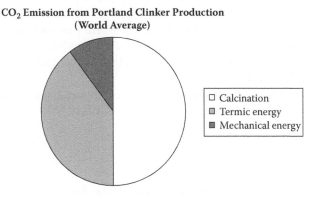

CO₂ Emission from Portland Clinker Production (World Average)

☐ Calcination
▨ Termic energy
■ Mechanical energy

Figure 3.25 CO₂ emission from Portland clinker production—simplified illustration of world average.

Percentage

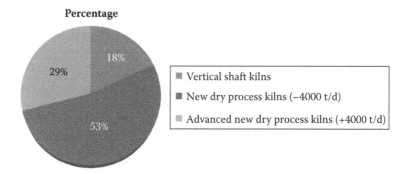

Figure 3.26 Variation in Chinese cement production types in 2008, according to Cheng.[94] (From Cheng X., Manufacture Building Materials by Using Industrial Waste Residue. Presented at 2008 International Expert Workshop on Cement and Concrete Technology for Sustainable Development, Beijing, October 7–12, 2008.)

Thanks to the China national policy on cement industry restructuring, there is continuous improvement in energy savings and emission mitigation, as presented in 2011 by Sui[198] on China's cement status. New dry process new suspension pre-heating and precalcining cement (NSP) cement output reached 1.3 billion tonnes, with the proportion of 72.25% in 2009. The comprehensive energy consumption (thermal and electrical) per tonne of cement in 2007 was 115 kg standard coal (ce), 4% lower than 2006 and 9% lower than 2005. A recent survey organised by the China Cement Association on 38 NSP cement lines found that the thermal consumption per tonne of clinker ranges from 2885 to 3980 MJ, depending on the kiln capacity. It is estimated by the Chinese Cement Association that in 2007, by using NSP to replace the backward vertical shaft kiln process, 3.3 million tonnes of SO_2, 37 million tonnes of CO_2, and 4.4 million tonnes of dust were reduced from the baseline of 2001.

In 2011, Kulkarni[4] claimed that for the world's second largest cement-producing country, India, by 2006, 96% of the production had shifted to dry process, and that the average consumption of thermal energy per kg of clinker was 725 kcal.

In a meeting in 2006, Galitsky gave the following energy consumptions for various types of kilns, in GJ per tonne of clinker:[199]

- Wet kiln: 5.9–7.0
- Lepol kiln: 3.6
- Long dry kiln: 4.2
- Short dry kiln—suspension preheater: 3.3–3.4
- Short dry kiln—preheater and precalciner: 2.9–3.2
- Shaft kiln: 3.7–6.6

The IPCC in 2000 and 2002 gave similar figures:[200,201]

- Wet process long kilns: 5.0–6.0
- Dry process long kilns: Up to 5.0
- Semidry/semiwet processes (Lepol kilns): 3.3–4.5
- Dry process kilns equipped with cyclone preheaters: 3.1–4.2
- Precalcined kilns and dry process rotary kilns equipped with multi-stage cyclone preheaters: About 3.0

Damtoft[62] claims that a theoretical minimum in energy consumptions in cement production is less than 3000 kJ per tonne of clinker. The energy consumptions in the countries that are the largest producers in the world are far from this limit. Referring to Table 3.13,[57] one can estimate that if the production in the world could be modernised down to 3 to 3.5 MJ/kg clinker, the reduction in CO_2 emission would be in the order of 0.10 kg CO_2 per kg of clinker, a very considerable reduction.

If, in addition, one-third of the thermal energy consumption could be converted to climate-neutral fuel, which is possible, then a more efficient cement production can represent a reduction of CO_2 emission of about 20%.

Table 3.13 Energy consumption in clinker production[57]

Cement industries' energy intensities by region and subregion					
Region energy intensities			Subregion energy intensities		
	MJ per kg clinker			MJ per kg clinker	
Region name	1990	2000	Subregion name	1990	2000
I. North America	5.47	5.45	1. United States	5.50	5.50
			2. Canada	5.20	4.95
II. Western Europe	4.14	4.04	3. Western Europe	4.14	4.04
III. Asia	4.75	4.50	4. Japan	3.10	3.10
			5. Australia and New Zealand	4.28	4.08
			6. China	5.20	4.71
			7. Southeast Asia	5.14	4.65
			8. Republic of Korea	4.47	4.05
			9. India	5.20	4.71
IV. Eastern Europe	5.58	5.42	10. Former Soviet Union	5.52	5.52
			11. Other Eastern Europe	5.74	5.20
V. South and Latin America	4.95	4.48	12. South and Latin America	4.95	4.48
VI. Middle East and Africa	5.08	4.83	13. Africa	5.00	4.75
			14. Middle East	5.17	4.92

Table 3.14 Energy consumption by the world cement industry in 2006

	Thermal energy GJ per tonne clinker	Electrical energy kWh per tonne cement
10% percentile (10% best in the class)	3.1	89
Global weighted average	3.7	111
90% percentile	4.4	130

The *Cement Technology Roadmap 2009* made by WBCSD and International Energy Agency (IEA)[151] claims that the figures shown in Table 3.14 were valid for the energy consumption by the world cement industry in 2006.

Further, with reference to the European Cement Research Academy (ECRA), they project that the average thermal energy for clinker manufacture will be reduced to 3.3 to 3.4 GJ/t clinker in 2030 and to 3.2 to 3.3 GJ/t clinker in 2050 (Figure 3.27).

As indicated above, there has been a considerable reduction in energy consumption in the cement industry over the last decades, but some places in the world still have very large improvement potentials, and of volume importance, in particular, are CIS and North America (CIS = Azerbaijan, Armenia, Belarus, Georgia, Kazakhstan, Kyrgyzstan, Moldova, Russia, Tajikistan, Turkmenistan, Uzbekistan, and Ukraine).

With reference to Locher (2006), the WBCSD claim that the theoretical minimum energy consumption (heat) for the chemical and mineralogical reactions is approximately 1.6 to 1.85 GJ/t. It is, however, important to

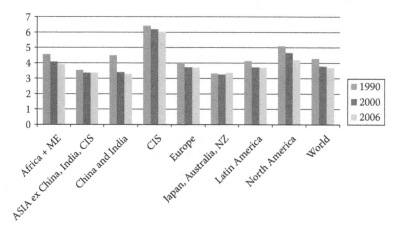

Figure 3.27 Development in energy consumption per MJ/tonne clinker.[151] (From WBCSD/IEA, *Cement Technology Roadmap 2009*, Report 2010, World Business Council for Sustainable Development and International Energy Agency, www.wbcssg.org and www.iea.org.)

remind us of the many practical obstacles in reaching such a value, for example, unavoidable conductive heat loss.

Canadian researcher Pierre-Claude Aitcin makes an understandable and simple calculation about the challenge of emissions from Portland cement clinker production in his fascinating book *Binders for Durable and Sustainable Concrete*.[116]

Portland cement clinker has an average lime content of between 60% and 70%. Aitcin uses 65% to simplify the calculation.

The decarbonation or calcinations process is $Ca\ CO_3 \rightarrow CaO + CO_2$.

Using the molecular weight, this is $100 \rightarrow 56 + 44$.

In order to make 1 tonne of clinker having 65% CaO, it is necessary to start with $65/56 = 1.16$ tonnes, which emits $65/56 \times 44/100 = 0.51$ tonnes of CO_2.*

In a modern plant, depending on the energy efficiency of the process, it is necessary to burn between 100 and 200 kg of "coal equivalent" to produce 1 tonne of clinker:

$$C + O_2 \rightarrow CO_2$$

giving

$$12 + 32 \rightarrow 44$$

Therefore, the amount of CO_2 generated during the process varies between:

$$44/12 \times 0.1 = 0.37 \text{ tonnes of } CO_2 \text{ and } 44/12 \times 0.2 = 0.74 \text{ tonnes of } CO_2$$

In addition to these figures comes the emission from the mechanical energy consumed.

Aitcin's simple calculation model helps us throw light on the challenges and possibilities we have for improvements.

The Japanese cement industry has for many years been a leading lady in energy efficiency and utilisation of waste materials in cement production. Noguchi reported the figures shown in Table 3.15 for utilisation of waste and by-products by the Japanese cement industry in 2000.[202]

3.5.11.1 Alternative fuels

One possibility, for places in the world that have the fortunate possibility, is to use natural gas as a replacement for coal. Referring to WBCSD, Tokheim

---

* In the IPCC *Good Practice Guidance and Uncertainty Management in National Greenhouse Gas Inventories*[201] it is recommended that we to use an emission factor of 0.51 tonne of CO_2 per tonne of clinker; if not, sufficient data on the CaO content of the clinker are available for use with other factors.

Table 3.15 Waste and by-products utilised by the Japanese cement industry, 2000[202]

Type	Uses	Amount (t)
Blast furnace slag	Raw material, additive	12,162,000
Fly ash	Raw material, additive	5,145,000
By-product gypsum	Additive	2,643,000
Sludge	Raw material	1,906,000
Nonferrous slag	Raw material	1,500,000
Steelmaking slag	Raw material	795,000
Ash dust	Raw material, fuel	734,000
Coal waste	Raw material, fuel	675,000
Casting sand	Raw material	477,000
Scrap tire	Fuel	323,000
Reclaimed oil	Fuel	239,000
Oil waste	Fuel	120,000
White clay waste	Raw material, fuel	106,000
Plastic waste	Fuel	102,000
Others		433,000
Total		27,359,000

and Brevik[203] claim that typically, coal has an energy-specific CO_2 emission of 96 kg/GJ, whereas the corresponding value for natural gas, which contains a considerable part of the chemical energy as hydrogen, is around 56 kg/GJ.

However, on global bases, a more realistic and very important aspect is the use of energy from waste fuels that otherwise would have been land-filled or incinerated under worse efficiency conditions. According to the Kyoto agreement, emissions from burning of biological waste/fuels will not be regarded as harmful.

When fossil waste, otherwise incinerated under other conditions, is used, the cement industry must take over the quota from the original source to be credited. In this respect, it might be worthy to remember in the future that even if this emission is not quoted as part of the cement production emission reduction, it is environmentally a gain, as we are reducing this emission from other sources.

There are also other important environmental advantages from this process. The cement production demands high temperatures over a long period. This makes the cement kiln ideal for the destruction of polychlorinated biphenyls (PCBs). Norwegian control shows a degree of destruction of more than 99.9999%.[189]

The UN Intergovernmental Panel on Climate Change (IPCC) guidelines say the following:[204]

- CO_2 from biomass fuels is considered climate-neutral, because emissions can be compensated by regrowth of biomass in the short term. CO_2 from biomass fuels is reported as a "memo item," but excluded from the national emission totals. The fact that biomass is only really climate-neutral if sustainably harvested is taken into account in the "land use change and forestry" section of the national inventories, where CO_2 emissions due to forest depletion are reported.
- CO_2 from fossil fuel-derived wastes (fossil AFR), in contrast, is not a priori climate-neutral fuel. According to IPCC guidelines, GHG emissions from industrial waste-to-energy conversion are reported in the "energy" source category of national inventories, while GHG emissions from conventional waste disposal (landfilling, incineration) are reported in the "waste management" category.

The alternative fuels most commonly utilised are:

- Waste tires
- Biomass
- Used solvents
- Sewage sludge
- Municipal solid waste
- Petroleum coke
- Other waste

There might be some disadvantages in using some of the mentioned fuel alternatives, varying from country to country in a longer perspective. As attention to environmental and climate change issues grows, alternatives like used tires might find other and "competing" sources for use/destruction.

As awareness of the advantages increases, it might also be possible that governments will give emission credits to destruction of the least hazardous waste fossil fuels in the future. Estimates of the advantages of using alternative fuels in cement production have been made, with emission reduction potentials ranging from 6 to 20%, and with a global average of 12% in 2020.[60]

As examples of increased use of climate-neutral fuel/waste, it is worth mentioning the following:

1. The Norwegian cement producer Norcem reports that in 2007 it was using more than 50% of its thermal energy at its Brevik plant as

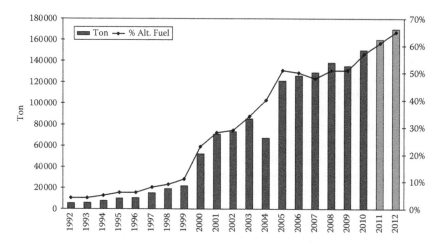

Figure 3.28 Use of alternative fuel, Norcem Dalen plant, Norway.[205] (From Brevik P., The CO$_2$ Challenge in Cement Production: Can CCS Be Part of the Solution? Presented at 2nd International Workshop on CO2: CCS and CCU in Germany, Norway, The Netherlands, Poland and Scotland—Challenges and Chances? Dusseldorf, November 10, 2011.)

an alternative fuel. The use of alternative fuel started in 1992, and has had steady growth since. In the period 1995–2004 the company invested 350 million NOK in new and more efficient equipment to achieve this.[205] In 2011, Norcem's environmental manager, Per Brevik, presented the results of the development, as seen in Figure 3.28 (see also Figure 3.29).[205]

2. The Australian cement producer Adelaide Brighton reports that in 2005 it started using shredded wood waste as an energy source in its kilns. In 2007, this resulted in a 25% reduction in its natural gas usage, which was its normal energy source.[206]

The WBCSD/IEA *Roadmap 2009*[151] mentioned earlier claims with reference to ECRA that the average use of alternative fuel in 2006 was 16% in "developed countries" and 5% in "developing countries." They predict that these figures will grow to 40 to 60% and 10 to 20%, respectively in 2030, and with unchanged percentages for "developed countries" and to 25 to 35% in "developing countries" in 2050. Perhaps the most important obstacle for a more aggressive goal and use from the cement industry is the limited future availability in some areas.

The Cement Sustainability Initiative (CSI) group, representing 20 to 25% of the cement production in the world in recent years, reports on an increase in biomass fuels, as shown in Figures 3.30 and 3.31.

Figure 3.29 Cement factory at Dalen, Norway. The first production started in 1916.[205] (From Brevik P., The CO_2 Challenge in Cement Production: Can CCS Be Part of the Solution? Presented at 2nd International Workshop on CO2: CCS and CCU in Germany, Norway, The Netherlands, Poland and Scotland—Challenges and Chances? Dusseldorf, November 10, 2011.)

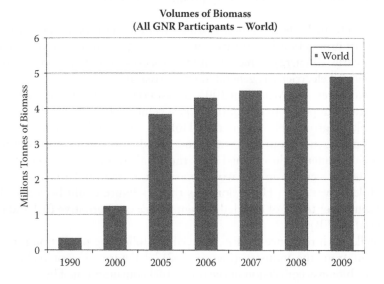

Figure 3.30 Development in use of biomass fuel among CSI members.[151] (From WBCSD/IEA, *Cement Technology Roadmap 2009*, Report 2010, World Business Council for Sustainable Development and International Energy Agency, www.wbcssg.org and www.iea.org.)

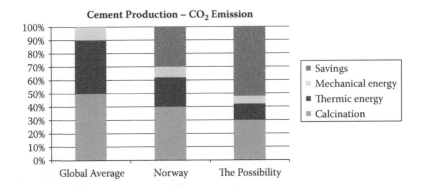

Figure 3.31 Cement production, possible improvement.

There are considerable differences between the emissions from the mechanical energy used in the cement production process, varying from over 10% of the total emission to 0. There are two major reasons for this difference:

- The efficiency of the process machinery
- The source for the electricity production utilised

As seen from Table 3.16, there is wide variation in the national figures for electricity production in various countries, ranging from countries where all or nearly all energy comes from hydroelectric power stations to countries were all electricity is generated from fossil fuel.

At a conference in Milano in 2003, Massimo Gelli gave a report from the environmental policy in one of the largest cement groups in the world[207]— the Holcim group. He stated that Holcim had building materials and cement plants in 75 countries with a total production volume of 75 million tonnes of cement. He claimed that the total average emission of CO_2 per tonne of cement in the world was 0.81 tonne, and that the figure for Holcim was 0.71. His vision was that its figure could be reduced to 0.40 tonne per tonne of blended cement by use of alternative fuel and raw materials (AFR).

Brevik, with reference to IEA, showed the following prediction for CO_2 emission from the industry (Figure 3.32).[205] Figures do not include the effect of biomass conversion in the use of fuel consumption. The prediction indicates a formidable improvement potential, but far from enough, looking at what would be a reasonable goal for binders for concrete production. To avoid that the rest of what is needed is taken in blending of supplementary materials at the site, hopefully the cement industry will do better than this forecast.

Table 3.16 CO_2 emission from electricity production in different countries[73] (1 kWh = 3.6×10^{-3} GJ)

Country	1990 Emission factor kg CO_2/MWh	1996 Emission factor kg CO_2/MWh	Country	1990 Emission factor kg CO_2/MWh	1996 Emission factor kg CO_2/MWh
Albania	228	19	Korea	317	297
Algeria	487	620	Kuwait	591	512
Argentina	320	301	Kyrgyzstan		106
Armenia		247	Latvia		172
Australia	777	791	Lebanon	1833	652
Austria	192	155	Libya	471	626
Azerbaijan		150	Lithuania		142
Bahrain	1014	767	Luxembourg		425
Bangladesh	604	540	Malaysia	664	594
Belarus		301	Mexico	523	502
Belgium	289	281	Moldova		535
Bolivia	286	269	Morocco	674	632
Bosnia-Herzegovina		943	Nepal		17
Brazil	26	32	Netherlands	516	435
Burma	1015	711	New Zealand	103	99
Bulgaria		419	Norway	1	2
Canada	189	163	Pakistan	410	438
Chile	274	318	Paraguay		133
China	710	772	Peru	6	14
Colombia	178	117	Poland	464	605
Croatia		217	Portugal	494	384
Cuba	629	654	Romania	473	304
Czech Republic	539	420	Russia		282
Denmark	454	446	Singapore	890	622
Ecuador	196	307	Slovak Republic	306	297
Egypt	546	561	South Africa	796	770
Estonia		747	Spain	408	322
Finland	202	249	Sri Lanka	3	205
France	57	40	Sweden	40	62
FYROM	609	825	Switzerland	8	3
Georgia		49	Syria	546	650
Germany	460	419	Tajikistan		68
Greece	971	812	Thailand	619	618

continued

Table 3.16 *(Continued)* CO_2 emission from electricity production in different countries[73] (1 kWh = 3.6×10^{-3} GJ)

Country	1990 Emission factor kg CO_2/MWh	1996 Emission factor kg CO_2/MWh	Country	1990 Emission factor kg CO_2/MWh	1996 Emission factor kg CO_2/MWh
Hungary	379	362	Tunisia	578	522
Iceland	2	1	Turkey	492	461
India	761	890	Turkmenistan		731
Iran	541	534	United Kingdom	632	477
Iraq	459	554	Ukraine		376
Ireland	724	716	United Arab Emirates	616	783
Israel	814	801	Uruguay	40	100
Italy	488	420	United States	546	503
Japan	720	791	Uzbekistan		432
Jordan	720	791	Venezuela	237	176
Kazakhstan		131			
World	489	466	Non-OECD Europe	496	420
Africa	660	663	Former USSR	417	328
Middle East	632	650	Latin America	184	164
			Asia (excluding China)	658	724

Source: Vanderborght B., Brodmann U., *The Cement CO_2 Protocol: CO_2 Emission Monitoring and Reporting for the Cement Industry, Guide to the Protocol,* Version 1.6, WBCSD Working Group Cement, October 19, 2001, http:/www.ghgprotocol.org.

3.5.12 New/other types of cement/binders

Modern research work with alternative binders has been going on for more than 100 years, and with increased interest over the last two to three decades. In a report from 2003, Jahren[56] refers to reports from more than 20 different countries that tell about projects on alternative binders.

The majority of these reports, however, are utilising supplementary cementing materials already mentioned as possible emission reduction tools.

As examples of the wide and versatile international interest in sustainability in general, and in this field in particular, we mention some examples of reports from various countries, in alphabetical order of countries.

Australia:
- In a paper from 2010, researchers from Australia and Canada reported on microstructure analysis of a binder based on fly ash

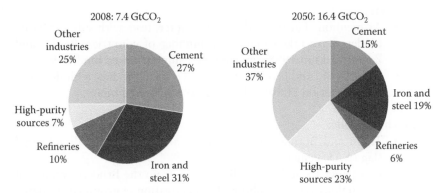

2008: 7.4 GtCO$_2$

Other industries 25%

Cement 27%

High-purity sources 7%

Refineries 10%

Iron and steel 31%

2050: 16.4 GtCO$_2$

Cement 15%

Other industries 37%

Iron and steel 19%

Refineries 6%

High-purity sources 23%

Note: Biomass conversion is not included in this figure.
Source: IEA analysis.

Figure 3.32 Forecast CO$_2$ emission from the cement industry.[205] (From Brevik P., The CO$_2$ Challenge in Cement Production: Can CCS Be Part of the Solution? Presented at 2nd International Workshop on CO2: CCS and CCU in Germany, Norway, The Netherlands, Poland and Scotland—Challenges and Chances? Dusseldorf, November 10, 2011.)

activated with sodium silicate and a mixture of sodium silicate and sodium hydroxide.[208]

- At the same conference in Jinan, China, van Deventer et al. reported on a commercialised product of alkali-activated fly ash and slag.[209] The product is called E-Crete, and the authors claimed that the savings in CO$_2$ emission compared to Portland cement concrete is 80%. They also claimed that the first solid alkali-activated binder comparable to Portland cement was first introduced by the Belgian civil engineer A.O. Purdon in 1940.

Belgium:
- At a RILEM conference in Jinan, China, in 2010, researchers from Belgium and the Netherlands reported on a fly ash-based geopolymer activated by different amounts of SiO$_2$ and Na$_2$O.[210] In their conclusion, they claim:
  - Increasing amounts of sodium oxide and silicate increased the setting time of the binder.
  - From different mixes the sample with the SiO$_2$:Na$_2$O ratio of 1:1.5 exhibited the highest compressive strength. Increasing the amount of SiO$_2$ or decreasing the amount of Na$_2$O in solution led to the decrease in compressive strength.
  - Tensile strength declined with more sodium oxide added in the system.

Canada:

- As mentioned at the end of this chapter, tests with sulphur-based binders have taken place in a number of countries. Poulin and Zmigrodzki in 1998 reported on a sulphur-based binder.[211]
- In 1995, Shi and Day[212] reported on tests to produce a binder from a combination of 80% slag and 20% hydrated lime.

Czech Republic:

- Tallin and Brandstetr reported on activity with alkali-activated slag (AAS) cement.[213]
- In a paper from 2010,[214] Bilec et al. from the Brno University of Technology reported about alkali activation of ground granulated blast slag of two types and modification with brown coal fly ash, a very fine fly ash powder (under 10 microns), metakaolin, and intergrinding of Portland cement concrete. In their conclusion, they claimed that the cost of alkali-activated concrete is higher, at best comparable with the cost of Portland cement concrete.

China:

- Tongbo Sui has, in a number of papers, reported on the Chinese development of belite-rich cement (for more about this, see Chapter 6).
- A group of researchers from Wuhan, China, in 1998 reported about a binder based on 48% slag, 32% fly ash, 18% other supplementary materials, and 2% "key" material.[215] They reported good concrete properties in general and compressive strengths up to 70 MPa.
- In an article in *ACI Materials Journal* in 2005,[216] Ding and Li report the advantages of addition of fly ash to a magnesium phosphate cement. They blended two types of magnesia with MgO contents of 71.5 and 89.51% with variable fly ash dosages. The strength of the mortars tested increased with increased fly ash content from 10 to 40%. They claimed that the optimum dosage was between 30 and 50%. For early strength, the 1-hour compressive strength had similar values for mortars with or without fly ash, while the mortars with fly ash had higher compressive strength after 4 hours. The aim for the binder was to find a good high-early-strength cement.
- In a RILEM conference in 2010, Zhu et al. reported on a binder based on 60% metakaolin, 40% slag, and a NaOH and water glass solution.[217] The binder was developed for cementing of oil and gas wells. A special retarder was developed to solve the problem of short thickening of the slurry.

- At the same RILEM conference, Zhang et al. from Xi'an University of Architecture and Technology, China,[218] reported on the synthesis of an alkali-activated binder based on slag and metakaolin as the main constituents.

- Lu et al. from Nanjing University of Technology[219] reported on the use of alkali activation by dolomite in a mixture of fly ash and metakaolin. They claimed that the addition of 10 to 25% dolomite powder was beneficial to the compressive strength.

- Another three papers from the same conference reported on different material properties testing and combinations of copolymers by use of blast furnace slag. Wang et al.[220] reported on shrinkage testing of a slag-based polymer combined with metakalin and activated with water glass. He et al., from Chongqing, reported on carbonation resistance of a binder based on slag activate with water glass,[221] and Shi and He from Hunan[222] reported on a test with various activators for a slag-based binder.

- Again at the same conference, Xiao and Luo[223] reported on a binder based on activation of magnesium slag tested with different activators.

- Finally, Pan and Yang from Nanjing gave an interesting state-of-the-art report on the topic,[224] claiming that research work in China on alkali-activated materials started in the 1980s. So far more than 50 organisations, most of them being universities and research institutes in China, are engaged in ongoing research activity. National conferences on the topic were held in Nanjing in 2004 and Xi'an in 2007, with a total of 75 research papers presented. From their report, we see that the main raw materials activated have been blast furnace slag, fly ash, phosphorous slags, and metakaolin, or combinations of these. In their report they mention several challenges that need further investigation:
  - Better flowability
  - Shorter setting time
  - Drying shrinkability
  - Brittleness
  - Efflorescence
  - Cost of activators
  - Range of activators

  However, they conclude that there are numerous examples of excellent performance regarding stabilising and immobilisation of heavy metals.

- In a parallel conference in Jinan, Xiao and Hu from Changchun reported on steam-cured bricks based on a binder of extracted aluminium fly ash with different types of activators: cement, lime, gypsum etc.[225]

- Liu et al.[226] reported on a binder based on manganese slag. Water-cooled manganese slag is a kind of granulated blast furnace slag with 5 to 18% manganese oxides. The authors claim that China has a net emission of 12 million tonnes of such slag each year (2004), with the exception of 30% that goes for cement, concrete, and recycled metal utilisation. They activated this slag in combination with 10, 20, and 30% substitution with metakaolin, and produced mortars with a strength in the order of 60 to 70 MPa after 28 days. They claim that the concrete is very erosion resistant, even in high concentrations of seawater.
- In Beijing, China, in 2012, Wang et al.[227] reported on a geopolymer based on slag and metakaolin activated by water glass. Of their interesting findings, we mention that the pH value increased from 1 to 3 days, then decreased from 3 to 7 days, while it rose again from 7 to 14 days, until it remained stable. The 28-day compressive strength of the geopolymer reached a maximum pH value of 10 to 11.

Finland:
- Tallin and Brandstetr reported on research with alkali-activated slag cement.[213]
- A special kind of alkali-activated slag was introduced in the 1970s, called F-cement. A report from Finnish-Swedish joint research project was presented at a conference in Trondheim, Norway, in 1989.[228]

France:
Several French researchers have reported about interest in high-belite cements:

- At a conference in Lyon in 2002,[229] Beauvent and Holard reported on using slurry from Carriers du Boulonnais aggregate production as a possible raw material source. From a production of 6 million tonnes of aggregate, 50% is washed, giving 0.35 million tonne of slurry waste.

Germany:
- Researchers from Leibniz University Hannover, Germany, reported on slag and fly ash-based geopolymeric cements incorporating 0, 10, 30, and 50% metakaolin.[230] The blends were activated by potassium hydroxide and potassium silicate solutions. They reported compressive strengths from 5 to 70 MPa. They concluded that the addition of metakaolin leads to densification with a corresponding significant reduction in porosity.

India:

- In 1990, a technology named FaL-G was developed in India. In a paper from Ottawa, Canada, in 1998, Bhanumatidas and Kalidas[231] reported on the technology, which is used for brick production. The binder is based on the fact that the fly ash–lime reaction is more active in tropical climates in combination with gypsum. The technology is aimed to produce bricks as replacement for clay bricks. The brick market in India was reported to be 180 billion units, giving a CO_2 emission of 146 million tonnes. As a consequence, clay bricks have been banned in some territories. The FaL-G organisation aimed to replace half of this market with its product, and thereby reduce emissions by more than 70 million tonnes of CO_2. The paper reported compressive strengths ranging from 8.0 to 47.0 MPa after 28 days, for various mix designs. The same authors presented a paper at a conference in Hyderabad, India, in 2005,[232] where they could report that the technology, developed in 1989, now had been used in full scale for 8 years, where they detailed usage, in addition to the main use in bricks, dams, ground and overhead water tanks, etc. They claim that the typical FaL-G concrete had strengths in the order 40 to 50% of a comparable Portland cement concrete after 28 days. However, tests with FaL-G concrete showed almost doubling of the compressive strength from 28 days to 180 days.
- Rai in 1992 reported on a somewhat similar technology.[233] The Central Building Research Institute (CBRI) in Roorkee, India, in 1968 started investigations on a hydrothermal reaction between fly ash and lime. This led to patenting a production process for fly ash–sand–lime brick in 1972. By 1986, two plants used this technology.

Japan:

- In 1983, Uomoto and Kobayashi reported on tests with a slag-gypsum cement, and also on some disadvantages with this binder.[234]
- In 2004, Hashimoto et al. reported on a binder based on three industrial by-products: fly ash, ground granulated blast furnace slag, and gypsum.[235] They reported a compressive strength of 35 MPa after 91 days, with a water/cementitious material ratio of 0.25, and the use of superplasticizer and vibration for compaction. They also claimed that of the 7.3 million tonnes of fly ash available in Japan each year, only 60% is used efficiently. The rest went for reclamation. The ratio in their mixes was fly ash:GGBFS:gypsum = 1.0:0.2:0.1.
- Noguchi reported on Eco-Cement,[236] made by using incineration ash from municipal solid waste and sewage sludge as its main and secondary materials, respectively, and the related Japan Industrial

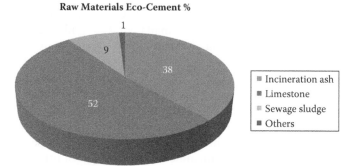

Figure 3.33 Raw material Eco-Cement in percent according to Noguchi.[236] (From Noguchi T., Sustainable Recycling of Concrete Structure, in *Concrete Sustainability through Innovative Materials and Techniques*, Roving National Seminars, Bangalore, January 10–14, 2011, pp. 86–97.)

Standard, JIS R 5214 (Eco-Cement), which was established in 2002. The production processes are basically the same as those for normal cement (Figure 3.33).

Iran:
- Researchers from Iran and the UK tested activation of natural pozzolans from Iran, with reported strengths in the 30 to 50 MPa range, and curing at elevated temperatures.[237]

Latvia:
- At a conference in Singapore in 1994, Dubin[238] reported on what he called a "new binder," although we would rather call it a type of blended cement. Fifty percent of the Portland cement binder was replaced by sand and expanded clay dust.

Netherlands:
- Bijen and Waltje in 1998[239] reported on tests with a binder based on 60% ground blast furnace slag and 40% fly ash, activated by 7% hydroxide. While other properties were favourable with the material, they claimed that carbonation resistance was leading to a decrease in strength compared to Portland cement.
- Buchwald and Wierex in 2010 reported on a technology named ASCEM, where raw materials like fly ash are thermally treated to get reactive glass, and then milled and mixed with fillers like fly ash as well as an activator.[240] The main purpose seems to be to get better control over a process based on variability in the raw material source.

Norway:

- In a report in 1989, Gjørv told about activation of Norwegian granulated blast furnace slag.[241] The tests demonstrated the importance of the fineness of the slag on strength development, and that strength development of the activated slag was comparable to standard blended cement with 10% fly ash.

Poland:

- A report from a Finnish/Swedish research project in 1998[228] mentioned tests in Crakow, Poland, with activation of slag, where old and new tests were compared.
- Deja and Malolepszy, from the Academy of Mining and Metallurgy in Cracow, in 1998 reported on durability tests with alkali-activated slag,[242] and mentioned that they had been working with AAS since 1981 and 1973, respectively.

Romania:

- In 1986, Ionescu and Ispas[243] reported on tests on durability with a binder based on slag and activated ashes. As activators, the following were tested: gypsum, phosphpogypsum, anhydrous limestone and semihydrated lime powder, calcium chloride, sodium carbonate, and liquid sodium silicate.
- At a conference in Poland in 2007, Georgescu et al. reported on durability testing of a binder consisting of slag activated with NaOH.[244] In their conclusion, they claimed that provided an adequate binder dosage of slag of about 400 kg/m^3, the concrete with NaOH-activated slag has good freeze-thaw resistance.

Russia:

- In 2001, Pavlenko et al. reported at a conference in San Francisco[245] on research at the Siberian State University of Industry on a new composite binder based on high-calcium fly ash and moulding sand (burnt sand) from a foundry. The fly ash contained up to 40% of SiO_2 and 36% CaO, including 15% free CaO, and the used sand above 90% SiO_2, mostly amorphous.
- Pavlenko et al. had previously reported on other combinations of "cementless binders" by the use of high-calcium fly ash (Istanbul, Turkey, 1992[246] and Bangkok, Thailand, 1998[247]) by utilising other waste products as silica fume, slag, etc. In the paper from 1998, Pavlenko explains that some of the background for the research is that the Pavlodarsky Tractor Plant in Kazakhstan was faced with the following problem: it needed to utilise waste products from steelmaking and the iron foundry process, but had a shortage in

housing construction for the workers due to lack of cement and aggregate—for both lightweight and heavy concrete. So, the "new binder" concept was truly a holistic sustainability effort.
- Tallin and Brandstetr reported on activity with alkali-activated slag cements.[213]

Serbia:
- In a paper from Seville, Spain, in 2009, researchers from Belgrade[248] reported on alkali activation of slag and fly ash with sodium silicate and water glass as alkali activators. They reported rather high compressive strengths with the slag mortar. They concluded that under similar conditions, the alkali-activated blast furnace slag performs better than the alkali-activated fly ash, and that this is closely connected with the nature and properties of the industrial products selected.

South Korea:
- Lee et al. presented a report in 2010[249] detailing interesting results with an alkali-activated fly ash binder. The tests were done on mortar cubes. The activator was a combination of NaOH and sodium silicate. The authors show how temperature is a key factor for the reaction product and the properties of the binder. While the compressive strength after 28 days with curing at 20°C was recorded to be below 2.0 MPa, the 28 days' compressive strengths cured at temperatures of 30, 60, and 90°C were 21.3, 42.8, and 42.0 MPa, respectively.

Spain:
- In a paper in 2010, Criado et al.[250] reported on corrosion tests on embedded reinforcement in a fly ash-activated binder compared with those in a Portland cement mortar. The fly ash was activated with an aqueous solution of NaOH and water glass. They claimed that the fly ash-activated binder passivated the steel reinforcement as rapidly and effectively as Portland cement mortar.
- In 2007, Guerrero and Goni[251] reported a test with a fly ash belite cement in a sulphate solution.

Sweden:
- A special kind of alkali-activated slag was introduced in the 1970s and called F-cement. A joint report from a Finnish Swedish research project was presented at a conference in Trondheim, Norway, in 1989.[228]

Turkey:
- Yeginbali[99] in 1998 told about research to produce a belite-rich binder from 30% oil shale and 70% phosphogypsum (by-product from the fertilizer industry).

Ukraine:
- In 2001, Krivenko and Kovalchok[252] reported on a special binder for lightweight gas concrete that they called geocement, based on metakaolin, sodium hydroxysilicate (soluble glass), and various additions. The material was particularly designed for thermal insulation of industrial equipment. They reported a unit weight of 400 to 600 kg/m^3, compressive strength from 2 to 5 MPa, and heat resistance up to 1000°C.
- The principle of alkali activation of slags has been known for more than three-quarters of a century. The research on AAS cements started in Kiew about 50 years ago.[213]
- A report by Tallin (Finland) and Brandstetr (Czechoslovakia)[213] mentioned a conference in Kiew, Ukraine, about alkali-activated slag, where more than 400 papers were presented.
- In 2010, Krivenko et al.[253] presented a paper in Jinan, China, reporting on tests with both alkali-activated slag and fly ash, with different activator combinations, and comparison with Portland cement concrete. Of the various findings we mention the following:
  - To modify the alkali-activated slag cement an alkaline ligno-sulphonate was used as a plasticiser.
  - They experienced a reasonable linear relationship between the w/c ratio and compressive strength for the activated slag binder, from about 60 MPa at w/c = 0.34 to about 32 MPa at w/c = 0.42 for 28 day tests, and from 32 MPa to 16 MPa for 2 days' testing, respectively.
- At the same conference, Kavalerova et al. reported on a new Ukrainian standard for alkali-activated cements.[254] After 50 years of experience Ukraine could finally present a new universal standard under which any aluminosilicate constituent, after its preliminary testing for suitability, can be used in the alkali-activated cements. Compressive strength testing is done after 2, 7, and 28 days of curing. The materials are classified in the following classes: slag alkali-activated cement, alkali-activated Portland cement, alkali-activated pozzolanic cement, alkaline slag Portland cement, and alkali-activated composite cement, which differ in combination of such aluminosilicate components as granulated blast furnace slag, OPC clinker, and ashes from coal combustion and basalt used in combination with the alkaline activator.

Compressive strength classes under the standard go from 300, in steps of 100, up to 1000. It is permissible to add to the cement during its grinding special water-reducing admixtures such as alkaline lignosulphonates (or of similar action) in quantities not to exceed 1%, and water-repellent surface-active substances in quantities not to exceed 0.1%, in order to intensify the grinding process and increase storage time.

United Kingdom:
- In 2007, Vlasopoulos and Cheeseman[255] reported on tests with magnesium oxide as a binder for the production of lightweight aggregate blocks. In their test they compared the MgO binder to Portland cement as a binder and with a 50% replacement. The aggregate was fly ash aggregate, Lytag. They were working with rather moderate compressive strengths of 5 MPa and less. They claim that replacement of Portland cement with MgO decreased the strength up to 28 days, but the samples had comparable strengths after 90 days. Further, they were of the opinion that the ongoing carbonation of $Mg(OH)_2$ might explain the strength gain between 28 and 90 days.
- In a paper at the same conference in Coventry, UK, in 2007, Vandeperre et al., from the University of Cambridge, reported on various tests with reactive magnesium oxide cements.[256] In a very interesting paper, the authors discussed a number of factors. To a large extent they have worked with blends with Portland cements and additions of fly ash, but mainly with blends where MgO is the main constituent, to reduce CO_2 emissions. They reported that the binder types are primarily recommended for moderate-strength concrete, for example, masonry products. They also claimed that the concrete will absorb considerable amounts of CO_2 through carbonation over its lifetime, possibly making the material carbon-neutral. Among their observations is that the MgO blend sets faster than pastes based on Portland cements and fly ash.

In addition to the above listed examples, we also like to mention that:
- High-alumina cement or "Ciment Fondu" was patented in 1908 and was produced in France for nearly 100 years, and can hardly be categorised as a new binder. This type of cement has been produced not only in France, but also in countries like Germany, Hungary, Czechoslovakia, Italy, Japan, Spain, and Russia.[257] Besides having a distinct restriction in its possible use, high-alumina cement does not have any advantages with respect to reduced CO_2 emission.

- Substudy 8: Climate Change[60] also mentioned several binder alternatives attempted to be commercialised.
- Lone Star Cement was an early commercial pioneer of geopolymers with hundreds of commercial applications and some market success for special applications. However, gaining customers' acceptance for new geopolymers over long-held preferences for conventional cement was difficult.
- The University of Illinois in the United States has been conducting research to co-dissolve nitric salts and colloidal silica in (poly) vinyl alcohol (PVA), creating a copolymer that, when dried and ground, can be calcinated at 70°C to form the key components of Portland cement. The product is extremely reactive, develops strength quickly during hydration, and offers the prospect of using reinforcement in concrete that usually corrodes with ordinary Portland cement.
- Ceramicrete is another binder in the experimental stage. This is a chemically bonded ceramic formed by mixing magnesium oxide powder and soluble phosphate powder with water, resulting in a nonporous material with compressive strength higher than that of conventional concrete.

Few of the above references give the volume impact of the various alternatives, and the commercialised effect experienced so far. In addition, most of the reported examples utilise resources from supplementary materials mentioned previously as tools to reduce the binder content in Portland cement: fly ash, slag, silica fume, metakaolin, and limestone powder. Consequently, from a resource point of view, they do not add to reduction in CO_2 emission compared to using the supplementary cementing materials as binder reduction in Portland cement concrete.

Substudy 8[60] claims that as a rule of thumb a new product needs a production level of more than 250 million tonnes per annum (5% market share) before a measurable positive impact on the overall unit-based CO_2 emission on the cement industry is achieved.

Of the more interesting cases not using supplementary materials is the Chinese high-belite cement (HBC) and other ongoing low energy and low CO_2 cement development headed by Tongbo Sui,[258] which is specifically explained in Chapter 6.

Similar developments are on the agenda for most of the major cement companies in the world. Hopefully this might lead to interesting developments in the future. However, we find it unlikely that the ongoing development we have mentioned will have a major impact on the total global emission amounts for cement in the next 10 to 15 years.

3.5.12.1 High-belite cement

At a conference in Lillehammer, Norway, in 2007, Gartner and Quillin gave a very thoughtful paper about the advantages of belite cements.[65] However, they stressed that developing new cements for large-scale use is an inherently slow process, and a great deal of work is required to:

- Confirm that such cements can be manufactured industrially, with the desired CO_2 emissions savings and low environmental impact, and at an acceptable cost
- Better establish the effects of compositional and processing variables on performance
- Optimise cements for performance in major use categories, in terms of physical properties, raw material availability, manufacturing parameters, etc.
- Develop the scientific understanding necessary to explain the observed performance, and thus to help predict the effect of variations in conditions on such performance
- Clearly establish the long-term performance of these materials in concrete and provide data that will give users and specifiers confidence in their durability
- Develop appropriate codes and standards for their use in construction

The main advantage environmentally of belite-based cements is that belite, C_2S, is formed rapidly above 1200°C, if the raw materials are sufficiently finely ground and well mixed, versus +1400°C in normal cement production. In addition, decarbonation is also less in the belite process. Gartner and Quillin[65] claim a reduction in CO_2 emission of 20 to 25% versus the ordinary Portland clinker process. However, a disadvantage is a somewhat slower strength development than for Portland cement clinker.

More details about these alternatives are found in Chapter 6.

3.5.12.2 Sulphur concrete

Sulphur as a by-product is relatively easily available at a relatively low cost. Several researchers have tested melted sulphur as a binder with low CO_2 emission. The properties of the final product might be interesting for several uses, but taking into consideration the heat involved and the health precaution necessary in the production process, we do not believe this alternative will have any significant effect on the reduction of CO_2 from the production of concrete. The earliest known use of sulphur in construction was in the seventeenth century in Latin America, where it was used to anchor metal to stone. In 1859, the cementing properties of sulphur were mentioned by Malhotra, referring to a U.S. patent.[259]

Sulphur might be an interesting binder, but it has some other environmental challenges concerning possible toxic and corrosive effects on reinforcing steel under wet and humid conditions. This fact makes sulphur concrete unfit for many applications and hardly a strong candidate to replace Portland clinker to promote a more sustainable development.

Successful development and the use of new binders might possibly have some effects on the market situation for the traditional Portland cement producers before 2030, but it is hard today to identify which alternative introduced so far will gain a leading role. In a shorter time span than 10 to 20 years it will not be easy to threaten the position of Portland clinker-based cements with respects to:

- Raw material availability
- Production cost
- Logistic organisation
- Accepted technology, codes, regulations, and standards
- Environmental friendliness
- Existing production equipment investments

Production equipment for cement production is expensive and investment-intensive, and is depreciated over a long time period. It is not a very drastic prophetic statement that considerable amounts of the cement delivered in 2050 will be produced with equipment already installed.

In a 10- to 20-year time span new binders globally will hardly have any strong significant effect on the CO_2 emission from concrete production in comparison with the use of available supplementary materials combined with Portland cement clinker. However, in a longer time span, some of the alternatives mentioned above and later in Chapter 6 might have promising potential to play a role in reducing global CO_2 emissions from concrete production. They may also be of significant importance in a shorter time span in a local or national context.

3.5.13 Increased carbonation

In Section 3.4 we tried to explain the absorption of CO_2 in concrete through carbonation. Increased carbonation is an interesting tool, but it also has some side effects.

We preferably only want increased normal carbonation in unreinforced concrete products or when we have noncorrosive reinforcement. This decreases the target area considerably.

Intentionally increasing carbonation can give a potential strength/quality increase that can be utilised for better durability or cement reduction, and consequently has a further positive effect on emissions.

3.5.14 Better energy efficiency in buildings

We have treated this item in Chapter 5, Section 5.3, but here want to mention some of the findings and statements in a report from the United Nations Environment Program in 2007:[8]

- Worldwide, 30 to 40% of all primary energy is used in buildings. While in high- and middle-income countries this is mostly achieved with fossil fuels, biomass is still the dominant energy source in low-income regions. In different ways, both patterns of energy consumption are environmentally intensive, contributing to global warming. Without proper policy inventions and technological improvements, these patterns are not expected to change in the near future.
- Generally speaking, the residential sector accounts for the major part of the energy consumed in buildings; in developing countries the share can be over 90%.
- It is clear that there are no universal solutions for improving the energy efficiency of buildings. General guidelines must be adjusted to the different climate, economic, and social conditions in different countries. The local availability of materials, products, services, and the local level of technological development must also be taken into account.
- In Organization for Economic Cooperation and Development (OECD) countries buildings are responsible for 25 to 40% of the total energy use. In Europe, buildings account for 40 to 45% of energy consumption in society, contributing to significant amounts of carbon dioxide emissions.

Concrete has a very good ability/capacity to store heat and a low heat transmission factor.

To better utilise this ability, it is of particular importance to reduce energy consumption/emissions from air-conditioned installations.

As an example, Biasioli and Øberg[260] show a practical example of a 2 to 7% difference in energy consumption between buildings. They indicate possibilities in reductions with best practice in the order of 30 to 40%.

In the literature (e.g., Sakai[57] and Biasioli and Øberg[260]) one often finds observations from LCA studies that document that the emission from the construction of a structure represents less than 10 to 20% of the total emissions, including the user phase.

The experience from this is that in a lifetime perspective, the savings from how we construct the building is more important than the actual materials used.

However, it is important to remember that this possible improvement has no or very little effect in a shorter time span.

3.5.15 Improved mixture design/packing technology/water reduction

There has been continuous activity in the development of chemical admixtures to concrete, as evidenced by published literature and patent applications. In the years 1985–1989 alone, nearly 600 patents were issued on concrete admixtures, and the enormous interest is demonstrated by the 242 patents in 1989 and 140 in 1992.[261]

When it comes to the possible effect of reduced CO_2 emission of increased use, we have concentrated our attention on ordinary plasticizers, water reducers (WRs), and the so-called superplasticisers, or high-range water reducers (HRWRs).

The effect on CO_2 emission comes from the ability to reduce water, and thereby act as a cement reducer. However, plasticizers are also used to make a more flowable concrete without reducing the cement content. Environmentally, this might be an important use in itself, as it may reduce energy consumption and make life easier for the concrete worker. But, it has little effect on the emission issue.

The plasticizers may, with normal dosage, have water-reducing effects of 6 to 10%,[262] while the superplasticizers may have water-reducing effects (dosage 1 to 5%) in the 20 to 40% range. We have seen superplasticizers with water-reducing capabilities above 45%.

Traditionally the most common WRs are sugars or sugar derivatives (gluconates, hydroxyl acids), often by-products recovered from the agricultural and food processing industries. Lignosulphonates (sodium and calcium salts) are by-products from the bisulphate wood pulping process. The lignins have a somewhat more complex molecular structure than the gluconates, and will include a greater product variety of possibilities and effects on water reduction.

The world market for lignin is in the excess of 1 million tonne, and use in concrete accounts for some 60% of this. Some 15 to 20% of the lignin market is for higher-value purposes than use in the concrete industry, and hence more interesting for the producers. There is potential for transferring 100,000 to 200,000 metric tonnes of lignins, presently used for low-value bulk market segments, to the construction area for use as a further CO_2 emission reduction. Furthermore, there is an annual growth in the lignin supply of 10 to 20,000 tonnes over the next decade or so, representing another potential for the concrete industry.

There is no doubt that the use of plasticizers and superplasticizers is a very positive contribution to making concrete a "greener" material. However, to what extent they will contribute considerably to reduce CO_2 emissions in the future is more dubious; there are several reasons for this:

- In some countries, for example, Norway, more than 90% of the concrete produced already contains water-reducing admixtures. Further, the effect on emission reduction due to increased use of admixtures will be negligible.
- More important than using water reducers for CO_2 emission in the future, from a sustainability point of view, will be to make more flowable concrete, like self-leveling concrete.
- Even if the price of water reducers is competitive to obtain special effects in concrete mixture design, there are doubts whether they are competitive for increasing cement reduction from the present level, compared to other alternatives.
- The raw materials for the water reducers are also used for other purposes, like surfactants and other plasticising challenges in the industry. Limitations in the raw material supply do not point in the direction of lower prices for water reducers in the future.

As mentioned earlier, there is an impressive activity in developing new chemical admixtures to use in cement and concrete production. It is therefore possible that the effect of this tool might be stronger in the future. However, experts seem to think that priorities will go in other directions.

The search for new, more cost-effective admixtures in this group will obviously continue, but the influence of new admixtures is likely to be less dramatic than the impact of the few key products that supported the early development, namely, PNS and PMS. Further work in this field is likely to yield more international benefits to cope with the shortcomings of current available admixtures, as well as the requirements from development in cement and concrete technology. Such products would typically:

- Have prolonged influence in concrete rheology
- Exert minimal effect on the cement hydration reaction and set
- Be tolerant toward variation in the binder composition (fly ash, slag, others)
- Be compatible with a maximum variety of other chemical admixtures
- Be applicable in a broad range of different types of concrete and applications

Given the rapid evolution of binder systems, the inherent composition variability of these systems, and the increasing variety of concrete applications, it can readily be predicted that a key requirement of the future chemical admixture will be its ability to perform under the widest possible range of parameters (binder composition, temperature, other admixtures), that is, tolerance, robustness.[262]

Reduction of the paste volume in concrete through increased use of water-reducing admixtures is an interesting tool. However, restrictions due to variation in availability, price, etc., might reduce the effect of this possibility. Theoretically, increased use of water-reducing admixtures and superplasticizers *alone* has the potential to bring the total global CO_2 emission down some 0.02 to 0.04 kg per kg of binder. However, water reducers are a vital tool to utilize the other alternative tools.

However, we believe that chemical admixtures will continue to be, possibly even more so in the future, an important tool to solve other sustainability challenges.

An interesting option that might increase the strength of the water-reducing tool as an emission reducer is if the cement companies starts to blend some plasticizers in the cement. Nearly all concrete could benefit from the effect, and this would give a w/c reduction and need for clinker, as well as possible other possible environmental effects. The profit in the cement industry is not based on volume, but on contribution margins, and there should be no doubt that the increased cost should be compensated in increased price. In addition to the savings in clinker and emission, this also has potential advantages in logistics and in simplifications in concrete production.

3.5.16 Increased building flexibility, and more sustainable design and recycling practice

A more conscious attitude regarding sustainability challenges at the planning stage may have considerable effects on the emission amounts for structures in general, including concrete construction. These effects might be very important from a lifetime perspective, but have no or minor effects in a shorter perspective.

The most important possible improvements regarding reduction in CO_2 emission, in addition to what was previously mentioned about increased utilisation of the thermal mass, seem to be:

- A better utilisation of the possibility to design concrete with a durability to withstand an optimised life expectancy specification. For many structures, in particular a wide range of typical infrastructure projects, increased durability might be an important contribution to reduce the emission over the life span of the structure, However, an undocumented analysis claims that well over two-thirds of all structures that are demolished, are demolished due to changes in socioeconomic demands, not due to lack of durability. Possibly we should sometimes also make concrete with lower-quality demands, and less binders for some of the structures that are not expected to last that long.
- A demand to make design structures with a more flexible usage perspective will reduce the increasing tendency to demolish structures that do

not meet modern usage demands. This will reduce the demand for new constructions, and thereby the cement consumption and CO_2 emission.

- Instituting a demand that a minimum part of any structure shall be reusable without destruction.
- Increased crushing of demolished concrete to increase the absorption through carbonation.
- Typical "public" buildings and structures that are expected to last a long time, like important bridges, harbours, libraries, theatres, operas, and religious meeting places like churches, mosques, temples, shrines, and synagogues. A best possible durability with life expectancy of 300 years would be reasonable.
- Increased use of lightweight concrete, which may reduce the emission footprints due to reduced deadweight loads and reduced material use, as well as being advantageous regarding energy consumption in buildings. However, care should be taken in the LCA studies, as the emission from the production of the aggregates may override the other emission advantages.

However, we firmly believe that lightweight aggregate might be an interesting contributor to a more sustainable design in a number of ways. In addition to possible savings in material consumption and corresponding reduction in transport burdens, lightweight concrete may contribute to longer spans and more flexible building structures for later different functional requirements. Lightweight aggregate has also been used as a growth improver on roof gardens, an interesting alternative, both to reduce the so-called heat island effect in densely populated areas and to create increased areas with CO_2-absorbing substances.

- Care in design, by optimising strength of concrete, longer span possibilities, and weight of structures through use of hollow blockouts in the floor structures to reduce weight. This has shown favourable LCA analysis figures. A very rough evaluation of concrete in columns with an average loading and length gives the following rule of thumb: by doubling the concrete strength, the column side is reduced by one-third, the concrete volume by 55%, the binder content by 18%, and the CO_2 emission by 20%.

In general, we are convinced that more active use of the LCA tool will result in reduction of emission over time. However, we also believe that a holistic attitude in the design will show that reduction in energy from optimising ventilation systems with respect to heating and cooling will have a greater impact on the emission figures than changes in the material specifications. It is not possible to give general views on this, as conditions vary with different climate and traditions.

One of the possibilities regarded as an important feature in some of the major rating systems for more sustainable buildings is the so-called heat island effect.

- Surface and air temperatures in urban and suburban areas tend to be higher than those in adjacent to rural areas. This effect, also called the heat island effect, is the result of many factors, but the solar energy absorbed by buildings is an important part of it. To reduce the absorption, or rather to increase the reflectance, might save energy for cooling, and consequently reduce emission of CO_2. A paper in *Concrete International* by Marceau and Vangeem[263] showed interesting results from testing of various materials in concrete. Regarding reflectance, based on a scale from 0 to 1, polished aluminium has an emittance factor of 0.1, and a black nonmetallic surface a factor of more than 0.9. The opposite of this is the solar reflection index (SRI); the index for "new typical grey" concrete has a value of 0.35, and 0.7 for a "new typical white concrete." In the mentioned paper they reported tests from 25-week-old concretes with white and grey cement, with SRI values of 0.68 to 0.77 and 0.41 to 0.52, respectively. From the conclusions in the paper we note that data from 45 concrete mixtures show that solar reflectance of the cementitious materials has more effect on the solar reflectance of the concrete than the other materials. The solar reflectance of the fine aggregate has a small effect, while the effect of the coarse aggregate does not have a significant effect.

A definite alternative is more innovative use of the possibilities in using high-performance concrete. Fidjestøl et al. gave a case study example in a paper in India in 2010.[264] The authors compared two different structural designs for a building in southern Norway, with a 1000 m² ground area in a four story building. They compared a conventional structure with a concrete quality of 45 MPa with a structure that utilised precast filigree components of 60 mm concrete as permanent formwork, plastic balls to make voids and reduce mass in the centre of the slab, and high-strength concrete of 85 MPa for load-bearing structures. The authors claimed to have reduced the CO_2 emission from the construction phase by 54.5% from the conventional structure to the more innovative solution. The reported case might be a little bit special, but it demonstrates quite clearly that there are very interesting potentials in an innovative design attitude.

We have seen several analyses indicating that use of precast concrete solutions might be favourable; however, the differences versus innovative *in situ* cast structures are seldom reported with more than 10% savings in emission on the construction stage. Probably more important in innovative design for less emission is, for example:

- Use of longer span to build more tolerant structures with respect to possible other utilisation in the future
- Use of high-performance concrete to reduce weight and increase durability

- More flexible use of concrete quality to optimise demand versus resource consumption (for example, walls often need a certain thickness to be an acceptable noise reduction barrier, but seldom have the same strength requirement as the slabs)
- More active use of the thermal mass potential for saving energy

3.5.17 Miscellaneous

3.5.17.1 Production restrictions

Some people have advocated that we must enforce production restrictions if the building industry in general, and the cement and concrete industry in particular, fail to develop technology to fulfil the emission goal that is necessary.

As mentioned earlier, we will in the future have restrictions, at least in Europe, not on the production of cement, but on the emission of CO_2 from the cement industry, in terms of a quota regime. But, Europe is not the future problem. The heavily populated countries in Asia represent the biggest growth in cement consumption. It is probably unlikely that they will accept a major restriction in their social expansion, but hopefully there might be some regulation of how this expansion will develop.

3.5.17.2 The testing regime

We have in the national and international codes and standards a well-documented and accepted regime for testing and specifying cement and concrete. There is probably time for looking at this regime to analyse if in some areas it has rules that are sustainability-unfriendly. An adjustment of the ages for testing and specifying, and the temperatures under which one tests, might be more realistic to real life, and not so unfavourable to new, slower, and more environmentally friendly binders.

With a rather conservative standardisation regime, it is not likely that such changes can take place in a short time frame. But hopefully it might be possible to see some modifications with alternative methods in a reasonable time.

3.5.18 Carbon capture and storage (CCS)

This is a technology that will have no impact from a 2020 perspective, but the development might be very important from a focus 2050 perspective.

As seen in Figure 3.34, there are activities going on in quite a few places in the world. The plan for the EU directive regarding CCS was scheduled for 2009–2010, and in 2015 it was expected that 10 to 12 full-scale projects would be started in Europe. However, by early 2012 plans have been reduced to eight.

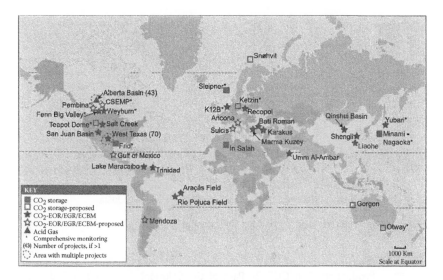

Figure 3.34 Location of sites where activities relevant to CO_2 storage are planned or under way (IPCC, 2005).[265] (From Bellona, *Carbon Dioxide Storage: Geological Security and Environmental Issues—Case Study on the Sleipner Gas Field in Norway*, Bellona Report. Oslo, Norway, May 2007.)

In December 2008, the EU decided on a budget of 1 billion euros for CO_2 handling before 2015.

The technology is expected to have its breakthrough in the middle of the 2030s. Estimates are that 25 to 40% of the global emission of CO_2 can be captured and stored, and 37% of the emissions will be captured and stored in 2050.[265]

The cement industry is of particular interest from the CCS perspective, as the industry represents very concentrated emissions.

An interesting thought for the future is that in some areas with a wide use of CCS, concrete might be a minus contributor to the CO_2 emission, taking into account the absorption through carbonation.

Brevik, referring to IEA, in 2011 showed the predictions given in Figure 3.35 for capture.[205]

CCS has two different sectors of challenge: capture and storage.

3.5.18.1 Capture

Quite a few technologies have been presented over the last decade or so. Among the specialists there seems to be little doubt that there soon will be a technology breakthrough in this sector, but still we have not seen any large-scale projects in operation, and it will be some years before we see projects that will be competitive. The projects are investment-intensive

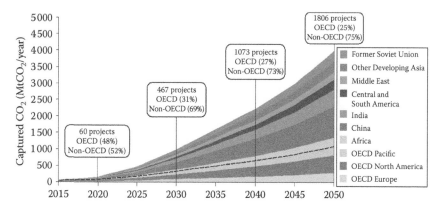

Note: The dashed line indicates separation of OECD/non-OECD groupings.
Source: IEA analysis.

Figure 3.35 Prediction for use of CCS[205] (From Brevik P., The CO$_2$ Challenge in Cement Production: Can CCS Be Part of the Solution? Presented at 2nd International Workshop on CO2: CCS and CCU in Germany, Norway, The Netherlands, Poland and Scotland—Challenges and Chances? Dusseldorf, November 10, 2011.)

with a definite need for governmental or intergovernmental subsidies for the first projects before they can be realised. Engineering companies are reluctant to make large test project plants unless they are paid for their work. The economical crisis has definitely not accelerated the development. Hopefully, and probably, we will see the first projects in half a decade. One important remark: Of the planned European projects, not one of them is for the cement industry. All are based on capture from the energy sector. The Norwegian cement company Norcem has completed a concept study and a preengineering study in 2011; Norwegian authorities are now giving positive reactions. Norcem believes that this is the first cement capture project in the world.[205]

3.5.18.2 *Storage*

When it comes to the challenge of getting rid of the CO$_2$ when it is captured, there are two principle options: industrial use of CO$_2$ as a raw material, and storage.

Media tells about a number of developments going on around the world for industrial use of CO$_2$ as a raw material, from use in new types of plastic to converting the gas to aggregate in concrete. Depending on the proposed use, a cleaning and concentration process might be needed. We have very briefly mentioned some of the ideas for use later in the chapter.

The other option, and the probably the most important, as under any circumstances it will be a must, is storage. There is still a lot of both public and scientific discussion about the alternative. We know from the oil industry in the North Sea that storage of gas in geological formations has been done over quite a few years (10 years at the Sleipner field), and seemingly with good results. Scientific proof of the non- or minimum leakage is still limited, and makes people in some countries sceptical and reluctant to explore the possibilities on a large scale. In several places in the world, geological structures have been identified that most probably will be of good quality for carbon dioxide storage, but there is also a logistic challenge with cost implications involved. The flue gas from the cement industry, or other industries, has to be concentrated from, for example, 16 to 18% to a transportable liquid concentration—and pipelines to possible storage structures have to be built. Even if large amounts of CO_2 might be utilised as an industrial raw material, the storage challenge still has to be solved, to balance the difference between the supply chain and the receiver end with variations in output and input, and market variations in different industries.

Of the "user" methods in development, we mention the following:

- According to Malhotra,[266] in January 2010, Carbon Sciences, Inc. of Santa Barbara, California, announced that it had developed a breakthrough technology that converts atmospheric CO_2 into commercial-grade gasoline. According to its information, this technology combines chemical and biological processes in a biocatalytic process that converts CO_2 into a cost-efficient energy source.
- Since 2008, Calera Corporation, in the United States, has been working with an innovative process capturing CO_2 emitted by coal- and gas-fired power plants and cement plants and converting it into calcium and magnesium carbonates for use in manufacturing carbon-negative products such as sand, aggregate, supplementary cementing materials, and cement.[267,268] As far as we understand, the process is still in a testing stage. The gasses from the power plants are transported through seawater. Calera claims that the process is the same as when coral reefs are formed, when calcium and magnesium in the seawater form calcium carbonate.

There is little doubt that CCS might be a very powerful tool for the cement and concrete industry sometime in the future, and possibly the final tool that enables the industry to reduce emission by more than 50%, in spite of a considerable volume increase by 2050. However, it is unrealistic to expect this tool to have any significant global effect for the cement and concrete industry before the end of 2030, at the earliest.

3.6 VARIATION IN FOCUS

Over the years we have seen partly contradictory evaluations and papers regarding emissions from cement and concrete. One reason might be the motive of the writer. Another important element in evaluating positive emission actions is whether one puts weight on comparison with other building materials, the political reduction scheme one has to consider, or future long-term general demands based on evaluations from the United Nations International Panel on Climate Change (IPCC).

Normally, using a lifetime expectation view will be a basic principle in sustainability evaluations, but in this case *this is not enough*. One also has to consider the political reality, and the possibility that catastrophic changes might occur in shorter time spans than a lifetime evaluation if we do not intensify emission reduction actions in shorter time spans.

To improve evaluation of status, needs, and possibilities, a triple-focus system for evaluations might be helpful, including:

- Lifetime expectancy perspective
- Society/political accepted reduction strategy
- Recommendations from IPCC

Why is it so important to have a triple focus?

- Because we need to handle all three challenges to have a holistic attitude to the situation
- Because the various tools we have in the toolbox will have different effects, depending on which challenge we are evaluating

3.6.1 Focus I: Lifetime expectancy perspective

Concrete is one of the most important pillars for a sustainable development. Development of infrastructure and modern living quarters in the former nonindustrialised world is the most important reason for the increase in cement consumption during the last 10 years or so.

Life cycle analysis based on a "from birth to grave" evaluation shows concrete to be a sustainable building material. However, it is doubtful if one always is using the material in the best sustainable manner, and there are interesting possibilities for improvements.

In more and more countries in the world, the practice of making analysis of the effect of a structure on the environment is getting more popular. Several methods have been developed for this, and they are normally referred to as life cycle assessment (LCA). The emission of greenhouse gases is one of the many effects that are evaluated. In Europe, a perspective of 60 years is normally a standardised time perspective, unless special circumstances make other age perspectives more reasonable.

There are examples of up to 300 years for special structures as foundations and harbours. Various analyses normally show concrete to be a very sustainable material in the birth-to-grave analysis, but they also show that there still is room for important improvements.

3.6.2 Focus 2: 2020

In January 2007, the EU Commission presented its climate and energy package solution. In a short and simplified version, this document stated that the EU will reduce its greenhouse gas emission by 20% by 2020 compared to 1990.[269]

Many of the various actions considered are still under discussion, but there is a high degree of acceptance of the final goal. The final decisions will set the framework for all the cement (and concrete) industry in Europe for the future years regarding CO_2 emission reduction.

The *minimum* 20% emission reduction by 2020 is about to become not only the target, but also a rule for working conditions for the cement and concrete industry far outside Europe.

In its December 13, 2008, meeting, the EU summit finally affirmed the protocol that states that EU will reduce its emission of CO_2 by 20% before 2020 from the 1990 figures. Further, the political leaders in Europe will use their political ability to try to get other nations to accept similar actions. Some countries have already accepted stricter targets.

There exist quite a few estimations of the development in global cement consumption.

Malhotra[97] estimates the cement consumption in 2020 to be in the order of 4 billion tonnes (Table 3.17).

Other estimates for the year 2050 seem to range from 2.9 to 4.5 billion tonnes. As mentioned earlier, the consumption in 2009 was 3.0 billion tonnes. For example, the WBCSD's 2009 estimates for 2020 are: low, 3.66 billion tonnes; high, 4.40 billion tonnes.[68]

Table 3.17 Cement consumption in millions tons[97]

	1985	2003	2020E
Developed countries	320	400	500
Developing countries	360	1200	3500
Total	680	1600	4000

Source: Malhotra V.M., Sustainability and Global Warming Issues, and Cement and Concrete Technology. Presented at 2008 International Expert Workshop on Cement and Concrete Technology for Sustainable Development, Beijing, October 7–12, 2008.

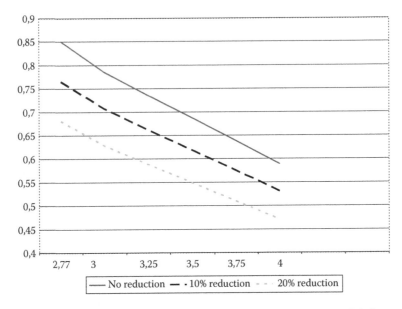

Figure 3.36 Emission in tonnes of CO_2 per tonne of cement, depending on global cement consumption in billion tonnes, and requested percent reduction factor.

Even with the economic chaos experienced lately, which undoubtedly will slow down the development somewhat, most analysts expect a considerable increase.

This increase must also be taken into consideration when evaluating the total emission factors from production of cement/concrete.

Today's present emission per tonne of cement of approximately 0.85 tonnes of CO_2, the result of a goal to reduce emission by 20% and at the same time increase the consumption from 3.0 to 4 billion tonnes, points against a future target emission amount per tonne of binder (Figure 3.36):

$$0.85 \times 0.8 \times 3.0/4 = 0.48$$

That is quite a challenge, and will probably need effects from most of the tools that we have available.

3.6.3 Focus 3: 2050

In his speech to the UN general assembly on September 24, 2007, the chairman of the IPCC panel said, "We have the time to 2015, if we want to stabilise emissions, then they have to be reduced considerably."

The IPCC published four climate reports in 2007. One of the most important conclusions from the reports, apart from us having to achieve control over the emissions by 2015, which seems to be taken care of by the political actions taken (Focus 2020), is that we have to reduce the human-created emissions by 50 to 80% by 2050.

It might be speculative to discuss a Focus 2050 and possible goals, as there are too many variables. However, what is certain is that the IPCC states that all human-made emissions must be reduced by 50 to 80% compared to the 1990 level. This statement is also valid for the cement and concrete industry. The most important uncertainty is what cement consumption will be in 2050.

Several authors have tried to give their guesstimates. WBCSD[60] has indicated four different scenarios with estimates for cement consumption in 2050, ranging from 3.766 to 5.5 billion tonnes. Several analyses also indicate that the consumption will flatten out around 2050 to 2060, after we have achieved the expected economical/social development in what was previously called development countries. Developed countries in Europe and America today have a rather stable consumption of 0.3 to 0.4 tonne of cement per capita. A world population of 8 billion people and a consumption of 0.5 tonne per capita point toward a consumption of 4 billion tonnes of cementitious material.

Simple evaluations shows that based on a worst-case analysis, even with utilising all the tools mentioned previously in this chapter, the IPCC goals will be very difficult to meet, unless new technologies are developed.

Based on historical developments, it is known that it takes at least 10 to 15 years for a new technology to be effective in full-scale use, before it has general acceptance and is in use on a global scale. With a capital investment-intensive industry like the cement industry, it is not reasonable to believe that changes will take place faster than this.

Likewise, it takes 5 to 10 years for a new technology to be utilised at a major pilot scale until full-scale utilisation takes place, and it takes 5 to 10 years from when a new technology is on the drawing board until a full-scale pilot plant can be fully opened.

Consequently, development of new technologies for cement and concrete production with an IPCC perspective must be one of the most important items on the research agenda for cement and concrete research institutions in the coming decade.

Carbon capture and storage (CCS) might be one technology that will be important for the cement industry in a 2050 perspective.

It is important to use the tools that are available now and not to wait until later to take action. Several researchers and economical analysts have said that the longer we wait to reduce emissions, the harder will be the demands for action in the future. The so-called 600-page Stern report to Britain's then-chancellor of the exchequer Gordon Brown from Sir Nicholas Stern on October 30, 2006, fully titled *Stern Review on the Economics of Climate Change*, gives indications that can be judged so that it might be 20 times more expensive per capita to repair the damages in 2050 than doing it now.

3.7 SOME CONCLUSIONS

There is hardly any doubt about the fact that some of the 17 tools mentioned previously will have greater global impact in the effort to reduce emissions from concrete production than others. It should also be obvious that we need all the mentioned tools in the toolbox of the cement and concrete industry. There are several reasons for this:

- Local and regional availability of the tools is different from one part of the world to the other.
- The optimum economical solution might differ from one part of the world to the other.
- Combinations of the different tools might lead to new and improved possibilities.
- The effect of the tools differs depending on what focus we have in a time context.
- There is no single tool that will be sufficient to meet the necessary goal.

Table 3.18 gives a rough indication of some of the most important tools/possible actions, depending on what focus one has for the evaluation.

Table 3.18 The effect of various actions versus focus

Possible actions/tools focus	Lifetime	2020	2050
1. Increased life/durability	Large	No	No
2. Increased building flexibility	Medium	No	Small
3. Demand for reuse of components	Medium/small	Medium	Medium
4. More optimised mix design/component dimension	Small	Medium	Medium
5. More energy-efficient buildings	Large	Small	Medium
6. More efficient clinker production/alternative fuels	Medium	Large	Large
7. More use of supplementary materials	Medium	Very large	Large
8. More efficient concrete production	Very small	Medium	Small
9. Sequestration technology	Possibly large	No	Possibly very large
10. New binders	Possibly large	No	Possibly large

Chapter 4

Recycling

The recycling question in concrete technology comes up in various aspects:

- Recycling of concrete
- Recycling of other materials as aggregate in concrete
- Recycling of other materials as reinforcement in concrete
- Recycling of other materials as binders in concrete
- Recycling of cement kiln dust

Later in this chapter we cite R.N. Swamy, who claims that *awareness* is one of the most important driving forces in the recycling of concrete. We firmly believe that just awareness of the sustainability issues in general, and emission challenges and recycling in particular, has been a very important inspiration to the present vitalisation we experience in concrete technology development. Awareness has inspired concrete technologists worldwide to try out new possibilities and solutions that have not been seen before, or have been forgotten in the modern industrial goal to streamline the concrete industry and material based on a too narrow line of material thinking and technology. The recycling possibilities and initiatives are too many to give the topic full attention in a single chapter. We have only tried to give a glimpse into the alternatives and the wide geographical spread in the ongoing development.

4.1 RECYCLING OF CONCRETE

The idea of recirculation in concrete technology is far from new. About 25 years BC, the Roman architect and engineer Vitruvius (i.e., Marcus Vitruvius Pollio) wrote his publication *De architectura*. He had many recommendations about the production of concrete. Among them were that the use of recycled burnt tile was better than sun-dried tiles as concrete aggregate, because it had already proven its durability (Figure 4.1).

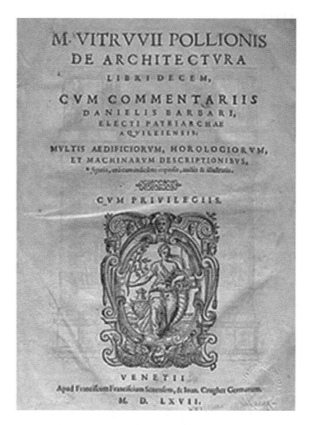

Figure 4.1 The front page of Daniele Barbaro's famous edition of Vitruvius's book of 1567.

In 1998, at a sustainability conference in Ottawa, Canada, R.N. Swamy[270] presented an interesting paper. He said the following about recycling:

> Everybody knows, we think, that recycling makes environmental sense. For example, around half of all the waste produced in our homes is made up of recyclable materials such as paper, textiles, glass, cans and plastic bottles. Yet, in the UK for example, currently only about 5% of all household waste is recycled—far less than the 40% needed to achieve sustainability. The keys to attain this level of recyclability are:
>
> Awareness
> Availability, and
> Accessibility

To achieve sustainable development in the cement and concrete industry, we need to understand and appreciate what has happened to the world during our lifetime. The century that we live in is almost at its end. The world at the end of this century is very different to the world that we inherited at the beginning of the century. There have been, particularly the last four to five decades, unprecedented social changes, unpredictable upheavals in world economy, uncompromising societal attitudes, and unacceptable pollution and damage to our natural environment.[270]

Awareness, availability, and accessibility, possibly combined with inventive minds and economical incentives, might be the key words when we discuss the recirculation challenge in the concrete industry.

With great variation from one part of the world to another, a certain amount of concrete is demolished every year, due to changes in the needs of society, durability problems, or disasters of different kinds. From a sustainability point of view it is important that as much as possible the values embedded in the original structure are recycled.

The ideal solution would be to recycle complete parts or components. This has also been done. The example shown in Figures 4.2 and 4.3 is from a storage building constructed at Løren in Oslo, Norway, in 1967–1968. In 1972 the owner decided on an extension, but 400 m away from the old building. The new building was 500 m², with standing double-T (DT)

Figure 4.2 Recirculation of double-T precast concrete façade panels; white panels are recirculated panels, gray panels are new, Løren, Oslo, 1972.

Figure 4.3 Recirculation of double-T precast concrete façade panels, Løren, Oslo, 1972.

panels in the façade, and saddled prestressed beams and DT components in the roof. The whole building was in precast concrete. In the picture we see the new building under construction, with a mixture of old recirculated and newly produced components.[271]

To further show the possibilities, we give an example from a report from the Norwegian Public Road Administration from its impressive recycling project. Over the years thousands of kilometers of concrete guard rails have been cast along Norwegian roads, often in a very mountainous landscape. Deteriorations have taken place over the years. A Norwegian contractor, NCC, through a daughter company, has specialised in a concept to recycle these guard rails on site through a specially developed slip forming machine. The old guard rails are equipped with reinforcement and given a new shape and size on site (Figures 4.4 and 4.5).

It would be a great step forward, in the sense of better recirculation and saving resources and reducing emission, if in the future we plan the structures we build so that certain parts or components of the structure can be recycled when this is due. It is a great step forward for sustainability that many rating systems for better sustainability honour designs that take this into consideration. Very few concrete structures that are built are of such complexity that not at least 10 to 20% of the structure can be recycled in full components when the structure is due for demolishing or rebuilding. This could lead to considerable advantages to the society.

However, it is the recycling of the raw materials we first and foremost think of when we discuss recycling of concrete. The fact that all material

Figure 4.4 On-site recycling of concrete guard rails at the scenic mountain serpentine road, Trollstigen, Norway.[272] (From Tangen D.A., *Utradisjonelle gjenbrukstiltak. Eksempelsamling*, Technology Report 2377, Norwegian Public Road Administration, Oslo, Norway, September 2006.)

from demolishing of concrete should be recycled should not be up for discussion, but it might be discussed how we recycle it.

When we recycle the reinforcement, hardly anyone claims that the new recycled steel coming from the smelter should go for new reinforcement. The new "old" steel might be for a car, construction profiles, reinforcement, or something else. When it comes to the aggregate, the thinking should be the same. The recycled aggregate should go to the use of aggregate that gives the most sustainable solution, sometimes for concrete, sometimes for something else (Figure 4.6).

Figure 4.5 On-site recycling of concrete guard rails at the scenic mountain serpentine road, Trollstigen, Norway.[272] (From Tangen D.A., *Utradisjonelle gjenbrukstiltak. Eksempelsamling*, Technology Report 2377, Norwegian Public Road Administration, Oslo, Norway, September 2006.)

The situation differs considerably from one part of the world to another. Let us mention a few cases with different conditions with respect to the demand and supply.

4.1.1 Norway

In 2005, Norway produced 53 million tonnes of aggregate. Nearly half of this went to other countries (export). Between 10 and 14 million tonnes was used as aggregate for concrete. The rest was utilised in roads, trenches, fillings along buildings, etc.

The total amount of debris from buildings in Norway was at the same time less than 1.3 million tonnes, and of this, less than 1 million tonne was from concrete (0.2 tonne per capita). Of this, less than one-third could practically

Figure 4.6 Recycled concrete aggregate for use in concrete.[273]

be recycled in concrete. Consequently, the amount of aggregate from demolishing was less than 1% of the total amount of aggregate produced in Norway.

4.1.2 Japan

According to Noguchi et al.[274] and Sakai,[275] Japan annually uses 500 to 700 million tonnes of aggregate, whereas two-thirds goes for concrete (Figure 4.7).

According to Noguchi et al.,[274] the Development Bank of Japan in 2002 anticipated the amount of building demolition waste to exceed 300 million tonnes in 2010 and 400 million tonnes in 2030. The last figure is triple that from 2000. More than half of this is concrete. The need for advanced recycling technology and a high proportion of recycled aggregate in new concrete is therefore obvious. Noguchi claims that between 2035 and 2045 the amount of recycled concrete from demolition waste will be higher than the need for aggregate in concrete. Noguchi and Fujimoto,[26] referring to the Ministry of Environment, claim that the total material input in Japan in 2000 was 2.0 to 2.2 billion tonnes, where concrete production accounts for 25%.

According to a paper from Dosho in 2007,[276] referring to the Ministry of Land, Infrastructure and Transport (MLIT), the amount of construction waste produced in Japan in 2002 was approximately 83 million tonnes per

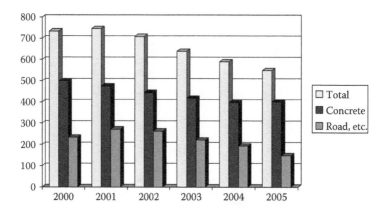

Figure 4.7 Aggregate consumption in Japan, according to Japan Crushed Stone Association.[275] (From Sakai K., Mail Correspondence, Numbers from Japan Crushed Stone Association, August 17, 2007.)

year, where concrete makes approximately 35 million tonnes. Ninety-eight percent of this is recycled, and most of it used for roadbed gravel. He reports on figures from the Tokyo Electric Power company, which in 2004 owned 5800 buildings, excluding 700 nuclear power facilities. As most of them were built more than 30 years ago, they will in the future generate a lot of waste, anticipated to be about 7.8 million tonnes, of which 7.6 million tonnes is concrete waste, 0.139 million tonne mixed waste, 29,000 tonnes woody waste, 64,800 tonnes plastic, and 27,400 tonnes glass and pottery.

4.1.3 The Netherlands

The Netherlands was among the first countries to systematically look into the possibility of aggregate recycling in later decades. In a paper from 1998, Pietersen et al.[277] reported that demonstration projects for the use of recycled concrete as aggregate in new projects have been carried out for more than 15 years. In an overview of the size of the challenge, they claimed that at least 200 to 300 tonnes of building and demolition waste was landfilled or applied in road construction within the EC countries. In the Netherlands, almost 15 million tonnes (MT) of building and demolition waste was produced in 1996, equivalent to 1 tonne per capita. Of this, about 8 MT was recycled, where:

- 1.0 MT was recycled concrete aggregate (+0.07 MT was applied in concrete)
- 0.9 MT was recycled masonry aggregate
- 5.3 MT was recycled mixed aggregate (+0.11 MT was applied in concrete)

Of this, 11 million tonnes is recycled, where the majority (80%) goes for "on the ground" applications. The total consumption of aggregate in the Dutch building industry is estimated to be 120 million tonnes per year, where 15% is imported.

The paper reported that roughly 80% of the annual concrete production of 16 million m³ in the Netherlands was used for ordinary concretes in the strength class of 25 MPa, and indicates that this is the primary focus for increased use of recycled aggregate in the production of concrete. The authors further report from surveys that a replacement level of 20% of recycled concrete or mixed aggregates, for common strength classes such as B15 or B25, is very well possible without any loss in performance. In 2001, Fraaij et al.[278] reported that a landfill ban for recyclable C&D waste in the Netherlands, effective from 1997, stimulated notably recycling, and recycling of construction and demolition (C&D) waste increased from about 60% in 1990 to about 95% in 1999. They also claim that R&D waste in Europe is equal to 0.6 to 0.9 tonne per capita per year, two to three times the domestic waste produced per capita per year.

4.1.4 Hong Kong

Poon and Chan[279] explain that Hong Kong, with its 6.8 million people on islands of mountainous landscape with limited land space, has a very special situation. The drive to create new floor space by demolishing old buildings and replacing them with new buildings has led to an increasing amount of construction and demolition waste, that in 2004 amounted to over 20 million tonnes over a rather limited area. In the past most of the waste has been used beneficially to form land for the fast-growing perpetual development. However, increasing environmental awareness has reduced this possibility considerably. Advanced crushing and sorting operations have been built, but due to reluctance to use an inferior quality material, the ease to find natural aggregate at an acceptable price, and the logistic challenge to have space and silos for different aggregate types in the production facilities, etc., very little of the demolition waste is recycled in concrete (Figure 4.8).

4.1.5 General

The World Business Council for Sustainable Development (WBCSD) mentions in a brochure an estimate saying that 900 million tonnes of demolition waste is generated each year in Europe, the United States, and Japan.[280] The annual construction waste from abandoned buildings in China reaches more than 100 million tonnes, in which concrete occupies 40%. Recycling of concrete demolition waste in concrete differs, as mentioned above, considerably from country to country and from region to region in many countries, for various reasons, probably mostly based on availability and market supply and demand, and the price of natural aggregate.

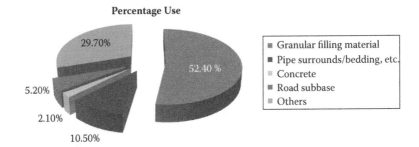

<figure>
Percentage Use

29.70%

52.40 %

5.20%

2.10%

10.50%

■ Granular filling material
■ Pipe surrounds/bedding, etc.
▪ Concrete
■ Road subbase
▪ Others
</figure>

Figure 4.8 Distribution of the use of recycled aggregate in Hong Kong.[279] (From Poon C.S., Chan D., A Review on the Use of Recycled Aggregate in Concrete in Hong Kong, in *International Conference on Sustainable Materials and Technologies*, Coventry, UK, June 11–13, 2007, pp. 144–155.)

We should probably see these figures in relation to the total amount of natural stone aggregate that is excavated each year in the world, about 20 billion tonnes. Consequently, the recycling challenge probably boils down to reducing the excavation with less than 10% and substituting this with demolition waste and recyclable materials—an effort that should not be too complicated, if the awareness is present.

According to WBCSD,[280] two countries in Europe have a consumption of recycled aggregate higher than 20%: Netherlands and UK. One country has a consumption percentage between 15 and 20%, Belgium, and two countries have consumption of recycled aggregate between 7 and 10%, Germany and Switzerland. All the other countries have lower recycling rates. In total, recycled aggregate accounts for approximately 6 to 8% of the concrete aggregate used in Europe.

Because the motives, partly accelerated by government incentives, and the need and inspiration for recycling concrete debris as aggregate in concrete are fundamentally different from one country to another, the methods and incentives for more advanced recycling technology will also be different. Japan, which has greater need for using recycled aggregate in concrete than most other countries, has naturally invented more advanced methods for recycling and "purifying" the recycled aggregate to a higher degree. Scrubbing, use of heat, grinding, etc., are tools in designing impressive recycling systems that might be very interesting and useful. However, it is not necessarily so that these methods, which also require environmental resources in their application, will be the perfect choice under other circumstances.

It is interesting to reflect on the development from a paper by Corinaldesi and Moriconi in 2003.[281] They reported that in the European Union there are annually more than 450 million tonnes of construction and demolition waste (C&DW), which constitutes the third largest waste stream, after mining and farm waste. If one excludes earth and excavated road

Table 4.1 Construction and demolition waste practice in EU countries, 1996, according to Corinaldesi and Moriconi,[281] referring to European Commission statistics of April 2000

EU member countries	C&WD MT	Reused or recycled %	Incinerated or landfills %
Germany	59	17	83
UK	30	45	55
France	24	15	85
Italy	20	9	91
Spain	13	5	95
The Netherlands	11	90	10
Belgium	7	87	13
Austria	5	41	59
Portugal	3	5	95
Denmark	3	81	19
Greece	2	5	95
Sweden	2	21	79
Finland	1	5	95

material, the amount of C&DW is estimated to be 180 million tonnes per year. Roughly 75% of the waste is disposed to landfills. However, some of EU countries, like Denmark, the Netherlands, and Belgium, are achieving recycling rates above 80% (Table 4.1).

Among the comments in the paper regarding testing of recycled material in concrete, the authors are of the opinion that the presence of masonry rubble is particularly detrimental to the mechanical performance and durability of recycled aggregate concrete, and that the same negative effect is detectable when natural sand is replaced by fine aggregate recycled aggregate fraction. However, they add that the strength losses can be counteracted by, for example, reduction of the water/cement ratio and addition of mineral and chemical admixtures.

At a conference in Lillehammer in 2008, Eddy Dean Smith from the U.S. Army Corps of Engineers gave an interesting glimpse into the challenges and possibilities when larger, concentrated amounts of concrete and other debris have to be taken care of.[282] He referred to Hurricane Katrina in 2005, where 100 million cubic yards of debris was generated, and to the U.S. Army, among the largest landowners in the world, owning some 93 million m² of real property, which plans to demolish old barracks, where a total of 26 million tonnes of demolition debris will be generated over a 15-year period. In such cases new and volume-related alternatives are coming to the surface, together with well-known alternatives. Among the alternatives mentioned in the paper, in addition to huge recycling of concrete as a road base, and as aggregate in new concrete, are reuse of precast components as shoreline protection and stabilisation, and stabilisation and protection against erosion of old mining operations. He also claimed that harbour and pier constructions are excellent candidates for recycling of large precast and prestressed concrete members from demolition.

Chen and Liau from Taiwan in 2001 presented an interesting test program trying to systematically find the effect of impurities from bricks and tiles in the recycled concrete aggregate.[283] In addition to control mixes with natural aggregate, they replaced the coarse natural aggregate with recycled concrete, and with 2/6, 3/6, 4/6, 5/6, and 6/6 replacement with bricks and tiles. They also tried out recycled aggregate from different parts of Taiwan, and the effects of changes in the w/c ratio. From their conclusion, we mention that they claim that when the proportion of bricks and tiles was lower than 60%, it caused only moderate reductions in the mechanical properties in recycled concrete. They also reported that when unwashed recycled aggregate was used, the strength properties were effected, and in particular for the lower w/c ratios. At a w/c ratio of 0.38, the compressive strength of recycled concrete was only 60% of that of normal concrete.

4.1.5.1 Processing technology

Various types of processing for recycling of aggregate have been introduced in different countries, and to a high degree in technology advanced processes, depending on market demands in the respective countries.

As mentioned previously, Japan has been a "leading lady" in developing some of the most advanced processing systems with heating and scrubbing devices in addition to crushers and sieves.

Noguchi[284] gives the illustration shown in Figure 4.9 of process flow in Japan depending on desired quality.

Sakai[28] summarises some of the more important Japanese recycling methods as:

- Heating and rubbing (Tateyasshiki et al., 2000)
- Eccentric-shaft rotor (Yanabashi et al., 1999)
- Mechanical grinding (Yoda et al., 2004)

In the heating and rubbing method, the concrete debris is heated at 300°C and the cement paste is weakened to remove mortar and cement paste from the aggregate. In the eccentric-shaft method, crushed concrete is passed downward between an outer cylinder and an inner cylinder that eccentrically rotates at a high speed to separate it into coarse aggregate and mortar through a grinding effect. In the mechanical grinding method, coarse and fine aggregate are separated by small holes in partition plates placed at an angle, with the centre shaft in a horizontal rotating drum where grinding is done by the aide of steel balls.

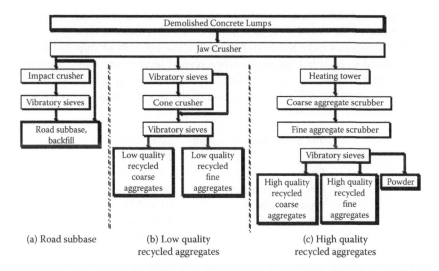

Figure 4.9 Process flow for production of recyclable aggregate depending on desired use.[284] (From Noguchi T., Optimum Concrete Recycling in Construction Industries—Japanese Experiences. Presented at 2008 International Expert Workshop on Cement and Concrete Technology for Sustainable Development, Beijing, October 7–12, 2008.)

Another leading lady in recycling of concrete is the Netherlands. In 2007, Mulder et al.[285] reported on a new advanced processing method. They called the concept closed-cycle construction and claim that the process is the first of its kind in the world. They claim that the processed materials are being reused on a higher-quality level, and that the quantity of the wastes that have to be disposed is minimised. They say that for concrete and masonry, the concept implies that the material cycle will be completely closed, and that the original constituents are recovered in a thermal process. One of the benefits of the concept is that the thermal process steps are fuelled with combustible fractions of the construction and demolishing waste itself. The demolition rubble is processed in separate waste streams, where concrete is one of them. The concrete rubble is thermally treated at a temperature of about 700°C to dehydrate the cement stone. The concrete rubble pieces disintegrate, and the original pieces are set free. After treatment, only 2% of the hardened cement paste is left on the sand and gravel grains.

There is reason to believe that the quality of the recycled aggregate is increasing with increasing number of steps in the production process, but it is hard from the literature to conclude what is an optimum process and what is "good enough," taking into account both the percentage replacement range and the type of concrete that the recycled aggregate is going to be used for. It is also evident that the more advanced methods have a higher demand for resources in terms of investments and energy consumption.

The optimum environmental choice therefore will be different from one part of the world to another, depending on local circumstances.

Most countries have made standards for the amount of impurities that are allowed in recycled aggregate when reused in concrete. Here are some typical approximate maximum values. The values are approximate, as the specifications might differ from one part of the world to another, and sometimes for different classes of concrete:

Tile, bricks, ceramics, asphalt	2.0 to 5.0%
Glass	0.5 to 1.0%
Plaster	0.1% (J)
Inorganic substances	0.1 to 0.5%
Plastics, insulation	0.1 to 0.5%
Wood, paper	0.1 to 1%
Total	**1.0 to 6.0%**

There is a wide range in requirements, probably based on results from both local testing and the differences we experience in building culture, and consequently in constituents in the demolition rubble. International standards might be interesting, but are not necessarily the most efficient tool in this case, due to the great variability on both the debris end and the user end (Figures 4.10 and 4.11).

In addition, most regulations have additional requirements on, for example, maximum water absorption, minimum density, toxic substances, etc.

4.1.5.2 Fines

One of the challenges in recycling crushed demolished concrete is the *fine fractions*. In use as both road base and recycled aggregate in concrete, the fine fractions create some limitations.

The fine fraction (less than 4 mm) might make as much as 15 to 25% of the crushed concrete,[287] and 30 to 60% of this might be cured cement paste, with some reactive properties (not only in recycling for concrete, but also in recycling as a road base the fines give headaches). The fine fraction might lead to clogging and carbonation in draining systems, and undesired stiffness of the road base.

Researchers from the German Cement Works Association[288] report on the use of fine crusher sand in cement clinker production. They also refer to similar work in the United States and Spain. In their research they looked at crusher sand from demolished concrete from various parts of Germany. The chemical composition compared to clinker is illustrated in Figure 4.12.

In their findings, they conclude that there is potential for use of such sand in the cement industry, but large differences in the chemical constituency of the sand exist. In particular, the variation in the content of silica and

Figure 4.10 Typical fracture planes in concrete with recycled aggregate. Impurities marked with darker spots.[286] (Photo courtesy Norwegian Public Road Administration. Tangen D.A., *Gjenbruk av betong*, Report 2479, Norwegian Public Road Administration Technology, Oslo, Norway, December 2007.)

Figure 4.11 Perfect concrete surface in concrete retaining wall with recycled aggregate.[286] (Photo courtesy Norwegian Public Road Administration. Tangen D.A., *Gjenbruk av betong*, Report 2479, Norwegian Public Road Administration Technology, Oslo, Norway, December 2007.)

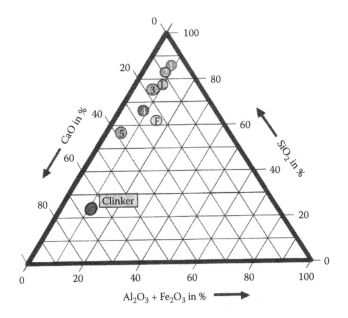

Figure 4.12 Ternary diagram over CaO-SiO₂-Al2O₃+Fe2O₃ with the composition of clinker and concrete crusher sands 1 to 5 and 1L and 1F according to Hauer and Klein.[288] (From Hauer B., Klein H., Recycling of Concrete Crusher Sand in Cement Clinker Production. Presented at *International Conference on Sustainability in the Cement and Concrete Industry*, Lillehammer, Norway, September 16–19, 2007.)

calcium oxides leads to different potentials that can limit the substitution. The potential substitution rates vary between 4% and almost 10% for the investigated samples.

In a sustainability context they claim that the use of crusher sand (0 to 2 mm) can preserve up to 10% of natural raw materials and up to almost 3% of fuel and energy.

Other recycling uses of the fine fraction of sand that have been recorded, besides use as landfill, are:

- Stabilisation of old mine shafts
- pH balance of acidic fertilisers in the soil in agriculture
- Fine fraction addition in self-compacted concrete

Of the many papers about ideas trying to utilise the fine fraction, we like to mention a very interesting test by Peng et al., presented in 2012.[289] They tested fines from 2-year old mortar, powder smaller than 2.36 mm from 20-year-old limestone concrete, and powder with a diameter smaller than

1.18 mm from 40-year-old concrete. They tried to activate the fine by a grinding activation method and by autoclaving. Then they cast blocks from the activated fines by mechanical pressure of 80 MPa. Their final product became a lightweight concrete with a density ranging from 1.5 to 1.8 g/cm³ and a compressive strength of 9 to 17 MPa. By refining the method further, the idea should be an interesting way of using fines for the production of building blocks.

In particular, over the last two decades numerous projects with research, testing, practical utilisation, and regulatory work have been done with recycled crushed concrete, in most countries in the world. Whether the crushed concrete is used for road bases or in new concrete or for other purposes, the rules for investigating the material should, in principle, not be different from testing of any other debris with respect to controlling it for possible harmful substances. The substances that most likely have to be checked with construction debris are lead, quicksilver, cadmium, zinc, chromium, arsine, nickel, copper, polychlorinated biphenyl (PCB), and polycyclic aromatic hydrocarbon (PAH). An important question, where the answer is somewhat different from one part of the world to another, and somewhat different from one type of use to another is: How clean is clean enough?

From being a rather rare item a few decades ago, use of recycled concrete aggregate in concrete has become a more normal alternative when the "best option" shall be chosen. In *Concrete International* in March 2010, a group of Canadian researchers even presented a new mixture proportioning method.[290] A particular feature with the method is to highlight the original coarse aggregate and the residual mortar (RM) within the recycled concrete aggregate as distinct phases. The RM phase is considered to contribute to the total mortar content in the recycled concrete aggregate concrete. The procedure developed for the method adds the RM to the volume of mortar in the natural concrete aggregate, and is therefore called the equivalent mortar volume (EMV) method. The method is developed from thorough testing as well as strength properties like durability of the concrete.

D. Bjegovic from Croatia has presented a number of interesting papers on the recycling challenge in her country at various international conferences and meetings over the last decade. In a paper at a conference in Jinan, China, Bjegovic et al.[291] claim that approximately 80% of construction waste, given certain recycling technologies, may be used as an economically valuable resource. However, still only 5% of the waste in Croatia is recycled. They recommend using a maximum of 30% recycled aggregate in concrete mixtures, because up to that amount, properties of concrete regarding compressive strength, free shrinkage, and water permeability would be satisfactory, or very near to the control concrete properties.

Numerous papers from all over the world have been reporting on the use of recycled concrete in new concrete. There are variations in the findings in the research and the conclusions, often depending on:

- What replacement levels/percentages that have been tested
- The contaminations (other building materials) in the aggregate
- The aggregate size tested
- The original quality of the concrete that is recycled
- The amount of paste left on the aggregate

As an example we mention an interesting paper by Yang et al. in *ACI Materials Journal* in 2008.[292] They tested various sizes of aggregate with 30, 50, and 100% replacement and tried to find relationships between different concrete properties and replacement levels and water absorption of the aggregate. They concluded that increasing water absorption of the recycled aggregate will slightly affect the initial slump, will decrease the rate of bleeding, and might have different effects on the early strength than on the long-term compressive strength, that the compressive strength, as well as the modulus of rupture and splitting strength will decrease, and that the long-term shrinkage strain will increase.

General impressions from the papers we have had access to are:

- Reduced quality of the new concrete will depend on the amount of paste residue on the coarse aggregate.
- Increased use or higher dosages of superplasticisers are often recommended or necessary.
- Increased shrinkage might be expected in using unwashed recycled aggregate in a higher replacement dosage.
- Higher substitution rates might reduce the freeze-thaw properties somewhat.
- There is less reduction in strength properties and other mechanical properties from recycled concrete of initially high quality (lower w/c ratios) than from low quality concrete.
- Substitution of *clean* recycled aggregate from higher-strength classes of concrete might sometimes give slightly better results than new aggregate for the strength and E-modulus properties. This has been explained with better bonding to recycled aggregate.
- Cleaned recycled aggregate from temperature treatment or similar does not show any quality reduction compared to new aggregate (slightly the opposite might be the case).
- Smaller substitutions of recycled aggregate (less than 15 to 30%) only show differences in quality from 100% new aggregate that might be characterised to only be of academic interest.

We have for many years seen examples of what we might call *in-house recycling* in the concrete products industry. When it comes to recycling of wash water, we have treated this subject in Section 5.9. However, it is not abnormal to register wastes in the concrete product industry in the order of 0.5 to 2% of the production volume, somewhat different from one type of production to another. Many producers crush this waste and reuse it in smaller substitution dosages. The most fascinating example of such recycling we have is from a small concrete product factory outside Dar es Salaam in Tanzania. The production plant produced pigmented red roof tiles and decorative exposed aggregate pavers. Wastes from the roof tile production were crushed in a simple crusher and screened once a month, and the appropriate fractions were utilised as decorative aggregate in some of the pavers. The rest was substituted for natural aggregate in the roof tile production. The owner and plant manager simply argued that he had to recycle his most valuable raw material, the red pigment.

As arguments for not using, or lesser use of, recycled aggregate as aggregate in concrete, we have summarised the following obstacles:

- Logistic challenges and costs for using several aggregate types in the production (more and special silos, etc.)
- The availability and price of competing natural aggregate
- Lower concrete quality at high replacement rates
- Requirements for increased costs in testing and control
- Unstable supply of recycled aggregate
- Unstable quality of recycled aggregate
- Conservatism, suspicion, and lack of technical knowledge on the buyer side, including the buyer's advisors and consultants
- Easier for the recycled aggregate supplier to find markets outside the concrete sector

Realistically, it will also in the future probably, in most cases, be more sustainable to recycle concrete for other uses than for use as aggregate in concrete, but there are reasons to believe that more will be used as concrete aggregate in the future. To overcome some of the hindrances mentioned above, we might need both carrots and sticks as advantage points for recycled aggregate in rating systems, and positive governmental regulations regarding use of recycled aggregate.

We also mention a rather special item for recycling. The Norwegian journal *Byggeindustrien* in 2007[293] reported that a contractor specialising in grinding and polishing concrete floors collected the dust from its grinding machines in special dust-collecting bags, and supplied the dust to a local ready-mixed operation, using it as fine raw material in self-compacting concrete.

There are probably still many new rational recycling possibilities not yet tried. In the introduction to this chapter we cited Swamy, who argued that awareness is one of the key factors for successful recycling strategies in the future.

However, an overall goal for concrete waste should be zero landfill, and the concrete industry should be a leading lady in finding the most sustainable recycling alternatives.

We are definitely in favour of a 100% recycling policy for demolished concrete, but from an environmental standpoint, we do not consider it natural that concrete necessarily should be recycled in concrete. When steel is recycled, hardly anyone would request that reinforcement should be recycled as reinforcement, or that steel from outdated cars should be recycled in cars, or that stainless steel forks should be recycled as forks. In the same way, we think that it is unnatural that aggregate from demolished concrete necessarily should be recycled in concrete. Recycled, yes, but to the purpose that requires the least environmental burden in a holistic evaluation.

4.2 RECYCLING OF OTHER MATERIALS AS AGGREGATE IN CONCRETE

There hardly seems to be any limit to the various materials that have been suggested to be recycled in concrete. Some of them give concrete special properties, for example:

- Granulated used tires for the production of paving stones, etc., to produce concrete with shock-absorbing effects (Figure 4.13)
- Building blocks from organic waste like coconut husk, sisal, newspaper, sawdust, etc., to make concrete that can be sawed and nailed with ordinary tools
- Decorative concrete from crushed glass (Figure 4.14)
- Recycling wastes to improve self-compacting concrete
- Recycling process waste to produce lightweight aggregate

Other initiatives have been just to produce concrete from an available abundant resource.

As mentioned previously, there seem to be few limits to the possibilities and innovations, and the research, development, and ideas come from many countries around the world. As examples, we mention some of the many alternatives shown so far.

4.2.1 Used rubber tires in concrete

Granulated or crumb rubber has been tried out in a number of countries. At a conference in Jinan, China, in 2010, three different groups of researchers from three different parts of China—Guangzhou, Henan, and

Figure 4.13 Concrete pavers with granulated rubber from old tires as aggregate.

Figure 4.14 Beautiful precast wall at the new concert hall in Stavanger, Norway, with crushed glass. (Producer, and photo from AS Betong, Sandnes, Norway.)

Dalian—discussed various aspects of this interesting topic. Zhang and Guan[294] claim that 100 million waste tires were stockpiled in China only in 2006, with a possible health and environmental risk. Liu and Wang,[295] having conducted a worldwide state-of-the-art study, claim that the number for the United States in recent years was 275 million waste tires each year, and that 290 million tires are stacked and not recycled.

A paper from Skripkiunas et al. from Kaunas, Lithuania, at a conference in Coventry, UK, in 2007[296] also mentions 275 million waste tires generated per year in the United States, and 180 million tires in the European Union. The authors report using crumbled waste rubber as fine aggregate in three fractions: 0 to 0.5 mm, 0 to 1 mm and 0 to 3 mm. In all tests they used 451 kg of cement, 380 kg of 0 to 8 mm crushed granite, 569 kg of 11 to 16 mm crushed granite, 821 kg of sand, the same amount of superplasticiser in all tests, and 158 kg of water, per cubic meter of concrete. In the rubber concrete specimens they used 22.8 kg of crumbled rubber. In their conclusion the authors claim:

- The finest waste rubber additives reduce the slump, while the coarse rubber particles increase it.
- Fine rubber particles increase the air content, while the coarse particles do not change it.
- The concrete strength decreases with use of rubber waste, and mostly with the medium-size particles.
- Rubber waste additives and their coarseness have no effect on concrete abrasion-resistance characteristics.
- The rubber particles have a positive effect on the concrete freeze-thaw resistance.

Tests around the world with granulated rubber have been with particles from sizes of more than 30 mm, down to rubber powder. The replacement of sand with rubber has been from 40%, down to less than 10%. The best material results are obtained with dosages in the lower range. The results are interesting and show possibilities of reduction of the elastic modulus, increased toughness and shock resistance, somewhat varying results regarding frost resistance in particular and durability in general, but generally good results, and tendencies to reduction in plastic shrinkage.

Of the many test reports available, we also mention a report from Iran,[297] where various percentages of the coarse aggregate were replaced by cut and shredded rubber particles 13 to 76 mm in size. The authors found that there was no reduction in the compressive strength of the concrete with substitution up to 5% by weight. They even found a minor increase. On higher dosages, the compressive strength was reduced. However, when 5 to 10% of coarse aggregate was replaced by shredded rubber, the tensile strength was reduced with about 2 to 12%, and the flexural strength by 30 to 60%. The modulus of elasticity was reduced by 17 to 25% for the same dosages.

Because the surface of the rubber crumbs is hydrophobic, while the cement matrix is hydrophilic, the binding of the matrix to the rubber aggregate is weak. The result of this, combined with the weaker material strength compared to natural sand, results in a comparatively higher reduction in compressive and flexural strength than might be expected. Washing and acid treatment of the rubber aggregate has shown improvements. In two of the mentioned papers from the Jinan conference,[294,298] interesting tests were done after special surface treatments of the rubber crumbs to improve bonding. Both papers show considerable improvements in the compressive strength. Zhang and Guan[294] also show an interesting improvement in the ratio between flexural and compressive strength.

Additional rubber crumb concrete information was given in two papers at a conference in Beijing in 2012. Researchers from Dalian, China, tested various cement and w/c ratios with a constant quantity of rubber crumb of 20 kg/m^3.[299] When the cement content increased from 288 kg with a w/c ratio of 0.66 to 528 kg and with a w/c ratio of 0.36, the compressive cube strength increased from 16.8 MPa to 49.0 MPa. In their paper they also, with reference to the World Sanitary Organization, claim that the number of accumulated used tires in the world has reached over 3 billion, and that this figure is growing by 1 billion tires every year. By the end of 2010, the used tires in China will exceed 200 million, to rank first in the world.

Researchers from Beijing[300] tested different dosages of rubber crumb as volume replacement of sand, in dosages of 0, 10, 15, 20, and 30%. Their base mix had a cement dosage of 350 kg/m^3 and w/c ratio of 0.34, with a compressive strength of 50 MPa. The testing showed a reasonable linear reduction in compressive strength to 28.63 MPa at 30% sand replacement as the rubber crumb dosage increased. They also concluded that the workability decreased with increased content of rubber. They claim that anticrack performance of rubber crumb concrete is much better than the benchmark, and that the cracks are both less and shorter. The antifatigue properties are also improved.

Wong and Ting presented a paper in *ACI Materials Journal* in 2009,[301] where they presented very interesting results from tropical conditions in Singapore. In principle, they tested the use of both substitution of rubber crumb for fine aggregate and rubber chips as substitution for coarse aggregate, and in both cases with 25% substitution. This was tested in both "normal" concrete and high-strength concrete. Various mix designs were tested, with Portland cement and high substation rates of cement with granulated blast furnace slag, rather high dosages of silica fume, and the use of steel fibres. The authors also reported that of scrap tires in Singapore, 88% were recycled and mostly incinerated. The compressive strengths of their different mixes with 25% substitution of rubber crumb or chips was in the range of 40 MPa for high-strength concrete and 10 to 20 MPa for normal concrete. In their conclusion they claim:

- Both types of concrete with 25% crumb rubber or rubber chips had lower slump and unit weight but higher air content than mixes without rubber.
- The use of rubber chips reduced the slump, but increased the unit weight compared to rubber crumb.
- Nonrubberised high-strength concrete experienced much more brittle failure than normal-strength concrete and the rubberised mix designs.

4.2.2 Aggregate manufactured from fines

Lupo et al. presented an interesting paper at a conference at Coventry, UK, in 2007,[302] where they reported on work to try to produce an aggregate from fines. However, many researchers from several countries have been working with the idea to try to utilise the silty materials normally deposited as waste that come from washing in aggregate production, wash water in concrete production, or when utilising recycled concrete. Ideas like sintering and production of building blocks and binding with Portland cement to produce low compressive strength blocks, etc., have been tried out, attempting to utilise this very large potential. The authors manufactured an aggregate with fine materials of particle sizes below 73 microns from five different UK sources. In a Hobard N50 mixer they supplied silty material and dry sand in a proportion of 23:57 dry sand:wet silt, and added a binder of ordinary Portland cement (OPC) and (poly)vinyl alcohol (PVA). The resulting mortar was processed in different ways and dried. They concluded that in this way it was possible to produce a concrete for backfills and low-strength applications.

4.2.3 Processed sugar cane ash

At a concrete conference in New Zealand in August 2011, researchers from Thailand reported positive tests with lightweight aggregate based on ash from sugar canes that were pelletised and sintered.[303]

4.2.4 Recycled plastic, e.g., bottles

At a sustainability conference in Milan, Italy, in 2003, researchers from the engineering faculty of Palermo, Italy, showed interesting tests with recycled plastic bottles (Figure 4.15).[304] The background for the investigation was the fact that Italy is the greater European consumer of polyethylene

From recycled plastic bottles....

.... to lightweight aggregate for concrete

Figure 4.15 Illustration on the front of the *Proceedings for an International Seminar on Sustainable Development in the Cement and Concrete Industries*, Milan, Italy, October 2003.

terephthalate (PET) bottles, with a yearly consumption of 400,000 tonnes. The quantity had increased 250,000 tonnes in 4 years. This recycled bottle cannot be employed to manufacture new bottles or precision mechanical pieces. The polymer itself has many good characteristics: versatility, lightness, hardness, low linear dilatation coefficient, good chemical and thermal resistance, low cost, and easy workability. Positive tests were reported, with compressed bottles as both aggregate varying in sizes from 10 to 20 mm and flakes as reinforcement.

From a conference in 2001, a paper was published by researchers from Belgrade in former Yugoslavia[305] looking at different types of demolition waste, including plastic, as aggregate in concrete. They stress the importance of a waste-sorting management in a war and demolition situation where unsorted particles of concrete, tiles, bricks, glass, plastic, etc., might be found, to be able to recycle as much as possible, and not only discard the materials into the landfill. In their report they tested and compared concrete produced from natural aggregate, recycled concrete, recycled plastic, and waste of chromium-magnesite bricks. Regarding the last two, here are some of their findings:

- There were two test series of recycled plastic—one with cement: plastic = 1:2 and w/c = 0.71 and one with cement: plastic = 1:3 and w/c = 0.55. The bulk density of the recycled plastic was 460 kg/m^3. The high w/c ratios were needed to achieve the same workability as their reference, because of the unsuitable particle size distribution of the plastic aggregate. The particle sizes were, with more than 90% of the particle, from 1 to 8 mm. The bulk densities of the plastic prisms were 1210 and 860 kg/m^3, respectively. The compressive strengths were recorded at 2.80 and 5.87 MPa, respectively, after 28 days. The authors conclude that the material might be used as thermo insulating blocks.
- The prisms produced with waste chromium-magnesite bricks had a cement:aggregate ratio of 1:3 and w/c ratio of 0.5. The prisms with this aggregate had the highest strength of the various materials tested, with a compressive strength of 62.62 MPa after 28 days, comparable to 51.24 MPa for the reference concrete of natural stone aggregate.

Expanded polystyrene, mostly in the form of beads, has for many years been utilised as aggregate to make lightweight concrete. Various practical technologies to prevent the light beads from floating to the top in the mixer have been developed in several places. Trussoni et al.[306] reported on tests comparing intentionally manufactured beads of expanded polystyrene with waste-expanded polystyrene. The waste polystyrene was expanded by the use of steam and an expansive agent, and later granulated. In their concrete tests, they did not use any special admixtures. In their test they worked with w/c ratios in the 0.5 range and unit weights in the 1860 to 1950 kg/m^3 range. Compressive strengths after 28 and 56 days were in the range of 16.5 to 20.7 MPa. There was not much difference in the concrete properties from the lightweight concretes with intentionally manufactured beads and the waste materials.

4.2.5 Hempcrete and other "straw concretes"

At a conference in Lillehammer, Norway, in 2007, researchers from Queen's University, Canada,[307] reported on tests with hemp as aggregate. Hemp fibres have been used in various manufacturing process in Canada since 1998. However, the woody centre of the hemp plant, making up 80% of the plant, is normally used for animal bedding or is discarded. To possibly find a use for this material was the background for the study. The study was performed with a very wide range of cement additions. The compressive strengths were rather low (e.g., 2.5 MPa), and possible uses are probably limited to building blocks, etc., but the deformability properties of the hemp mortar were interesting, with typical, very high strain after critical load.

At the same conference in Lillehammer, researchers from France[308] reported on research with a building material-based hemp and lime, which they called hemp lime concrete (HLC). The initiative was not so much a recycling effort as it was to create a material with low carbon footprints. The material has in France been used to:

- Fill load-bearing timber frame structures (wall application)
- Cover masonry walls (coat application)
- Insulate roofs (roof application)
- Insulate floors

The material can be sprayed or tampered, but has a very moderate compressive strength, typically below 0.2 MPa.

At a conference in Jinan, China, in 2010, Chinese researchers from Yancheng, Nanjing, and Suzhou reported that the total output of crop straw in China is more than 700 million tonnes. After the annual summer and fall seasons, the burning of straw is causing serious environmental pollution. On the other hand, the average porosity of straw is up to 83.5%, which makes it a potential thermal insulation and sound absorption material. In the mentioned paper,[309] they tested the effect of wheat straw, rice straw, cotton stalk, and rice husk on properties of concrete. As the sugar content of the straws has impact on the setting performance and retarding effect of concrete, increased knowledge on this item was the main purpose of the investigation. The main finding in the research was that the inhibitory effect on the hydration of cement gives the following order with respect to type of straw: wheat straw, rice straw, cotton stalks, rice husk. With a mortar containing 5% of straw they received 28-day strengths of 16.7 MPa with rice husk, 17.3 MPa with cotton stalks, 25.2 MPa with rice straw, and 25.9% MPa with wheat straw. These were 49.5, 51.3, 74.7, and 76.8% of the strength of the control mix, respectively.

4.2.6 Papercrete

A number of inventors and researchers in many countries have been working with recycled newspapers to make lightweight concrete, often in the shape of insulating, saw-able, and nail-able building blocks.

A paper by researchers from Finland and Portugal in 2011[310] reported on 28-day compressive strengths in the order of 1 to 13 MPa and thermal conductivity in the order of 150 to 420 mW/m°C.

4.2.7 Oil palm shell lightweight concrete

A 2007 paper from Sabah, Malaysia, reported on tests using oil palm shells as aggregate together with crushed stone to produce a lighter concrete.[311] The researchers used oil palm shell together with crushed stone, and produced a concrete with a dry density of 2000 kg/m^3 and recorded compressive strengths from 34.76 to 39.53 MPa after 28 days, depending on the curing conditions. However, they also recorded that the cylinder compressive strength was 56% of the cube compressive strength.

4.2.8 Glass concrete

The use of glass as aggregate in concrete has been reported in quite a few countries (Figure 4.16).

The use of glass as aggregate seems to have been initiated for at least four reasons:

- To find recycling possibilities for an important quantity of waste
- To use the properties of glass as an architectural feature
- To try to utilise the possible pozzolanic properties of finely ground glass
- To produce lightweight aggregate from crushed and ground glass

Figure 4.16 Weidemann Hall in Steinkjær Town Hall, Norway, with crushed glass as aggregate.

Use of crushed coloured glass as exposed aggregate finish in precast concrete façade panels was tried out in the UK as early as in the 1960s. Some claim problems such as a pedestrian on the street below might get pieces of glass in his or her head from the top panels on sunny days.

An inspiring fact with post-consumer glass is the huge quantity. Naik[312] claims that the quantity is about 10 million tonnes in the United States each year.

In an article in *Concrete International* in June 2003, Meyer[313] reported on the effort, research, and testing at Columbia University in New York. He explained that waste glass constitutes approximately 6% by weight of the solid waste in New York, and that the secondary market for recycling is virtually nonexistent. One of the challenges they were afraid of in their research was increased alkali silica reaction (ASR). Of the results they recorded was increased ASR, but with a clear size effect, and also ASR was much stronger with clear glass than coloured glass. Green glass did not seem to cause ASR-induced expansion, and they believed this to stem from the chromium oxide manufacturers add to the glass to obtain its green colour. They reported several ways to avoid ARS or its damaging effects using waste glass as concrete aggregate:

- The glass may be ground to pass at least mesh size 5.
- Mineral and pozzolanic admixtures could be added.
- The glass could be made alkali resistant by coating it with zirconium.

As mentioned later in this chapter, waste glass has also been tested as a binder substitute.

In a report at a SINTEF Norwegian Concrete Information Day in 2005, Kåre Johansen told about the development story of producing concrete lightweight aggregate blocks based on aggregate originating from recycled glass.[314] A national glass recirculation company, Norsk Glassgjenvinning AS (NGG), annually receives 80,000 tonnes of recirculated glass from 4000 deposit containers around the country. In its advanced processing plant, it takes out metals and sorts the glass in 30 different colours and particle sizes. Some of the glass is ground and transported to a company that expands the glass to a porous foam glass product that can be used as aggregate in lightweight concrete products.

The lightweight aggregate concrete blocks have been named Glasopor and are produced in thicknesses of 15, 20, and 25 cm, with a height of 19 cm and length of 50 cm.

Use of foam glass as a building product, however, is not new (Figure 4.17). Among others, foam glass insulation was used as insulation on the roof of what at the time was one of the highest buildings in the world, Cape Kennedy Space Flight Centre, in 1966.

Figure 4.17 Lightweight aggregate foam glass.

Researchers from SINTEF in Trondheim, Norway, have for many years been working together with the national glass recirculation company NGG to find uses for the recirculated glass in concrete. Per Arne Dahl[315] has made several reports about this. In a paper at Norwegian Concrete Day in 2000, he claimed:

- The single concrete aggregates are typically 3 mm thick, with smooth surfaces on two sides.
- Compressive strengths tested in cube samples are typically 20% lower than comparable normal concrete in cube testing, and 40% lower tested in cylinder samples.
- Unless special mix designs are used, most mix variants are very alkali reactive.
- Drying shrinkage is typically half that of comparable normal concrete.

Recycling of a related product to glass was reported by researchers from Nagoya in Japan in 2001.[316] Silica sand is the primary raw material to produce glass, and 7 million tonnes of silica sand was used in Japan per year, and processed in a refining process. In this refining process about 1.5 million tonnes of this comes out as an industrial by-product, while 0.6 million tonnes is silicious powdered industrial waste, as sludge with particle sizes from 5 to 100 microns. The sludge is partly recycled, but 80% remains unused. The purpose of the tests reported by the researchers was to find a recycling alternative for this in concrete. The silicious powdered waste typically contains 80 to 85% of quartz and 10 to 15% feldspar, where silica accounts for 93.2% of the total volume. The particles are typically coarser than slag powder, limestone powder, and fly ash. The discharged sludge has a moisture content of 30 to 33%. The authors with a positive result tested out use of the sludge in production of high-fluidity concrete to improve segregation resistance, as a partial replacement of fine aggregate.

4.2.9 Paper mill ash for self-compacting concrete (SCC)

Corinaldesi et al. reported on very interesting tests with paper mill ash used to produce SCC.[317] To guarantee stability (avoid segregation and bleeding) and durability in SCC, it is normal to use a higher amount of fines or to combine the use of superplasticisers with the use of a viscosity modifier. An excessive amount of cement is negative to heat of hydration, creep, and shrinkage properties of the concrete. The authors claim that in their market, the increased use of blended cements since the introduction of the cement norm EN 197-1 in 1996 has resulted in a shortage of fine mineral fillers.

In their research, they tested the use of ground and unground paper mill ash in SCC compared to use of limestone filler and a viscosity modifier. They claim that paper mill ash, in particular after grinding, can find proper use as a mineral addition in the production of SCC. The paper mill ash gives good stability to fresh concrete, and as a consequence, the viscosity stabiliser can be avoided. However, a higher dosage of superplasticisers is necessary to guarantee a high slump flow value without modifying the water dosage. The mechanical properties as compressive strength are benefited from the paper mill ash; however, shrinkage might increase moderately, in particular for the ground ash.

4.2.10 Slag

Anashkin and Pavlenko, from the Siberian State University of Industry, Russia, in 2007 reported on the use of 0.14 to 5 mm particle size sand from open-hearth furnace slag as aggregate.[318] They also explained the establishment of a "state standard" for the use of such sand in concrete, and the production process and the quality control process to ensure adequate quality. At the same conference, Lukhanin and Pavlenko[319] reported on the design of heat-resistant concrete by blending ferrochrome slag, quartzite, and activation of the waste materials with water glass. They claimed that the new binder performed better than high-alumina cement.

4.2.11 Recycling of "doubtful" waste as aggregate

Doubtful wastes from industrial production have to a far degree been deposited in nature as landfills or in the sea. Increased attention of the environment and the possible negative effect on public health has led to increased efforts and research to find other, nonharmful, or less doubtful solutions. Recycling in one way or another is obviously one of the answers. Recycling as aggregate in concrete is a natural option to investigate.

A paper from an international sustainability conference in Lillehammer, Norway, 2007,[320] reported on the use of Pb slag as a fine aggregate in concrete. The authors concluded:

- Slag generated from Pb production can be used as fine aggregate with natural fine aggregate. The optimum replaceable amount of slag for production of mortar is about 25%. Utilisation of slag up to this amount as fine aggregate does not alter the strength properties (compressive strength, bending strength, and E-modulus) of the mortars. Addition of slag as fine aggregate lowers the water requirement of mortar mixtures to obtain a consistency similar to that of normal Portland cement–natural aggregate mortar mixtures.
- The toxic element contents in the leachates generated from the slag-containing mortar bars obtained after a diffusion leaching test are far below the standard requirements, although some of the toxic element contents in the leachates of slag-containing mortar are considerably higher than those in the leachate generated from mortar prepared without addition of slag. The leaching of elements from the mortar cubes can be related to the pH of the leachate as well as probably the porosity of the mortars.

The report is interesting, as it gives promising results in the context that we have both a shortage of natural aggregates and an excess of doubtful wastes in many places in the world. However, it must be noted that the report is based on a rather low number of tests, on mortar tests and not concrete tests, and reports from a rather limited range of concrete strengths (about 40 to 70 MPa, which has a reasonable acceptable porosity).

Collepardi et al. reported on the replacement of 20% of fine aggregate with slag from the production of nonferrous metallic slag.[164] They compared both types with Portland cement concrete, a ground granulated blast furnace slag (GGBFS) cement concrete, and a concrete with 15% replacement of cement with ground slag. The compressive strength development of the concrete with the ground nonferrous slag was almost the same as that of the corresponding concrete with GGBFS. When the ground slag was used as aggregate to replace the sand, there was no significant effect on the compressive strength. The leaching by water of heavy metals from the hardened specimens was negligible, and the immobilisation of heavy metals was very effective, particularly when the unferrous slag was used to replace Portland cement only, or to replace natural sand only.

4.2.12 Iron mine mill waste (mill tailings)

Technologists from several countries have made concrete with mill waste as aggregate. In some or even many connections, mill tailings might also be classified as doubtful waste.

Tailings, also called mine dumps, slimes, tails, refuse, leach residue, or slickens, are the minerals left over after the process of separating the valuable fraction from the uneconomic fraction of an ore. Tailings are distinct from the overburden or waste rock, which are materials overlaying an ore of a mineral body that are displaced during mining without being processed. The extraction of minerals from ore can be done two ways: placer mining, which uses water and gravity to extract the valuable minerals, or hard rock mining, which uses pulverisation of rock, and then chemicals. In the latter, the extraction of minerals from ore requires that the ore be ground into fine particles, so tailings are typically small and range from the size of a grain of sand to a few micrometers. Mine tailings are usually produced from the mill in slurry form.[321]

As mill tailings represent one of the most important environmental challenges, with both the possible environmental hazards they represent and the enormous areas they occupy, they are of particular recirculation interest. Use in concrete might be a possible solution to reduce some of them.

Sometimes iron mine waste is discarded in nature by building a protection dam, and pumping the waste behind the dam for permanent storage. In a paper from researchers from Hebei, China,[322] the disaster in Shanxi Province in 2008 is referred to, where a mill waste dam burst and killed 276 people. Accidents like this seem to happen about once a year somewhere in the world.[321] Some of the same researchers, together with researchers from Tangshan, claim, with reference to the Ministry of Land and Resources, that the total amount of quarried ore in China was 8205 million tonnes in 2008, and that 0.909 million ha was covered by piles of waste.[323]

Cai et al.[322] in 2011 reported on tests where they used the fines from aggregate manufacturing as aggregate from mill tailings. They tested various concretes with mill tailing as fine aggregate, and additions of microfines from the manufacturing process. They worked with w/c ratios in the range of 0.37 to 0.43, and cement substitutions with fly ash and slag in the order of 35%. Their compressive strengths were in the 35 to 45 MPa range. They concluded that good concretes could be made with these material combinations.

4.2.13 Bauxite residue/red sand

Another mining residue is the bauxite from aluminium production. Davoodi et al.[324] claim that from 1 tonne of aluminium produced after the Bayer process, 1 to 2.5 tonnes of residue is generated. As one of many alternatives in efforts to recycle this residue, they tested the replacement of sand as aggregate in concrete. The bauxite residue (red sand) was first processed by neutralisation (carbonated) and washed. They claim that the tests show that the sand replacement had no significant negative impact on concrete compressive strength. The inclusion of red mud, however, had an adverse effect on other properties of concrete except the strength, and it reduces the workability of concrete dramatically.

4.2.14 Copper slag

Researchers from Taiwan have reported on the use of copper slag as a substitute for sand.[325] They tested 0, 20, 40, 60, 80, and 100% substitution of the sand, and concluded that substitution rates lower than 80% have only small effects on the strength properties. Of the 1000 tonnes of copper slag produced in Taiwan annually, 75% is utilised, while the rest is discarded off the seashore.

4.2.15 Other materials

One of the persons, and his institution, that has been impressively active and rich in initiative in trying new alternatives is Tarun R. Naik from the UWM Centre for By-Product Utilization, University of Wisconsin–Milwaukee. He has been testing a number of industrial by-products in concrete, both as aggregate and as substitutes for cement, and has presented a large number of papers about his work in many countries in the world. In a paper from Milan, Italy, in 2003,[312] Naik reported on tests with:

- Foundry sand and cupola slag as aggregate in concrete
- Use of post-consumer glass as aggregate, and to some extent as a pozzolan
- Wood ash, in combination with fly ash as a cement replacement material
- Pulp and paper mill residual solids as a reinforcing tool

These are some of the many possibilities they have been working with. The mentioned report gives interesting facts and findings about do's and don'ts utilising the enormous waste potential from other industries. Without going too far into technical findings, we mention a few of the most interesting results, or mentioned facts in the report:

- The foundry industry generates approximately 15 million tonnes of by-products annually, where sand is the main constituent. Based on work from the UMW Centre for By-Product Utilization, concrete bricks with replacement of ordinary sand up to 35% have been produced.
- Combination of wood ash and "normal" fly ash seems very interesting with substitutions up to 35% cement replacement.

Figure 4.18 Foundry sand, Norway, particles larger than 4 mm.[326] (From Haugen M., SkjølvoldO.,*LaboratorieundersøkelseravformsandfraUlefossNV,*Report3D0592.02, SINTEF, Trondheim, Norway, 2009.)

Use of foundry sand in concrete has also been tried in quite a few other countries. From Norwegian testing and practical testing in normal concrete and foamed concrete we mention a few experiences (Figure 4.18):

- The coarse particles might give separation tendencies in high slump and foam concrete mixtures.
- The sand in moderate substitutions has moderate effects on strength and water demand, but might affect the efficiency of some chemical admixtures, as air-entraining and foaming admixtures.
- Tests and trial mixes are definitely recommended.

In another report,[174] Naik claims that concrete with improved penetration to chloride ions and improved freezing and thawing resistance can be made with pulp and paper mill residuals, without loss of strength characteristics. Pulp and paper mill residuals, as well as de-ink solids from paper recycling plants, should be properly dispersed in water, preferably using hot water, before using such sludges in making structural-grade Portland cement concrete.[174]

4.2.16 Waste latex paint

In a paper in *ACI Materials Journal* in 2008,[327] Canadian researchers report on a test where waste latex paint was used as a substitute for some of the water in concrete. Waste latex paint is classified as a hazardous household waste (HHW) in North America and is collected by municipalities. Waste latex paint represents 25 to 30% of the total HHW, which makes up considerable quantities. The paper explains that waste latex paints are often stored in containers, and thereby achieve good homogenisation. Tested were 10, 15, 20, and 25% replacements of the mixing water. From the tests the researchers concluded that the substitution with latex paints enhanced the workability of concrete, did not cause problems with excessive air entrainment, did not negatively affect the setting time, led to considerable increase in flexural strength, but a decreasing effect of the compressive strength, increased resistance to rapid chloride penetrability, and experienced negligible emission of toxic metals.

A full-scale sidewalk was produced with 15% substitution of water with waste latex paint. This substitution rate was from the test results regarded as optimal.

4.2.17 Fillers for self-compacting concrete

Since the beginning of the 1990s, self-compacting concrete has achieved growing acceptance in many parts of the world. Self-compacting concrete is in itself an important step toward increased sustainability in the production of concrete, as it reduces the use of vibrators that can be a health burden to many workers. In addition, self-compacting concrete can be important in improving concrete durability and aesthetic appearance. To produce self-compacting concrete, the use of fillers, or the finest aggregate fraction, is of uttermost importance to avoid segregation and ensure a satisfactory result.

Sometimes, in many places in the world, it might be difficult to find the right fillers at an acceptable price to make self-compacting concrete a competitive alternative. In the mineral industry, from the production of clean aggregate for concrete, to rock quarries for other purposes, there is in many places an excess of "dust" or fine material. Many of these sources, earlier characterised as waste, have found interesting use as recycled fillers for self-compacting concrete.

In the very large variety of research in recycling, we also mention a paper from researchers from the University of Michigan[328] who tried to find an academic method to evaluate the many potential alternatives (Table 4.2). They called their approach engineered cementitious composites (ECCs), a kind of fibre-modified cement invented by Prof. V.C. Li, University of Michigan, specifically designed for increasing the ductility of cement-based materials up to the order of 3 to 7×10^{-2} with its ultimate tensile strain and normally comprised of a mixture of cement, fly ash–GGBS, micro silica sand–limestone powder, PVA fibre, and chemical admixtures et al.[329] As not all their examples normally would be registered as binder substitutes,

Table 4.2 Evaluation of various recycling alternatives according to Lepech et al.[328]

Material	Substituting material	Outcome	Reason
Fly ash	Cement	Passed	
Cement kiln dust	Cement	Passed	
GBS (slag)	Cement	Failed	Poor grain size distribution
Rice husk ash	Cement	Failed	Poor hydration
Solid municipal waste ash	Cement	Failed	Inconsistent chemistry
Foundry green sand	Sand	Passed	
Wastewater sludge	Sand	Failed	Inconsistent chemistry
Expanded polystyrene beads	Sand	Passed	Micromechanical synergy
Pot lining	Sand	Failed	Chemical incompatibility
Post-consumer carpet fibre	Fibre	Passed	Micromechanical synergy
Banana fibre	Fibre	Failed	Low strength

we have chosen to mention their approach in this chapter. They tried to develop their own toolbox to evaluate recycling alternatives. Their approach is interesting, even if some of their findings might deviate somewhat from research and experience in other parts of the world.

Another type of general evaluation of alternatives for recycling was presented by Brandstetr and Havlica from the Czech Republic at a conference in Bangkok, Thailand in 1998.[330] Their main concern was to utilise the ash and residue from fluidised bed coal combustion plants. In this process the combustion process has a temperature of 850°C, and the raw materials are often somewhat different than in a conventional coal-fired electricity work. Consequently, the residue or ash is also different. In the paper the authors discuss various alternatives for recirculation of the waste product: from using the material as it is as backfill for mines and pits, in reclamation of land, for soil amendment, for stabilisation of dumping sites, for road and dam construction, in solidification of municipal wastes, etc., to use as aggregate in concretes, heavyweight and lightweight, and to use as an independent binder. The paper is interesting, especially in the holistic thinking based on a philosophy that all the waste should be recycled. Then it is important to have a representative number of alternatives to choose from to find the sustainability-oriented best alternative, taking all factors into consideration.

A somewhat similar view on recycling is found in a paper by Jahren, originally published for an international symposium in 1998, and later published in a modified version in a special sustainability issue of Concrete International in 2002[331] and other places. In the paper are mentioned recycling efforts of wastes or secondary products from the mineral industry in Norway.

In Norway, many of the competitive mineral companies are located alongside a deep-sheltered fjord, which permits economical and efficient product transport to other parts of the world. Traditionally, the deep fjords have been used as deposit areas for the waste from the mineral activity. As awareness of the effects of such pollution has grown, deposits into the sea have been reduced, and often forbidden. Typically, the next step is to deposit the waste on land instead. Acceptable deposit areas have diminished, and taxation on deposits has produced stronger awareness and efforts to convert wastes to secondary products. Ecological awareness and stronger economical incentives have accelerated this process. In the paper are mentioned some examples of possible recycling of such secondary products in concrete, from three independent companies, located less than 10 km apart, with annual quantities of secondary products from 15,000 tonnes to well over 100,000 tonnes.[332]

One of the mentioned companies had a recycling challenge with a filter cake. The filter cake mainly consisted of $CaCO_3$ (87% of the solid content), with two main pollution components, 9.7% of the solid of free carbon, mainly as graphite, and about 28% water. In addition, there were 1.2% SiO_2 and various smaller quantities of other substances.

In the analysis of possible use of this filter cake, 14 different possible options were evaluated. One of these options was use as filler in machine-produced concrete production of blocks, pavers, etc. This was also successfully tested, where the special filler was both a good cement substitute in dosages of 10 to 15% and an efficient tool to improve standing properties of the products with direct demoulding.

However, the use in concrete production was not necessarily the best recycling option. When recycling alternatives are evaluated, it is important to:

- Conduct thorough work in finding as many realistic possibilities to choose from as possible
- Have a systematic and holistic approach in evaluating advantages and disadvantages, before the type of recycling is decided

Typical factors to be evaluated are:

- Possible economical and environmental effects of processing the waste to a more refined product (e.g., drying of the filter cake mentioned above, or splitting it in different chemical fractions)
- Cost and consequences of intermediate storage of the product
- Cost and consequences of type of reasonable packing, transport, and delivery systems
- Storage and handling by end user
- Volume quantity involved by usage alternatives compared to available quantity

- Concentration or spread of possible users, and logistics involved
- Testing and documentation required and resources to be spent on the evaluated alternative
- Marketing resources required
- Possible value for the end user (price)
- Technical adaptability
- Investment efforts involved with the alternative

In the case of the mentioned filter cake, in addition to the successful testing in concrete production, we mention two other alternatives that also were successfully tested in full scale:

1. Soil neutraliser, an "early spring tool" for farmers: To neutralise the effect of fertilisers, it is common to add limestone powder to the soil. Use of 1:5 to 1:6 with limestone powder:fertiliser is a normal dosage. When adding the neutraliser on the snow in the late winter and the early spring, the snow melts faster, and the farmer can bring his products to the market earlier in the season, often when the prices are the best.
2. Snow melting tool on mountain roads: In Norway we have a number of roads in the mountain that are closed during the winter because of the heavy snowfall. Spreading the filter cake on the snow increases the effect of the melting from the sun. The carbon content in the filter cake makes it even more effective than other alternatives.

Concrete is a very interesting alternative for recycling of various materials or wastes, because concrete represents a large-volume alternative, but it is important also to compare this in a sustainable manner to other alternatives.

4.3 RECYCLING OF OTHER MATERIALS AS REINFORCEMENT IN CONCRETE

To use other materials than steel as reinforcement in concrete is not new. We know that straw was used as reinforcement several thousand years ago. We also know that nails were tried out as a strength increaser at a bridge work in New York more than 100 years ago. In modern times the use of various wastes as concrete improvement alternatives has come up as a possible recycling alternative. In a paper in *ACI Materials Journal* in 2003,[333] researchers from Michigan State University reported on interesting tests with virgin cellulose, recycled paper, and various types of recycled plastics. In addition to strength tests, they tested impact resistance, abrasion resistance, permeability, and deicing scaling of concretes with these fibres. The results differed considerably between the different types of fibres for some of the tests. However, all the different fibres, with properly selected dosages, yielded important interesting gains for many of the various tests. Only for the abrasion resistance test was there little, no, or negative gain.

As mentioned in the introduction, use of organic fibres as reinforcement in concrete is not new, and research in this area was initiated long before this became a recycling topic. However, such research has gotten increased importance, taking into account the large amount of material in this sector that is naturally available. The possibilities seem extremely interesting. As an example, we mention a paper from *ACI Materials Journal* in 2011[334] that from very thorough testing claims that in comparison with alkali-resistant (AR) glass fabric-reinforced composites, a sisal fibre-reinforced cement composite represents a more ductile behaviour under impact loading.

4.4 RECYCLING OF OTHER MATERIALS AS BINDERS IN CONCRETE

The recycling of materials as cement supplementary materials is treated further in Section 3.5. In the following, we only mention a few further alternatives, where the aim is more to find recycling alternatives than emission reduction of significance.

4.4.1 Waste glass

In a paper on a sustainability symposium in Ottawa, Canada, in 1998,[335] Canadian researchers reported on tests with crushed glass as a pozzolanic material. They concluded that waste glass, if ground finer than 38 µm, exhibited pozzolanic behaviour. The compressive strength of lime-glass cubes exceeded 4.1 MPa, a minimum value specified by ASTM C 593. The strength activity indexes of the concrete with 30% by volume of cement replaced by 38 µm glass were 91, 84, and 96% at 3, 7, and 28 days, respectively.

Glass of 150 µm with a particle size larger than 75 µm and smaller than 150 µm was too coarse to be qualified as pozzolanic material. However, even if glass had a major constituent of SiO_2 (but in different phases), it should be noted that silica fume exhibits superior performance to ground glass in all respects.

For more about these possibilities, see Section 3.5.10.

4.4.2 Recycling of fluid catalytic cracking catalysts

At a conference in Warsaw, Poland, in 2007, researchers from Spain[336] reported on a study on the utilisation of two spent fluid catalytic cracking catalysts as pozzolanic materials in concrete. The reason for trying to recycle the catalysts was partly due to the interest in finding suitable recycling alternatives, and partly due to the fact that the catalysts' composition was mainly silica and alumina that showed pozzolanic activity. The pozzolanic activity of the tested material showed pozzolanic activity similar to that of other products usually used in commercial cement production.

4.5 RECYCLING OF CEMENT KILN DUST (CKD)

Recycling of CKD has been an interesting object for studies for many years. The chemistry and particle size of the dust can vary considerably from one cement production process to another, but the dust often has interesting properties for one type of recycling or the other.

At the sustainability conference at Lillehammer in 2007, Jacobsen presented an interesting overview of the area[337] and gave the resume shown in Table 4.3 of the composition of 36 different CKDs published in papers in the period 1982–2004.

Modern wet process plants typically produce less CKD than older wet process plants, and with more reactive CaO and less calcinated raw feed, giving higher particle size.

The quantities of CKD are considerable. Khanna et al.[338] claims that the quantity of CKD varies from 0 to 15% by mass of the Portland clinker produced. They say that the United States had 92 cement plants producing 68.8 million tonnes of clinker in 2000, and that the CKD from these plants

Table 4.3 Constituents of 36 different CKD according to Jacobsen[337]

Constituents and physical properties	All			Dry			Wet		
	Mean (%)	CV (%)	n	Mean (%)	CV (%)	n	Mean (%)	CV (%)	n
CaO—total	46.7	24	36	54.1	14	7	42.8	21	9
CaO—free	12.7	93	12	26.1	37	4	6.4	23	3
Ca(OH)$_2$	14.2	63	4						
SiO$_2$	14.8	40	36	12.4	43	7	16.6	27	9
K$_2$O	5.8	89	32	5.3	36	6	5.5	23	7
Na$_2$O	1.2	95	30	1.1	48	6	1.7	100	7
Fe$_2$O$_3$	2.4	35	36	2.2	25	7	2.6	29	9
Al$_2$O$_3$	4.6	66	36	3.4	40	7	7.1	72	9
SO$_3$	7.2	72	36	9.6	73	7	6.0	40	9
MgO	1.5	48	36	1.6	43	7	1.7	46	9
Cl	2.7	113	25	2.2	77	6	1.2	80	2
LOI	18.1	50	24	14.3	95	5	19.6	22	8
Blaine (cm^2/g)	8372	75	6	7205	21	2	5410	—	1
Particle density (g/cm^3)	2.73	7	11	2.78	7	2	2.55	6	3

was 2.8 million tonnes. Of this, with reference to Hawkins, they say that 2.2 million tonnes was landfilled and 0.6 million tonnes was beneficially reused. With reference to Dyer et al., they claim that 30 million tonnes of CKD was generated per year in 1999.

Recycling the CKD in cement would be the most desirable alternative, but the chemical composition in various CKDs, and the alkalis in particular, puts strong limitations on this alternative. The alkalis (potassium and sodium) normally occur as alkali sulphates in CKD. As most standards and regulations have strong limitations to the alkali content, recycling in cement can only be done in limited quantities. A considerable amount of research has been done worldwide to solve the challenge, among others, to produce high-alkali cements for special purposes, or to blend Portland clinker CKD and supplementary cementing materials as fly ash to produce a binder with less alkali problems, or to point out areas in the use of concrete, where the use of high-alkali cement is acceptable.

Typical recycling areas of CKD are, in addition to small amounts in cement, are as:

- Solidification and stabilisation tool with clay
- Filler material in the asphalt industry
- Supplementary cementing material in special precast concrete products
- Stabilising agent in the technical chemical industry
- Binder in a controlled low-strength material (CLSM)

In a paper in *Concrete International* in 2009, Lachemi et al.[339] tells an interesting story about the use of CKD in CLSM. The purpose of this special concrete is mainly for intermediate filling of trenches in streets, etc. The CLSM shall quickly acquire sufficient strength for possibly an intermediate period of time, and thereafter be easy to excavate. Flowabilty of the fresh concrete to avoid mechanical vibration is important. Typically they were looking for a maximum compressive strength of 2.1 MPa after 28 days, and a minimum strength of 0.7 MPa after 28 days. Tests with CKD as a binder came out very well: high flow ability, heavy trucks could transport the area after 18 hours, and the material was easily diggable with a hand spade after 7 days.

In a paper on high-belite cement at a conference in Ottawa, Canada, in 1998, Sui et al. briefly mention the use of CKD for producing a chemical fertilizer.[340]

Many researchers have tested CKD as a binder, in combination with Portland cement or alone. An example of this is a paper at a RILEM conference in Jinan, China, in 2010, where Ge and Wang[66] report on tests with the effects of sodium silica solution on hydration of nonclinker-made binders with CKD and Class F fly ash. They claim that the results indicated that the major hydration products of the binder were calcium aluminium oxide-related hydrates and aluminosilicate/C-S-H gel. The amount of gel formation was greatly influenced by the CKD–fly ash ratio and modulus of sodium silicate. Not only the amount, but also the microstructure of the binder hydration products might change with curing time. They tested various combinations of CKD and fly ash from 100:0 to 15:85, with various dosages of sodium silicate.

The Norwegian cement company Norcem tells about a success story regarding recycling of its CKD. In 2004, a working group proposed to identify areas for potential commercial use of its CKD, where the chemical properties could be utilised.[341] The typical market areas were as mentioned earlier in this chapter. From the cement plant in Brevik, 12,000 to 16,000 tonnes of CKD was collected each year. Early in 2012 the company could tell that for the last 6 years, all of its CKD had been sold to commercially acceptable recycling purposes, mainly for stabilisation of unstable clays. From 2012 on, the logistics, silos, etc., are ready so that Norcem can also market the CKD from its other cement plant, in Kjøpvik in northern Norway, which has a CKD capacity of another 5000 to 8000 tonnes per year. In addition to the positive recycling success, this also has a positive CO_2 emission aspect. The CKD replaces lime powder as neutraliser in soil stabilisation, approximately in the ratio 1:1. To burn the CaO-rich lime powder, approximately 0.8 tonne per 1 tonne of CO_2 is emitted for each tonne produced. So, the recycling of the CKD adds to reduction in greenhouse gas (GHG) emission (Figures 4.19 and 4.20).

Typical examples of blending ratios for soil stabilisation, depending on type of soil or clay, might be:

SP:FAC:CKD:L

75:0:25:0, 50:0:50:0, 50:0:0:50, 0:50:0:50, 0:50:30:20

(SP = standard Portland cement, FAC = Portland–fly ash cement, L = limestone powder.)

The typical strength of the stabilised soils/clays might vary from 150 to more than 400 kPa.

Figure 4.19 Stabilisation with CKD in the mix. (Photo courtesy of Norcem.)

Figure 4.20 Stabilisation with CKD, puddle equipment. (Photo courtesy of Norcem.)

Chapter 5

The environmental challenges—other items

To give an objective rating between the many other environmental challenges is not possible. We have chosen to treat them in alphabetic order:

Aggregate shortage
Durability/longevity
Energy savings
Health
Leakage
Noise pollution
Radiation
Safety
Water
Waste

As emissions and absorption and recycling have been most emphasised in the concrete industry, these items have also been given the most space, and separate chapters in the book.

5.1 AGGREGATE SHORTAGE?

Ordinary concrete contains about 80% of aggregate by mass, and the formidable consumption each year, probably in the order of 8 to 10 km^3, will in several areas of the globe have effects on the available resources. In addition to the possible general resource challenge, probably more environmentally important is how we collect this resource. Dredging of aggregate from riverbeds and the sea might in many areas be more negative to our environmental balance than the consumption itself. In other areas, river dredging might be positive to the flood balance.

However, talking about aggregate, concrete normally consumes one-third to one-half of the aggregate excavated from natural deposits or from the crushing of rock. The rest of the aggregate produced is consumed in various infrastructure requirements and developments, such as roads, backfills for buildings, pipelines, trenches, etc. In some areas of the world, technologists have been too worried about machine-produced aggregate in general and machine-produced sand in particular, as the angular particles might have a higher water demand. In other areas, where such aggregate for years has been a natural choice as raw material for concrete, the use of chemical and mineral admixtures, modified mixture designs, and the advantage of improved bonding between cement paste and aggregate have broken down old traditions and undesired barriers.

In Chapter 4, Section 4.2, we have tried to give examples of how recycling from various sources might both ease the pressure on possible aggregate shortage and reduce the negative impact of dumping useful resources in nature.

Aggregate shortage, especially high-quality aggregate, is probably not a global challenge, but it might definitely be a serious challenge in some areas of the world, both locally and regionally.

Taking China's case as an example, presented by Sui during the Second Asian Concrete Federation (ACF) Sustainability Forum in Seoul, Korea,[342] the current cement output is more than 2 billion tonnes in 2011. The total volume of concrete used can be estimated at over 3 billion m³, with 60% of cement being used for concrete making and average cement consumption of 333kg per m³ of concrete. We then can imagine how huge the consumption of aggregate is in China, as given in Figure 5.1.

The fact is indeed that natural aggregate is getting less, and the quality is getting lower. Therefore, artificial aggregates like crushed rock, recycled aggregate, etc., have to be used to meet the increasing demand of concrete making and the growing challenge of resources scarcity.

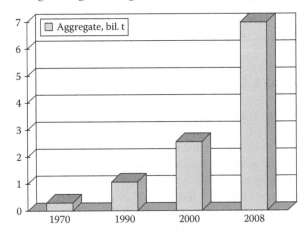

Figure 5.1 Concrete aggregate consumption in China.[342] (From Sui T., Current Situation of China Cement and Concrete Industry on Sustainability. Presented at Second ACF Sustainability Forum, Seoul, November 3–4, 2011.)

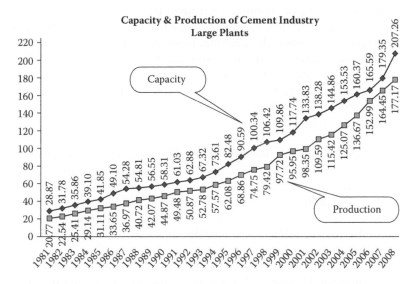

Figure 5.2 Cement production India.[4] (From Kulkarni V.R., Concrete Sustainability: Current Status in India and Crucial Issues for the Future. Presented at Concrete Sustainability through Innovative Materials and Techniques, Roving National Seminars, Bangalore, January 10–14, 2011.)

From 2008 on, India is the second largest producer and consumer of cement in the world, and has the largest predicted growth. The cement production reached 216 million tonnes in 2010, and the expected consumption in 2020 is 425 million tonnes, and in 2030 is 860 million tonnes (Figure 5.2).[4]

At a seminar in January 2011, Vijay Kulkarni, the president of the Indian Concrete Institute,[4] raised the question that shortage of aggregate might be an important obstacle to this growth—a nearly 10-fold growth in the use of concrete in 30 years. The aggregate shortage challenge differs considerably from one part of this big country to another. Kulkarni[4] expressed some worries about lack of resources. He said:

India has been lucky to have good reserves of basalt, granite, limestone etc. However, recent years, as the concrete volume have picked up, the concrete industry is feeling the pinch of shortage of aggregate. In fact, it might not be an exaggeration to state that the specter of aggregate due to the scarcity is haunting many parts of India. The problem is exacerbated in respect to river sand in view of the dredging restrictions enforced by local authorities. Reports emanating from different parts of the country indicate that shortage of river sand is widespread, extending from parts of Kerala and Karnataka to those of Maharashtra and Gujarat. In northern India, especially in the Indo-Gangetic plains, good river sand is generally available in abundance; however, in view of the very nature of all alluvial terrain of this region coarse aggregates are not easily available, forcing concrete producers to fetch them from long distances. Generally speaking, in most of the urban centers, leads for fetching aggregates are increasing and hence their cost.

With the slated increase in concrete volumes in the coming years, aggregate will be required in large quantity. Unfortunately, no estimates are available regarding the actual reserves of concrete-grade aggregate in the country.

In his paper, Kulkarni also looked at other resources for concrete production (see also Section 5.9). Regarding resources for production of cement, he said that the estimated gross reserves of cement-grade limestone stood at 97,430 million tonnes as of March 31, 2006, and that the proven reserves are only 22,931 million tonnes. Since 50% of the proven reserves are located in inaccessible or ecologically sensitive areas, the life cycle of cement plants will have some limitations.

Later in his speech, Kulkarni advocated for establishing a national security plan for natural resources for concrete. Our reaction is that such plans should be a natural planning tool in every country.

The raw materials for concrete are mined or quarried from the earth's surface. Consequently, in addition to resource consumption, the quarrying or mining operation also has other environmental challenges to be taken care of than the use in concrete:

- Operational and final impacts on the landscape.
- Some natural aggregate resources are positioned under valuable agricultural areas.
- Noise and dust.
- Traffic effects on the local roads.
- Groundwater protection.
- Effect on local flora and fauna.
- Some resources are positioned on or near urban areas where industrial activity might be unacceptable or give complications.

Utilisation of the second best alternatives might lead to:

- Higher cost than desirable
- Longer transport
- More water demand in the mixture design
- Secondary effects like increased alkali silica reactions (ASRs)

Most of these challenges can be solved, but they need attention, proper planning, knowledge transfer, and probably also considerations in research and development activity.

In developed countries we have rules and regulations for most of these questions, and the rules and regulations also give natural restrictions on the resources that can be excavated. While we in some areas in the world experience proper after-use of the landscape changes from quarrying and mining in terms of replanting, landscape restorations, water sport facilities, new apartment and house building areas, new land for industrial activity, etc., this is unfortunately not always the case for other places, where the effect of aggregate excavation has unreasonable and unnecessary burdens on the landscape and environment.

At a conference on possible shortage of natural resources (a conference initiated by the Norwegian Minister of the Environment at the time, Mrs. Gro Harlem Bruntland, who at different times functioned as Norwegian prime minister and chairman of the Bruntland Commission) in Trondheim, Norway, in 1977, Jahren tried in a paper to answer the resource shortage challenges regarding aggregate.[343] He concluded that a possible shortage problem was a definition problem, completely dependent on whether this was defined as a local, regional, national, or international problem—and the environmental restrictions at any time for the mining or excavation of aggregate.

The answer regarding the possible shortage of aggregate is probably the same today as it was some 35 years ago. The shortage possibility is a definition problem. It is a bit like being in the middle of the ocean asking for water, but meaning drinking water. We have no doubt that there is enough aggregate around. The question is rather: How can we best utilise the possible available aggregate resources in an optimum environmental fashion?

In many places in the world, collection of aggregate cannot be as done previously; in the future expect to excavate just outside our doorsteps. Restrictions in how we excavate the earth might lead to the necessity to utilise possibilities both with longer transport distances and with other origins—and with consequences for the cost of aggregate. There are some serious doubts if there will still exist some real shortage of aggregate if we utilise all possible alternatives. Some people might, however, rightfully claim that many places in the world have a shortage of traditional aggregate. If it is a real shortage, or it is shortage of areas that will be allowed use for quarrying, both will be a matter of opinion, and differ quite markedly from one country to another, and within areas of the countries. As an example, an article from 1978[344] reported on the growing potential shortage of natural aggregate within a radius of 80 km from the large urban areas in Canada, which comprise less than 10% of the Canadian land mass. The paper referred to similar situations in the United States.

The environmental challenges undoubtedly challenge the aggregate industry. In a overview paper about the sustainability challenges in the concrete aggregate industry in 2007, Danielsen[345] summarised the many challenges for the European aggregate industry. He claimed that the aggregate industry is not a favourite among the public and environmentalists.

The industry produces noise and dust, sites are often unsightly, changes to land are nonreversible, and high volumes of lorry traffic are associated with the industry. "It may be said that, in some regard, the aggregate industry is facing an image problem all over Europe." He claimed that many countries have expressed concerns about the sustainability of the aggregate resource, in terms of both tonnage remaining and the land use planning issues, due to the nonrenewable character of natural aggregate resources. This is especially pronounced in regions facing a shortage of adequate local materials. As an example of the marked change from the use of natural aggregate to crushed rock aggregate, he mentions that in the 1980s, natural aggregate made up 50 to 60% of the total aggregate production in Norway. In 2007, this figure was about 20% and is steadily decreasing. We have found reports about similar developments in other countries.

It should be mentioned that Norway is rather fortunate regarding aggregate, and exports about half of the total volume of aggregate excavated.

As another example, a business newspaper in Seattle, Washington, states in an article in 1998 that the director of the local aggregate organisation claims that there is no aggregate shortage yet, but there might be in the future.[346] The key problem seems to be difference in opinions regarding permission processes between the industry and the environmental community.

Adequate national, regional, and local planning is a necessity to secure adequate and sustainably optimised aggregate resources for the future. As an example of the importance of such planning documents, we refer to a planning document from the state of California,[347] claiming that construction aggregate is the leading nonfuel mineral commodity produced in California, as well as in the nation. Valued at 1.63 billion USD, the aggregate made up about 44% of California's 3.72 billion USD nonfuel mineral production in 2005. California is the nation's leading producer of construction aggregate, with a total production of 235 million tonnes in 2005. The study from California evaluated the resource situation in 31 aggregate areas. In total, the 50-year aggregate demand was estimated to 13,436 million tonnes, while the permitted aggregate resources were estimated to be 4343 million tonnes, with great variation in the percent of permitted aggregate resources compared to the 50-year demand from one area to another. Four areas had less than 10 years of permitted aggregate resources remaining.

We borrow the concrete aggregate from nature, and a sustainable rehabilitation of the landscape after use should be obvious.

We show an example of rehabilitation of a crushed stone aggregate operation after 80 years of service in one of the suburbs of Oslo, Norway, close to a popular skiing, hiking, and recreation area. Approval for rehabilitation for housing was given in 1988. The building operation started in 1996 (Figure 5.3).

(A)

Figure 5.3 (A) An overview.

Norway has traditionally had an abundance of glaciofluvial deposits of sand and gravel. One such deposit is found at Oreid, about 5 km east of the town of Halden. Aggregate was taken out over an area of about 270 da (1 da = 100 m²). Aggregate was produced over a period of more than half a century. To the west and south, the aggregate deposit was close to roads and local roads. On the other side were forest and agricultural areas. A rehabilitation program with replanting started toward the end of the 1970s (Figures 5.4 to 5.7).

5.2 DURABILITY/LONGEVITY

There are numerous examples of the potential longevity of concrete structures, from the testimonies we have of concrete cast many thousand years ago, through the Roman concrete époque 2000 years ago. Illustrations of the Pantheon temple in Rome are probably among the most commonly used in numerous concrete publications. The examples shown in Figures 5.8 to 5.10 are from more recent concrete history.

Many authors have claimed that making concrete more durable is one of the strongest actions that can be taken by the concrete industry to make concrete more sustainable. We will not argue this claim, but must remind readers of the fact that probably more than two-thirds of the concrete that

(B)

Figure 5.3 (Continued) (B) Another overview.

is demolished is not demolished due to lack of durability, but to changes in the society. Consequently, until we start a new practice in higher reuse of demolished components, the optimum solution would be to design the concrete more according to the desired life expectancy—or even more sustainable, to design concrete with good durability, in combination with

(C)

(D)

Figure 5.3 (Continued) (C) The terraced apartments. (D) Popular living area.

Figure 5.4 Old aerial view of the Oreid aggregate plant.

flexibility and robustness in design to enable functional changes of the structure over its life span, without major demolishing.

We have seen examples of "icon structures" for libraries, operas, religious buildings, etc., that have been designed with concrete expected to last at least 300 years, because it seems unlikely that the need for demolishing should come earlier. These are important examples in showing and proving what can be done, but in volume, these efforts in themselves are too small

Figure 5.5 Old pictures from the aggregate operations in the 1980s.

Figure 5.6 Old pictures from the aggregate operations in the 1980s.

Figure 5.7 Oreid in March 2012. (Photo by Inge Richard Eeg.)

to have any significant effect on the total GHG emission from concrete building activity.

There is much evidence that shows good economic reasons for society to put more emphasis on the durability question. One of the strongest, knowledgeable, and forceful advocates for increased sustainability through improved durability in the last few decades is P. Kumar Mehta. In a number of articles on conferences and in technical journals he has given advice and examples on how he thinks this should be done. He advises that deteriorations are closely linked to cracks, microcracks, and porosity of concrete. In an article in 2001, together with Burrows,[348] he raises the question of whether the high-speed construction practice might be negative to better durability. He is of the opinion that high early strength concrete mixtures are more crack-prone and faster deteriorating in corrosive environments. He believes that codes should be amended to stress this point adequately.

Durability is a topic with many challenges, and some of them even have some double-sided aspects with respect to environmental challenges. In Section 3.4 we discussed the advantages of carbonation and CO_2 absorption in concrete. These advantages, however, have a strong negative side as well, at least when it comes to concrete reinforced with corrodible steel. Lack of durability in concrete due to quick carbonation and damages from

Figure 5.8 John Smeaton's Eddystone Lighthouse, completed in 1759, in service until 1759, and later restored outside Plymouth.

corrosion of the reinforcement is a considerable area of research and experience and too voluminous to go into detail in this publication. We thus refer readers to an interesting paper by Chinese researchers from Baotou in Inner Mongolia, Shenzhen, and Xi'an,[349] who in a very analytic analysis remind us that sustainable technology definitely needs a holistic approach: increased emission of CO_2, and an increased amount of acid rain, which leads to increased carbonation and destruction of concrete through corrosion of reinforcement, and again is part of the total CO_2 balance regarding the CO_2 emission from concrete production.

In another of the many thoughtful papers by P. Kumar Mehta[350] from a symposium in Hyderabad, India, in 2005, he again stresses the important of reduction of microcracks in the concrete of the 21st century. He also stresses the importance of the fact that deterioration of concrete seldom

Figure 5.9 The Glenfinnan Viaduct, UK, built from July 1897 to October 1898, 30 m
high, 300 m long, with 21 spans, with concrete mix designs varying from
5:1 to 4:1.

occurs due to a single cause. When two or more causes of deterioration
are at work simultaneously, their interactions are overlooked by the current methods of durability enhancement. The strength-driven design and
construction practice in the concrete industry pay more attention to the
strength-limiting wide (less than 0.2 mm) and deep cracks than to the discontinuous and narrow microcracks in the 0.01 to 0.1 mm range, which
are invisible to the naked eye. In his conclusion, he stresses the three root
causes for lack of durability:

- A high degree of water saturation is one of the three root causes for
  durability problems because in almost every case, it precedes the
  occurrence of any visible damage to the concrete structure. Water is
  not only the primary vehicle for transport of ions and gases into concrete, but it is also implicated in expansion and cracking mechanisms
  in solids containing poorly crystalline materials.
- Microcracks in concrete are the second root cause of durability problems because their growth and connectivity with macrocracks and
  voids mark the point in time when the stage of "no visible damage"
  ends and the stage of "visible damage" begins as a result of loss of

Figure 5.10 The world's first skyscraper in reinforced concrete, the Ingall's Building, Cincinnati, Ohio, built in 1902. The 16 floors of the building are still in use.

watertightness. Being discontinuous and invisible to the human eye, the microcracks are generally ignored by the strength-driven design and construction practices in the concrete industry.

- The number one root cause of concrete durability problems is the presence of inhomogeneities in hydrated cement paste (e.g., large capillary voids and oriented layers of crystalline hydroxide). These inhomogeneities serve as potential sites for microcrack formation when concrete is subjected to tensile stresses generated by both environmental and mechanical loading conditions.

By properly selecting the materials and mix proportions, the microstructural inhomogeneities in concrete can be considerably reduced, and thus the durability of structures can be radically enhanced.

There is no doubt that increased durability of concrete is an extremely important factor for improved sustainability. The amount of money spent each year around the world in concrete repair and maintenance is close to a shameful waste of resources for the society. However, durability of concrete is much more than better understanding and optimising of concrete mixture design. Probably equally important is that we can find the reinforcement where it is supposed to be. From our own experience in repair analysis in the 1980s and 1990s, more than two-thirds of all corrosion damage problems investigated could be traced back to bad workmanship in placing the reinforcement too close to the surface. Yet, another cause for inadequate durability is improper curing and sloppiness in the curing stage—again, a cause that can be traced back to bad workmanship, sloppy management at the site, and improper education at many levels.

However, the effect of increased durability as a way to increased sustainability is a far less powerful tool than it could be, if we do not at the same time increase the understanding and ability to utilise this important advantage with concrete.

You do not need a higher university education to understand that increasing the lifetime of a concrete structure from 50 to 100 years, and even in cases to 200 or 300 years, will reduce the pressure on the resources in the world.

There exist a number of possible powerful, partly nonconcrete technological tools to increase the focus on increased life and durability, but only a few of them are still used:

- Life cycle cost (LCC) analysis, where anticipated (reduced) repair and maintenance are capitalised
- Specifications, taking into account the LCC
- Increased lifetime expectancy requested in specifications
- Combinations of investment cost and maintenance responsibility in investment contracts
- More relevant and flexible design that makes it possible to utilise structures even when social demands change
- More emphasis on the residual value of a structure after the expected life period, and the expected/specified return on investment period

In modern concrete technologies we have powerful tools/levers to enhance durability, such as use of superplasticisers, use of secondary cementing materials, use of synthetic fibres, etc.

It is unfortunately a fact in modern political planning that too often investments and maintenance and daily running costs are found in two different budgets and decision processes. In addition, return on investment requirement is based on rather short planning horizons. This is disadvantageous in a holistic sustainability context in general, and for concrete material in particular.

With increased general public emphasis and understanding of sustainability issues and climate change realities, we hope that this practice might be modified somewhat in the future.

Increased popularity of environmental rating systems might be a forceful tool toward increased longevity as the market value of a sustainable designed structure might increase. Most buildings today have a 50- or 60-year service life as a design criterion. Hopefully, sometime in the future we will get more advanced criteria, for example: the service life of the structure shall be 300 years, with a minimum of 50 years' service life without major maintenance and need for functional modifications.

There is no doubt that properly designed concrete can last for decades and centuries, with moderate maintenance. Research has shown that the proper use of high-performance concrete can even make durable concrete with considerable longevity in cases where concrete is subject to severe environmental impacts like frost cycles, coastal chloride attacks, water and wind attack, abrasion, sulphated soils, etc.

We like to repeat that sustainability requires holistic thinking in the design of durability on all levels to achieve the goal.

5.3 ENERGY SAVINGS

One of the most important advantages in the use of concrete in a sustainability context is the ability for a concrete building to reduce energy consumption when correctly designed.

Somewhat varying from one part of the world to another, one-half to three-quarters of the cement consumed goes for building structures. Measured in expenditure, the figures for the various sectors in Europe in 2009 were:

- New house building: 18%
- Nonresidential buildings: 31%
- Rehabilitation and maintenance: 29%
- Civil engineering: 22%

The energy consumption in buildings of the total national energy consumption varies somewhat, but in most industrialised countries is close to 40%, and it seems like the figure is increasing as the standard of living increases. No other single sector has grown in energy consumption over the late decades like energy in buildings. The energy consumption in buildings is mainly from the use of the buildings, and it is here we find the largest savings potential, and a unique potential for saving in energy consumption

Figure 5.11 Cliff houses in Mesa Verde National Park, Colorado.

and CO_2 emission. One of the saving potentials is more active use of the thermal mass in the building.

However, to utilise the thermal mass of heavy building materials is not a new invention. This has been used to stabilise the temperature in structures where people have been living for thousands of years (Figure 5.11).

The reasons for the energy consumptions in buildings might differ somewhat, depending on, for example:

- Type of building
- Climate
- Traditions/culture
- Standard of living

A typical split on various types of energy consumption in a developed country might be as shown in Figures 5.12 and 5.13.

Philosophically, we can discuss if 40% is a high or low percentage versus other consumption sectors. This is 40% more than, for example, the energy consumption in transport. Taking into consideration the time we spend and the importance of our buildings, we should probably not be too surprised that the energy spent for heating, cooling, cooking, and lighting is higher than other sectors. However, probably the best thing with this high figure is that through the increased attention that this challenge has been given over, in particular, the last two decades, we have learned that there is considerable potential in improvement in the way we spend energy in buildings.

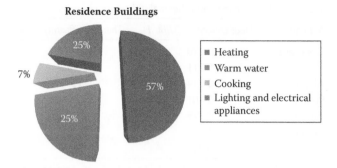

Figure 5.12 Typical residence building energy use.[351] (From Betongelementforeningen, *Betong for energieffective bygninger–fordelene med termisk masse*, Precast Concrete Association, BEF, Oslo, Norway. www.betonelement.no.)

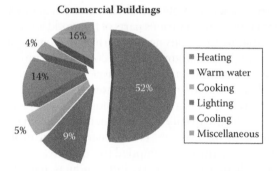

Figure 5.13 Typical commercial building energy use.[351] (From Betongelementforeningen, *Betong for energieffective bygninger–fordelene med termisk masse*, Precast Concrete Association, BEF, Oslo, Norway. www.betonelement.no.)

In the European Union (EU), more than 40% of the energy consumption is related to cooling, heating, and lighting of buildings. This is some of the background for the European Performance of Buildings Directive (EPBD), created in 2002, and coming into force in 2006. Every building with an area above 1000 m³ is covered by EPBD, and has requirements in radical improvements in the energy use, and consequently also the CO_2 emission. It is claimed that the buildings that the directive covers have a saving potential of up to 82 million tonnes of CO_2, and if the directive were extended to all existing buildings with a surface below 1000 m², typically to single family houses, an additional savings of 316 million tonnes per year in CO_2 emission could be obtained.[352]

Maher and Kramer,[16] referring to the U.S. Green Building Council, claim that buildings in the United States account for 36% of the total energy use, 65% of the electricity consumption, 30% of the greenhouse gas emission, 30% of the waste output, and 12% of the potable water consumption.

Table 5.1 Heat capacity and thermal diffusivity comparison[29]

Material	Heat capacity, BTU/(lb°F)	Thermal diffusivity, m²/s
Concrete	0.210	5.38–7.53
Iron	0.111	172
Aluminum	0.214	1270
Plaster	0.201	3.77–6.46
Wood	0.779	1.18
Water	1.21	1.51
Air	0.288	2.37

Source: Schokker A.J., *The Sustainable Concrete Guide, Strategies and Examples*, U.S. Green Concrete Council, ACI 2010.

Utilising the thermal mass of the heavy building materials as an energy reservoir is one of the most important keys to potential savings. Concrete and masonry are particular preferred choices in such actions.

Many research reports from various parts of the world with variable climatic conditions have told us that utilising the thermal mass as a passive stabiliser and energy saver might be marginal when it comes to heating, but it is nearly always an interesting stabiliser and energy saver when it comes to cooling.

Mineral wool and similar insulation materials are a slow thermal conductor (low diffusivity), and can therefore not be used as a reservoir.

Steel has a high heat capacity and conducts transformation of temperature very well (high diffusivity). The transformation properties are too good as the reservoir is filled and emptied too quickly.

Wood has a high thermal capacity, but conducts temperature changes too slowly (low diffusivity) for a regulation according to the daily temperature cycles.

Concrete and masonry have a high temperature capacity and moderate temperature conductivity (moderate diffusivity). This combination allows the energy reservoir to be filled and emptied according to the daily cycles (Table 5.1).

There are three basic principles that have to be present to utilise this savings potential:

- The thermal mass must be exposed, either against the room that needs cooling or in the ventilation channels to the rooms that shall be acclimatised.
- The insulation barriers must be outside the thermal mass that shall be utilised.
- The thermal mass must be in contact with cool night air through natural or mechanical ventilation.

Typically a 20 cm thick wall has the heat capacity to regulate the temperature changes over a 24-hour temperature variation cycle.

In a low-energy building with heavy materials, the thermal mass can be utilised:

- In the summer to reduce the need for room cooling: The heat load is absorbed by the thermal mass during the day, and is released during the nighttime through night ventilation.
- In the summer to avoid overheating: The thermal mass absorbs the heat load and softens the temperature variations in the building.
- In the spring to reduce heating: The thermal mass absorbs the heat from the sun during the daytime from the low spring sun radiation and releases it during the evening when the external temperature sinks. This is also partly true during the winter regarding heating from lighting and internal heating.
- In the winter to reduce the need for internal heating: The thermal mass is heated during the daytime, and gives heat back during the nighttime so that the total heat load is reduced.

These heat-stabilising effects have been known since ancient times. A number of cultures have built their houses into cliffs in the mountains or rocks to get a more pleasant and constant climate, in particular in warmer climates. The stabilising effects of the thermal mass have also for thousands of years been used in food and wine storage in cellars. We still are calling the storage rooms in vineyards in wine cellars even if many of them today are aboveground.

Tommy Kleiven, in an article in 2009,[353] gave some interesting advice, or rather a checklist, for planning to utilise the thermal mass potential. Here is some of his advice (freely translated and edited by the authors):

1. Where do we expose the thermal mass? In channels, the floors, the walls, or the roofs?
2. Strategy for cooling the thermal mass? Natural or mechanical air ventilation or water circulation in pipes in the heavy building materials?
3. Efficient sun protection on the sun-exposed façades.
4. Do not place rooms with high internal heat loads (meeting rooms, etc.) near the most sun-exposed façades, as this will reduce the cooling need.
5. Utilisation of natural light is better for the internal environment than artificial light, and reduces the heat load and need for cooling.

In a paper from 2007, Punkki et al.[58] show the difference between passive and active ventilation in utilising the thermal mass of concrete in a building in Finland. In the case of passive utilisation, the massive structures absorb additional heat from the ambient air, and later when the air temperature lowers, the structure emits the heat back to the air. The problem of passive utilisation is the relatively slow heat transfer from air to massive structures. In the passive utilisation of thermal mass the massive structures absorb and release energy without any control system. In active utilisation in the same kind of structure, the whole system is controlled. Therefore, a larger amount of heat can be stored in the structure and a larger savings in the energy consumption can be achieved. They claim from literature that from 3 to 15% of the energy savings has been reported for heating energy consumption, and 10 to 100% energy savings from the cooling energy consumption, using passive ventilation. In the paper they show calculations from a seven-floor office building in Finland, comparing concrete with lighter structures. The calculations show a 5% energy savings using massive (concrete) structures and passive ventilation. By using active ventilation they show an additional 6 to 23% energy savings, depending on what ventilation concept is utilised.

Biasioli and Øberg[354] refer to investigations by Cembureau/BIBM/ERMCO and claim that a solid residential building requires 2 to 7% less bought energy for heating than a building of lightweight construction. Where cooling is required, the energy savings are even larger, and cooling facilities can be avoided altogether in many heavy buildings. They have also calculated possible energy savings in various dwelling buildings (heating and cooling), from a light to a heavy building, in various countries; Sweden, four buildings, 3.7 to 9.7% savings; Germany, two buildings, 2.4 to 3.5% savings; Denmark, one building, 1.2% savings.

Finally, in this chapter, we must mention that the sustainable effect in this environmental subject has been questioned. In a climate research magazine we found a paper on the subject in 2005.[355] The author refers to the British House of Lords, which would like to delay actions to save energy due to what he calls the "return effect." An example of the return effect is that by saving money on less cost in energy consumption, the consumer will use his new money to buy other things that also consume energy. In addition, by saving energy, energy will be cheaper, and the consumer then increases his energy services compared to other consumer goods. In addition to this, increased energy efficiency will probably lead to technical progress that leads to economical growth, which again leads to increased use of energy and greenhouse gas emission. The climate change debate is not easy! However, there should not be any doubt about the goals for the concrete industry in this case—to reduce the burden on our common environment in all respects of concrete production. Other industries and functions in society must then take care of their own cleaning up.

As a summary from various publications, it looks like, based on European conditions, that an average of 7 to 10% savings in heating resources are possible, and in special cases, over 15% can be achieved in actively using the thermal mass in concrete. When it comes to cooling, some researchers have claimed that savings up to 30% are possible.

Figure 5.14 The heat island effect.

5.3.1 Cooling and heat island effects

In densely populated urban areas, temperatures up to 5 to 6°C higher than in the surrounding nondeveloped areas have been reported (Figure 5.14).

The U.S. Environmental Protection Agency claims that a city with 1 million people or more might experience temperatures 1.8 to 5.4°F (1 to 3°C) warmer than the surroundings. In the evening the difference might be as high as 22°F (12°C).[356] Wikipedia, referring to the IPCC, mentions, for example, that Barcelona, Spain, is 0.2°C cooler for daily maxima and 2.9°C warmer for minima than a nearby rural station.[357]

The heat island effect is more distinct regarding the rise of nighttime temperatures than daytime temperatures.

The heat island effect (also called the urban heat island) might be particularly distinct on clear summer days or evenings. The major reason for this effect is reduced reflection of sun radiation from darker surfaces in roofs and pavements compared to a natural environment. Other important factors are increased energy use, intensified traffic, blocking of winds, industry, etc. The heat island effect has multiple negative environmental consequences. The higher temperatures increase the need for cooling with consequences for energy consumption and CO_2 emission. The higher temperatures also increase the ozone production and the evaporation of VOCs.

> Volatile organic compounds (VOCs) are organic chemicals that have a high vapor pressure at ordinary, room-temperature conditions. Their high vapor pressure results from a low boiling point, which causes large numbers of molecules to evaporate or sublimate from the liquid or solid form of the compound and enter the surrounding air. An example is formaldehyde, with a boiling point of –19°C and fresh paint getting into the air.
>
> Many VOCs are dangerous to human health or cause harm to the environment. VOCs are numerous, varied, and ubiquitous. They include both the man-made and naturally occurring compounds. VOCs play an important role in communication between plants. Anthropogenic VOCs are regulated by law, especially indoors, where concentrations are the highest. VOCs are typically not acutely toxic, but instead have compounding long-term health effects. Because the concentrations are usually low and the symptoms slow to develop, research into VOCs and their effects is difficult.[357]

The heat island effect also leads to increased energy consumption for cooling. It is estimated that the heat island effect costs Los Angeles 100 million USD per year in energy.[358] This increases the importance of utilisation of the thermal mass in heavy building materials.

The normal best measures against the heat island effect are:

- Light-coloured materials
- Porous materials
- Increased vegetation

Light-coloured materials reflect the radiation better. Porous materials, for example, pervious concrete, have lower heat storage capacity and reduce heating through moisture evaporation. The increased vegetation adds to the absorption of CO_2.

The environmental rating systems gives credit to measures using environment-friendly solutions in concrete design, for example:

- Light façades
- Green rooftops
- Pervious concrete in parking lots

Wikipedia, referring to the fourth IPCC report in 2007, claims that the heat island effect itself does not have any effect on global warming.[357] "Studies that have looked at hemispheric and global scales that any urban-related trend is an order of magnitude smaller than decadal and longer-scale trends evident in the series."

Not all cities have a distinct urban heat island.

An important and large area in concrete technology that we have touched upon in this chapter is the many alternatives of utilising special insulating options of concrete, like lightweight aggregate concrete, porous concrete, aerated concrete etc., for saving of energy.

5.4 HEALTH

We do not know about reports showing health hazards regarding living in a concrete structure, but there are some reports that discuss possible health risks working with concrete. The risk of illness or injury depends on the level and length of exposure and the sensitivity of the individual. However, there are ways to prevent or control negative health effects when working with concrete and cement. First, dress for protection. Wear alkali-resistant gloves, long sleeves, and pants to reduce skin exposure to concrete or cement dust, and waterproof boots that are taller than the concrete is deep.

In addition, wear safety glasses, avoid wearing contact lenses, and when dust cannot be avoided, use respiratory protection.

The adverse health effects from working with concrete and cement are generally the result of exposure through skin contact, eye contact, or inhalation:

Skin contact: Getting cement dust or wet concrete on the skin can cause burns, rashes, or skin irritations. Sometimes workers become allergic if they have skin contact over a long period of time. Washing with soap and running water seems to be the best general remedy.

Eye contact: Getting cement or concrete dust in the eyes may cause immediate or delayed irritation of the eyes. Depending on length of exposure, the effect may range from redness to painful chemical burns.

Inhalation: Inhaling cement dust may, for example, occur when emptying cement bags to make concrete. Dust might also occur when sandblasting, grinding, cutting, or drilling concrete. The exposure might cause nose or throat irritation. Long-term exposure to concrete dust containing crystalline silica can lead to a disabling lung disease called silicosis.

As in other industries, in cement and concrete production, more important than anything else is the leadership attitude toward zero accidents, policies, training, and legal compliance.

We will claim that health safety in concrete production is more a management challenge than a product problem.

One of the organisations taking this seriously is the Cement Sustainability Initiative (CSI), a group of 22 major cement companies (2012) representing, in later years, 20 to 25% of the production volume in the world (Figure 5.15).

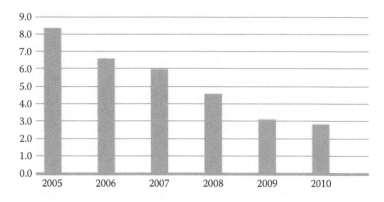

Figure 5.15 Development in employee accident frequency, CSI.[359] (From WBCSD, http://www.wbcsdcement.org/index.php?option?com (accessed February 2, 2012).)

One of the tools to improve health safety is documentation such as safety data sheets to inform about product data, responsible persons and organisation, possible first aid recommendations, handling advice, control measures, toxicity relations, possible allergy reactions, etc. In Figure 5.16, we have shown what a safety data sheet might look like for ready-mixed concrete.

5.4.1 Skin burn

Kjellsen et al. in 2007 reported on five separate cases of severe skin burn from handling of paving stone from one Norwegian producer.[361] The cases happened in the fall (October) of 2004 and 2005, and were thoroughly investigated, and the findings are of importance to the handling of concrete. In all cases the work with the paving stones was done during rain. In the third case the work was done in dry weather, but the surface of the paving stones on the pallets was wet from previous rain. In the findings it was found highly likely that the primary cause of the skin burn reported was high pH, probably around 13.3 in the water that wetted the workers as they carried the stones from the storage to the installation site, or wetted the knee region. The recommendations from the findings were:

- The material safety information was revised for more information about the potential risk for skin burn.
- Waterproof clothing was recommended.
- Pallets of concrete products should be protected at the top in rainy periods.

5.4.2 The chromium challenge

Several decades ago the cement industry became aware that the hexavalent chromium that could be found in cement was a strong sensitising agent that might form water-soluble compounds and have the capacity to penetrate the human skin. In contrast, Cr(III) occurs in water-insoluble compounds in cement. Much effort has been made to get rid of hexavalent chromium. One of the remedies was to utilise ferrous sulphate added to the cement to reduce the sensitivity. Then the hexavalent chromium is transformed to trivalent chromium. This method has been standard practice in many countries the last three to four decades. The leaching of Cr(VI) from cement and concrete is also discussed in Section 5.5.2.

In a paper by Thomassen et al.,[362] they report a case of reactive airways dysfunction syndrome (RADS) with overexposure to ferrous sulphate. They recommend keeping all workers that come in contact with iron sulphate in the cement factory properly informed about the possible hazard, and say that standard equipment in the handling includes overpressure mask,

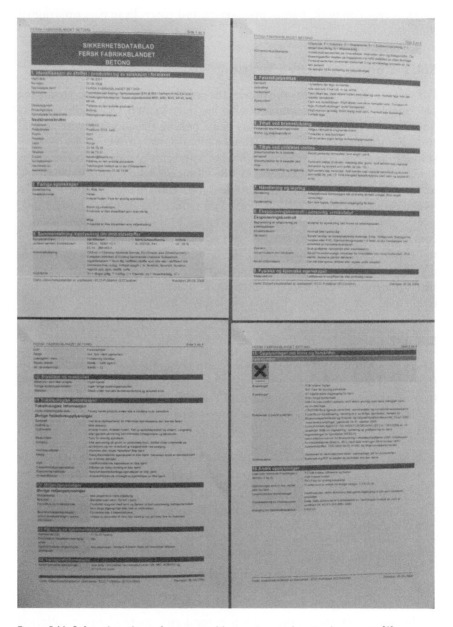

Figure 5.16 Safety data sheet, four pages, Norwegian ready-mixed concrete.[360]

barrier cream on neck, hands, and wrists, and coveralls in paper outside of the working clothes.

5.4.3 Compaction by vibration

For several thousand years the importance of compaction of concrete during placement to achieve acceptable quality has been known. Originally this was done by stamping with heavy boots or by poking the concrete with special rods. For the last century compaction by vibrators has to a large degree been standard practice. The irritating noise must and can be reduced to an acceptable level by ear protection devices. The vibration from handheld vibrators, however, has been recorded as a problem for a rather large number of sensitive persons. The use of special gloves can reduce the problem considerably, but the real breakthrough came in solving the problem with self-compacting concrete. Self-compacting concrete has had a considerable increase in use in most industrialised countries over the last decade or so, since it was introduced as an industrial alternative in Japan in the beginning of the 1990s. In addition to its health-improving possibilities, self-compacting concrete might also increase productivity and quality considerably.

5.4.4 Dust

Also see Section 5.9.3, where we mention the collection of grinding dust from the building site, to be recycled as filler in self-compacted concrete by a nearby ready-mixed company, and the use of cement kiln dust (CKD).

The dust challenge might appear quite a few times through the life cycle of a concrete structure:

- When producing cement
- In aggregate production, and in transport and storage of aggregate
- In concrete production; from the silos, the mixer, and in concrete product production, and when cleaning the equipment and production area
- In transport to the building site
- On the building site, in particular during drilling, grinding, and sweeping/cleaning
- As noncollected dust left in structures that might influence ventilated air
- In demolishing of concrete

The dust from cement production in modern cement plants is collected in advanced filters, and often recycled to useful purposes (see Chapter 4, Section 4.5).

Wind might be a troublesome factor in unwashed aggregate heaps, and in many places in the world, water-spraying equipment is required to avoid unpleasant and troublesome dusting to the surroundings.

Washing equipment for concrete trucks, and permanent surfaces on roads and approach aprons might be required in some areas to reduce traffic dust.

Most work situations with concrete have governmental or local regulations to protect against unreasonable exposure to dust, like demand for filters on silos, etc., use of dust masks and protective clothing, etc.

International studies have been done among workers in cement plants, concrete production, and on the building site.[363] They conclude that the highest exposure for cement dust was among the workers involved in cleaning in cement plants, and among the labour force involved in repair work at the building site. The lowest exposure was among workers involved in formwork and filling/placing of concrete. Which operation that was done and the level of protective measures decided the degree of exposure. As an example, cleaning with a brush increased exposure, while outdoor work reduced the exposure. All the tests from the cleaners at cement plants exceeded the German norm for allowable values for dust to be inhaled (10 mmg/m^3). However, all the workers involved in this type of work were utilising protective measures.

The amount of cement dust in the measured dust varied quite a bit from one type of work to another, from the excess of 60% in the cement plants to less than 30% to the people working with installations, etc. The referred study included 180 measurements, with an average exposure time of about 8 hours.

The health risk from cement dust exposure is somewhat uncertain. Some researchers have claimed an increased number of cancer cases in the lung, gut, head, and neck from workers exposed to cement dust, compared to the normal population. However, other studies could not find such a relationship.[363]

However, we know that exposure to dust is generally very unpleasant, and might cause allergic reactions for some people, and might also cause skin burn if protection and washing advice is not followed properly.

Some health and safety advice gives special attention to exposure and the need for protection during the use of concrete cutoff saws. The exposure might have particularly high concentrations.

Another health-wise difficult operation is sandblasting with silica sand. In some countries, such operations are banned, unless performed under very controlled production with strong protective measures. Exposure to dust from crystalline silica sand is a possible cause for silicosis, a very troublesome disease.

5.4.5 Emission and moisture in concrete

In an information brochure from the cement company Cementa AB[364] we find some interesting information about how to ensure production of healthy floors. Concrete itself hardly gives off any unhealthy gases, but moist, or

Table 5.2 Relative drying capability of concrete depending on construction thickness and water/cement ratio[364]

Thickness	0.4	0.5	0.6	0.7
10 cm	0.4	0.5	0.4	0.4
13 cm	0.8	0.8	0.8	0.7
18 cm	1.0	1.0	1.0	1.0
20 cm	1.1	1.1	1.1	1.2
25 cm	1.3	1.4	1.5	1.8

Source: Cementa, *Healthy Floors*, Information brochure, Cementa AB, Danderyd, Sweden. www.cementa.se.

rather, not properly dried out, concrete before use can result in emission of volatile substances from glue from floor covers, etc., that might be unpleasant to sensitive individuals. The emissions from the building materials themselves are hardly measurable, but the moist alkaline concrete might lead to emission of gases from some glues. Proper drying of the concrete before using glue or a similar substance on the surface is therefore important.

The general rules are:

- Thinner concrete sections dry faster.
- High water/cement ratios have higher porosity and will more easily absorb moisture from curing water, rain, etc., and will thereby take longer time to dry out.
- Low water/cement ratios might use "internal drying out" and bind water chemically.

Table 5.2 gives some indications of how the construction thickness and the water/cement ratio change the relative drying capability.

One special case that is worth mentioning is the emission of ammonia from concrete due to the use of certain nitrogen-bearing concrete chemical admixtures. Urea, an accelerating agent and freezing point-reducing agent, for example, is sometimes used separately as an accelerator or in combination with a water reducer for minus temperatures winter concreting. A typical case in China for the emission of ammonia happened in a basement of a concrete structure where the ventilation was poor, which received a lot of complaints from the residents due to the unhealthy and unpleasant emission. Investigation indicated that the basement was constructed in the winter using concrete containing urea as an ingredient of antifreeze, which reacts and emits ammonia in high-alkaline pore solutions. The Chinese national *Standard Limit of Ammonia Emission from Concrete Admixtures* (GB 18588-2001) was therefore enforced for limiting the ammonia emission from concrete admixtures to less than 0.1 wt%.

5.4.6 Form oil

Traditionally, a considerable amount of mineral oils has been used both on the building site and in precast production. The mineral oils had negative health aspects for the workers and were poorly biodegradable. Modern concrete production utilises to a higher degree vegetable oils, sometimes modified with, for example, marine ester and medical white oil. The modern form oils are easily biodegradable, and it should be a demand from the suppliers to supply the user with certificates that tests have shown that the product is health-wise harmless.

5.4.7 NOx-absorbing concrete

5.4.7.1 General

The air around us mainly consists of 80% nitrogen (N_2) and 20% oxygen (O_2). When chemical processes, for example, in the combustion in a car engine, use the gases, these basically harmless gases can become harmful/poisonous gases when they come out as a combination, for example, NO and NO_2.

These gases are called NOx. Nitrogen oxides (NOx) are important harmful pollutions, also called photochemical smog. In particular, NO_2 is also a catalyst in the formation of ozone. NOx pollution is a particular problem in heavy traffic areas, and will in particular be formed in strong sunlight, and when layers of air are colder than the upper ones. The pollution is dangerous to health, in particular for sensitive persons.

NOx are not only a health problem, but also greenhouse gases. However, absorption effects that are mentioned in the following have little effect in a greenhouse gas reduction context.

5.4.7.2 Principle of reaction

A very fine powder, titanium dioxide (TiO_2), with photocatalytic properties is included in the surface layer of the concrete. At the surface the TiO_2 particles will establish a coating of active oxygen, when the concrete is exposed to ultraviolet light. This active oxygen transforms NOx to NO_3^- ions that are absorbed in the concrete. NO is transformed to NO_2, and NO_2 to NO_3. NO_3^- ions in the concrete are later washed away with rainwater, forming highly diluted nitric acid (Figure 5.17).

5.4.7.3 The catalyst

The catalyst that is used in the process consists of very fine particles/powder of TiO_2. The available information is a bit contradictory regarding:

- Particle size
- Purity
- Special treatment

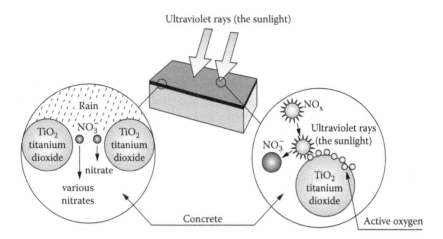

Figure 5.17 Reaction principle of NOx-reducing concrete.[365] (From Jahren P., NOx-reduserende betong, *Betong industrien*. Report 1-2001, Oslo, Norway, 2001.)

It seems like there are at least two sources of information confusion:

- Needs to protect technology.
- The researchers wish for reaction maximising, independent of economical optimising. SEM pictures of the catalyst show particle sizes in the order of 0.06 to 0.18 micron. The particles are agglomerated to secondary particles in the order of 0.5 to 10 micron.

5.4.7.4 The effects

NOx transformation increases with light intensity (Figure 5.18). An absorption in excess of 80% is expected at an intensity of the ultraviolet light of 0.1 mW/cm^2. This is similar to the light of an overcast day. Some tests show that this is particular to the transformation of NO to NO_2, and in a lesser degree of NO_2 to NO_3^- ion.

Absorption of NOx is very high, and reasonably constant up to a concentration of about 1 ppm (Figure 5.19).[365]

Normally the concentration of NOx near traffic varies from 0.1 to 1.0 ppm. In tunnels, registrations of 5 ppm are not unusual. Measurements up to 10 ppm have been registered. The absorption has a tendency to be reduced with increasing relative humidity.

The TiO_2 powder is stable and is not consumed in the process.

The absorption is relatively constant over time (some tests indicate a certain reduction in the reaction during the first 3 months).

Testing indicates that the reaction pattern of NO and NO_2 is different, and has different effects, depending on the light intensity. The transformation

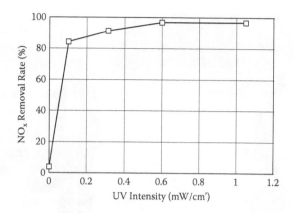

Figure 5.18 NOx absorption depending on light intensity.[365] (From Jahren P., NOx-reduserende betong, *Betong industrien*. Report 1-2001, Oslo, Norway, 2001.)

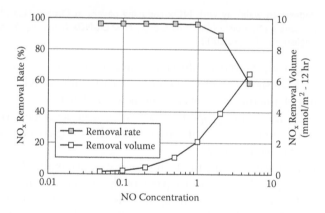

Figure 5.19 NOx reduction depending on NO concentration.[365] (From Jahren P., NOx-reduserende betong, *Betong industrien*. Report 1-2001, Oslo, Norway, 2001.)

of NO to NO_2 and further to NO_3^- ions depends on the light intensity. However, the picture is a bit complicated for the transformation of NO_2 to NO_3^- ions. Here other reaction factors might be present, as it is recorded that this reaction also takes place during the nighttime.

5.4.7.5 Concrete—product areas

In December 1999, the Japanese cement producers Ube and Mitsubishi Materials established a producer group covering all Japan that produces paving and interlocking paving stones, under the product name NOXER. The producers established an organisation called Photocatalytic Concretes Association.

Figure 5.20 Pedestrian walkway with NOx-reducing pavers in Omiya, Japan.[366] (From Brochure Mitsubishi Materials Corp., *NOXER-NOx Removing Paving Blocks*, English ed., Tokyo, 2000.)

The paving products were produced in a large number of colours and formats. Figure 5.20 shows an area in Omiya, Japan, with NOXER products on both sides of a main street several kilometers in length.

Special NOx-absorbing properties have also been developed for sound barriers, etc., around heavy traffic areas. NOx-absorbing concrete has also been used in sprayed concrete.

As far as we have learned, the technology was originally developed in Japan, but has spread also to other parts of the world. In Europe, among others, the Heidelberg Cement Group has been working with the opportunity. Figures 5.21 and 5.22 come from a PowerPoint presentation from Åsa Nilson in Sweden, showing examples from Italy and Sweden.[367] She explains that the tests from Bergamo, Italy, show more than 40% reduction in the NOx concentration compared to asphalt.

5.4.7.6 Other experiences

Most NOx-absorbing concrete products are produced with a special surface layer of 5 to 10 mm in thickness containing the photocatalytic admixture. The reason for using only a thin surface layer is economical. The admixture is expensive. However, it might be rational to combine the layer with NOx-absorbing properties with a concrete of high abrasion resistance.

NOx-reducing concrete is particularly of interest where we have a combination of high concentrations of both traffic and people (pedestrians).

Figure 5.21 Trial área, Via Borgo Palazzo, Bergamo, Itália.[367] (From Nilsson Å., TiOmix–
Fotokatalytiske møyligheter, Cementa, asa.nilsson cementa.se.)

The product developed by the Heidelberg Cement Group is called
TiOMix61. In Sweden also it has tested out the technology in precast concrete façade components.

The future will show if kiln dust from the production of TiO_2 can also be
recycled for this interesting opportunity.

A somewhat similar product was described in *Concrete International*
in 2009,[368] where the authors report a special Portland cement called TX
active having photocatalytic properties that will both clean itself and reduce
the NOx levels by 20 to 80%, depending on atmospheric conditions. They
further claim that the pollution from traffic could be 50% cleaner if just
15% of the buildings and roads in the surrounding area were resurfaced
with photocatalytic cement products.

5.4.7.7 Climate change and health

The Norwegian Government Climate and Pollution Directorate claims that
the climate change affects our health.[369]

Figure 5.22 Trial area, Malmø, Sweden.[367] (From Nilsson Å., TiOmix–Fotokatalytiske
møyligheter, Cementa, asa.nilsson cementa.se.)

The EU has anticipated that the mortality rate will increase by 1 to 4%
per degree in temperature increase, with the highest effect in Central and
Southern Europe. The climate change will increase the spread of water-,
food-, and vector-carried illnesses in Europe. One of the factors is the
increase in spread of illness through insects. Changes in air quality and
airborne dust, as well as increased ozone production, will have negative
effects on our breathing systems.

5.5 LEAKAGE

5.5.1 General

Traditionally concrete leakage is discussed and dealt with when cracking
develops due to a number of factors, such as inappropriate concreting,
changes in atmospheric temperature or humidity, differential settlement

due to load change, occurrence of chemical corrosion, and unexpected external load, like earthquake force, etc. A related topic to handle the challenge of course lies in the solution to making impermeable concrete for use in storing liquids and gas, or to making the concrete strong enough to withstand the undesired load. There should be considerable experience to learn from in this field to reduce various possible leakages to the environment.

As an example, we mention a paper by researchers from Teheran, Iran, and Arizona State University that studied gas permeability of air-entrained concrete[370] and found that entrained air decreased significantly the oxygen gas permeability of concrete. In the reported tests they used various mixes with a cement content of 350 kg/m^3 and w/c ratio of 0.5, and dosages of air entrainment admixtures of dosages from 0 to 2.5% of weight of cement. Even small dosages reduced gas permeability to about one-third, while there was little difference in the permeability between the various dosage rates.

Another related topic that is normally referred to is the leaching of lime compounds in the concrete that leads to deposits of salts on the surface of concrete, known as efflorescence. This is found when water percolates through poorly compacted concrete or through cracks or joints, and when evaporation can take place at the surface of the concrete. Calcium carbonate is formed through a carbonation reaction and deposited as a white deposit. The efflorescence can relatively easily be taken away by a diluted hydrochloric acid (1:10 to 1:20), followed by a wash-down with water quickly afterwards. When using pigments in concrete the efflorescence might appear as a shading/lighting of the colour, which can be brightened up by the same type of acid washing.

The leakage questions mentioned above of course have a relationship to some degree with concrete sustainability. This chapter, however, will mostly focus on the topics regarding concrete in a sustainability context that arise in two different ways:

- Leakage or release to the environment from concrete and its raw materials
- Concrete as a tool to reduce these leakages to the environment in the deposits of hazardous waste or in protection of hazardous substances

5.5.2 Leakage of pollutants from cement and concrete

This issue is causing increasingly greater significance with the intensive development of environmentally sound management of co-processing of various solid wastes and hazardous wastes as alternative fuels and raw materials such as artificial fish reefs in cement kilns, and is discussed hereafter in two aspects: cement manufacture process and use of the resultant cement, i.e., concrete.

5.5.2.1 Leakage from the cement manufacture process

First, from the point of view of the manufacture of cement, besides the fact that the process is associated with impacts of resource extraction (fossil fuel, limestone, and other minerals) upon environmental quality, biodiversity, landscape aesthetics, and the depletion of nonrenewable or slowly renewable resources, clinker burning is the most important phase of the production process in terms of the environmental impact.

Cement production causes emissions to air and waste emissions to land. The key pollutants released to air are particulates, nitrogen oxides (NOx), and sulphur dioxide (SO_2). Dust control has been well done so far in keeping the concentration less than 50 mg/Nm^3 and even below 10 mg/Nm^3. SO_2 emission normally is not a big concern for cement manufacture because the unique characteristics of the preheating and precalcining system can maintain the SO_2 balanced in the kiln system. While the removal of NOx to the designated level (less than 500 mg/Nm^3, for example, normalised as NO_2 based on 10% O_2) can be achieved through proven technology, such as use of a low NOx burner, staged combustion technology, and selective noncatalytic reduction (SNCR) technology.[371]

The waste generally emitted to the land is cement kiln dust (CKD) and bypass dust. In the cement manufacturing sector of the European Union in general, reuse of collected particulate matters in the process, wherever practicable, or utilisation of these dusts in other commercial products, when possible, is encouraged by the European Integrated Pollution Prevention and Control Bureau (EIPPCB).[372] Where appropriate, CKD not returned to the production process may be recovered in various types of commercial applications, including agricultural soil enhancement, base stabilisation for pavements, wastewater treatment, waste remediation, low-strength backfill, and municipal landfill cover, depending primarily on the chemical and physical characteristics of the CKD.[373]

With the dynamic deployment of co-processing with cement kilns, especially in the past 10 years, as a sustainable development concept based on the principles of industrial ecology to provide an environmentally sound resource recovery option preferable to landfilling and incineration, greater concern on the possible increased release of other pollutants arises and has to be dealt with. These other pollutants include VOC, polychlorinated dibenzo para dioxins (PCDDs) and polychlorinated dibenzofurans (PCDFs), CO, HCl, HF, ammonia (NH_3), benzene, toluene, ethylbenzene, xylene, polycyclic aromatic hydrocarbons (PAHs), heavy metals, and their compounds, as described by the EIPPCB in 2010.[372] Under some circumstances, emissions may also include chlorobenzenes and PCBs.[374]

Cement kilns co-processing hazardous wastes should comply with an emission limit for PCDDs/PCDFs of 0.1 ng I-TEQ/Nm^3 (corrected to standard conditions: dry gas, 273 K, 101.3 kPa, and 10% O_2).[374] For other pollutants, pertinent national legislation should apply for specific countries. Best-available technologies (BATs) or control emissions were introduced by Karstensen[375] in 2008 and summarised by the EIPPCB in 2010 with the associated emission levels in the European Union.

It should be also noticed that not all wastes are suitable for co-processing. Wastes such as radioactive or nuclear waste, electrical and electronic waste (e-waste), whole batteries, corrosive waste including mineral acids, explosives, cyanide-bearing waste, asbestos-containing waste, infectious medical waste, chemical or biological weapons destined to destruction, waste consisting of mercury contamination, and waste of unknown or unpredictable composition, including unsorted municipal waste, etc., are not recommended because of health and safety concerns and potentially negative impacts on kiln operation, clinker quality, and air emissions. Only waste of known composition, energy, and mineral value is suitable for co-processing in cement kilns. Co-processing shall be applied only if all tangible preconditions and requirements of environmental, health and safety, social, economic, and operational criteria are fulfilled.

5.5.2.2 Leakage from concrete

For better understanding of the leakage of heavy metals (or trace elements) from concrete, it is first necessary to know the existence of these elements contained in the clinker and the resultant cement.

In recent years, more investigations and works have been made on the behaviours of heavy metals in cement and concrete, especially for the concerns resulting from the co-processing with cement kilns.

GTZ and Holcim's joint investigation results indicate that the effect of wastes on the heavy metals content of clinker is marginal on a statistical basis, the one exception being the bulk use of tires, which will raise zinc levels.[376]

Achternbosch and his team[377] gave the results based on a detailed study of the German cement industry:

- Secondary input materials contribute to the total trace element concentration of cement, although primary raw materials represent the most important input pathway.
- Trace element concentrations of the cements are slightly higher for 2001 than for 1999. For cadmium, antimony, thallium, and zinc the increase amounts to up to 12%. The contribution to this increase mainly results from the fact that the amount of wastes used as input materials in 2001 was higher than in 1999.
- The release of trace elements from concrete elements is negligibly small during the phase of use. An increased release of trace elements is possible under special assumptions after demolition.

Similar observations on an increase in the content of trace elements such as heavy metals in cement have also been found in other countries or sources.[378,379] The change of the heavy metals in cement depends mainly on the primary raw materials, as well as types and fraction of alternative fuels and raw materials used for the manufacture process. For example, use of secondary fuels such as fractions of industrial, commercial, and municipal wastes, used tires, sewage sludge, etc., can result in a higher contribution of trace element concentration to cement than that of meat and bone meal.

The presence of trace elements derived from raw materials or fuels can lead to varying effects on the clinkering process and performance of the resultant cement and concrete.

First, during the clinkerisation, trace elements can induce changes in phase stability by incorporating into clinker phases.[380-383] Several consequences can arise with respect to technical properties of the cement, such as its setting times and compressive strengths.[384-386]

Second, during the hydration of cement, trace elements are released into the pore solution having a basic pH and can thereafter react. They are likely to be absorbed on some hydrates such as the calcium silicate hydrate gel (C-S-H) or form new compounds such as hydroxides.[380-383]

The environmental impact of such cement pastes containing trace elements has to be considered. Indeed, according to the immobilisation nature of trace elements in the cement paste, the kinetics of their leaching will be very different and probably very contradictory due to the complexity of the heavy metals and their behaviours in cement, as well as the different scenarios and testing procedures applied.

Studies by EIPPCB in 2010[372] show that metal emissions from concrete and mortar are low, and comprehensive tests have confirmed that metals are firmly incorporated in the cement paste matrix. In addition, dry-packed concrete offers high diffusion resistance, which further counteracts the release of metals. Tests on concrete and mortar have shown that the metal concentrations in the elutes are noticeably below those prescribed, for instance, by national legislation. Moreover, storage under different and partly extreme conditions has not led to any environmentally relevant releases, which also holds true when the sample material is crushed or comminuted prior to the leaching tests.

Further study by GTZ/Holcim[376] on leachable trace elements in cement and concrete shows that heavy metals are present in all feed materials, conventional and otherwise. However, under certain test conditions, leached concentrations from concrete of other metals besides chromium may approach drinking water standards. The main results of leaching studies done to assess the environmental impacts of heavy metals embedded in concrete are as follows:

- The leached amounts of all trace elements from monolithic concrete (service life and recycling) are below or close to the detection limits of the most sensitive analytical methods.
- No significant differences in leaching behaviour of trace elements have been observed between different types of cements produced with or without alternative fuels and raw materials.
- The leaching behaviour of concrete made with different cement types is similar.
- Leached concentrations of some elements, such as chromium, aluminium, and barium, may, under certain test conditions, come close to limits given in drinking water standards; hexavalent chromium in cement is water soluble and may be leached from concrete at a level higher than other metals, so chromium inputs to cement and concrete should be as limited as possible.
- Laboratory tests and field studies have demonstrated that applicable limit values, for example, groundwater or drinking water specifications, are not exceeded as long as the concrete structure remains intact, for example, in primary or service life applications.
- Certain metals, such as arsenic, chromium, vanadium, antimony, or molybdenum, may have a more mobile leaching behaviour, especially when the mortar or concrete structure is crushed or comminuted (for example, in recycling stages such as use as aggregates in road foundations, or in landfilling).
- As there are no simple and consistent relations between the leached amounts of trace elements and their total concentrations in concrete or in cement, the trace element content of cements cannot be used as environmental criteria.

A study in the Netherlands also focused on the effect of chromium (Cr).[387] The main sources of it in cement came from raw materials, refractory bricks in the kiln, and chromium steel grinders. The relative contribution of these factors may vary depending on the chromium content of the raw materials and the manufacturing conditions.

Sui et al.[388] studied water-soluble Cr(VI) content in cement in China and concluded that it varies upon the type of cement, specifically on the type of mineral additives blended in the cement. Portland slag cement normally has the lowest content compared with other types of cement due to the reducing elements, such as Fe(II) or S^{2-}, contained in slag, which can react with the reactive Cr(VI).

Leaching of chromium from concrete debris may be more prevalent than leaching of other metals. Limestone, sand, and clay contain chromium, making its content in cement both unavoidable and highly variable. Where there is a possibility of contact with the skin, cement and cement-containing preparations may not be used or placed on the market in the European

Union, if they contain, when hydrated, more than 0.0002% water-soluble chromium (VI) of the total dry weight of the cement.[372] As the main chromate source is from the raw material, a reduction in chromium levels (VI) in cement requires that a reducing agent is added to the finished product. The main reducing agents used in Europe are ferrous sulphate and tin sulphate.

Assessments of the environmental quality of cement and concrete are typically based on the leaching characteristics of heavy metals to water and soil. Various exposure scenarios need to be considered:[376]

• Exposure of concrete structures in direct contact with groundwater (primary applications)
• Exposure of mortar or concrete to drinking water in distribution (concrete pipes) or storage (concrete tanks) systems (service life applications)
• Reuse of demolished and recycled concrete debris in new aggregates, road constructions, dam fillings, etc. (secondary or recycling applications)
• Dumping of demolished concrete debris in landfills (end-of-life applications)

There are principally two different ways of testing concrete for leakage:

• By grinding concrete to a fine material, and then testing the powder
• By testing a sample of concrete.

The second type of testing method gives a more realistic way of analysing what can be expected in leaking from concrete. However, even this method might be regarded as only indicative, as it might be difficult to achieve a sample that is large enough to be representative of real life, and to be given the same conditions as a structure might be given in real life.

The methods of testing crushed and grinded concrete might give false results for the behaviour of a concrete structure, in particular for a concrete with low porosity. However, the method might be useful in analysing the effect on the environment of the concrete structure when at the end of its life cycle it will be demolished, or possible recycled.

In recent years, leaching tests for construction materials and wastes have been developed with an emphasis on using them as tools to predict release over a long term. These leaching tests include CEN/TS 14405 (2004)[389]and CEN/TS 14429 (2005)[390](European Union); NEN 7371,[391] NEN 7373,[392] and NEN 7375[393](The Netherlands); and DEV-S4[394] (Germany).

A typical pH dependence test specified by CEN/TS 14429 provides information on the pH sensitivity of the leaching behaviour of the material. The test consists of a number of parallel extractions of a material at a liquid/solid (L/S) ratio of 10 over 24 h at a series of preset pH values, depending on a specific environment situation. Normally a pH value of 3.0 to 5.0 is selected for an acidic rain environment.

A diffusion leaching test is another important procedure to be implemented. The NEN 7375 tank leach test is a procedure for evaluating release from monolithic materials predominantly by diffusion control (e.g., exposure of structures to external influences). Continuous tank leaching (i.e., without a storage period) and intermittent tank leaching (i.e., leaching with interspersed periods of storage) shall be carried out on concrete cubes to determine the release of heavy metals as a function of intermittent wetting. The leachate pH was set depending on specific environment situation.

Release mechanism determination can be conducted based on NEN 7375 (2005). On the basis of the concentration factors and slopes calculated, the leaching mechanism(s) involved in the release of different components from the test piece can be determined.

It is important to notice that the leaching parameters should fit the particular scenario to reflect the actual release behaviour of heavy metal, and the scenario and environmental exposure factors may vary from country to country.

With pavement scenarios such as road washouts and the immersion of roads by rainfall as an example, the heavy metals in the concrete pavement might leach out and be transferred to soil and groundwater. Leaching pH is one of the significant factors that can influence the release of heavy metals in concrete. Therefore, rainfall pH is an important parameter in pavement scenarios. In addition, concrete pavements undergo intermittent wetting, and the carbonation caused by intermittent wetting as CO_2 dissolves in concrete pore water during dry periods results in decreased pore water alkalinity, formation of calcium carbonate, and reduction in the Ca/Si ratio of the C-S-H gel.[395,396] So, intermittent wetting can also influence the release of heavy metals.

In a paper from 2001, Brameshuber et al.,[397] from the Institute for Building Materials Research, at Aachen University of Technology, Germany, give interesting views on the difficulty of finding adequate testing methods in the laboratory that might access the possible burden on soil and groundwater, accompanied by some results from methods tried out. They also distinguished between the leaching of fresh concrete in contact with groundwater and soil and that of hardened concrete. They claimed that the moisture content of the fresh concrete provides for a relatively unhampered diffusion of environmentally relevant substances into the surrounding medium. Apart from diffusion, the wash-off effect is also a dominant leaching mechanism. In fresh concrete, only partial fixation of environmentally relevant substances in the cement matrix has taken place. Fine and ultra-fine particles, in which or to which environmentally relevant substances may be incorporated or attached, are easily washed off. All these effects as a whole may lead to higher release rates during the fresh concrete phase. The authors have developed two different testing methods, which they have named dynamic test method and static test method. In the static method, leaching is tested into a partially saturated soil as well as with a water-saturated soil with water flow.

In their verification tests for the methods, they tested concrete with both normal Portland cement and blast furnace slag cement, and substitution with fly ash. The water/cement ratios tested were relatively high (all above 0.66), with w/sc (water/slag cement) ratios above 0.53. The tested concrete did not seem to have particularly low porosities. The number of samples tested was also moderate. However, it is interesting to note that they conclude that most heavy metals were released in very small quantities. Only chromium was released in detectable quantities, and at no point were the German limits for the total chromium content in water exceeded. (The analysis was carried out at 20°C. The following substances were included in the tests: arsenic, cadmium, chromium, copper, lead, nickel, and zinc.)

In a paper from 2007, Sugiyama et al. gave an interesting report about the present test methods for testing of leaching of concrete in Japan.[398] They explained that test methods for leaching behaviour of trace elements so far have been according to Notifications No. 13 (1973) and No. 46 (1991) from the Environmental Agency of Japan. In these methods, concrete specimens have to be crushed and ground to a particle size of less than 5 mm. However, the Research Committee on Concrete in Japan Society of Civil Engineers proposed a tank leaching test in 2006. For concrete testing a specimen of Ø100 × 200 mm is cast in the laboratory in a plastic mould without use of form release agents. Concentrations of boron, fluorine, hexavalent chromium, arsenic, selenium, cadmium, mercury, and lead are measured in the specified leachate taken from the first to fourth cycles of tank immersion. The authors, after testing the method, seem to be satisfied with the method.

In a paper from a conference in Coventry, UK, in 2007, Fava et al.[399] gave some interesting views after testing leaching behaviour and environmental impact of concrete manufactured with the addition of biomass ash from a combined heat and power plant. They produced test samples with 400 kg/m^3 of Portland cement, and with 100 kg/m^3 of the biomass fly ash from seven different sources. The w/c− ratios varied from 0.47 to 0.56 and the w/c+ ash ratios from 0.38 to 0.45. They report that the concrete specimens manufactured with biomass ash containing heavy metals did not manifest a release higher than the Italian Standards, and the pH leachate remained alkaline through the testing period, which shows that the concrete can be disposed of safely.

An interesting point is that the tests showed a linear relationship between the environmental load and the compressive strength as well as the loss in ignition (LOI) of the ashes. The authors therefore postulate the possibility to describe the environmental load through the compressive strength, and the LOI of the biomass ashes.

In a paper from 2004, Marion et al. from CRIC in Bruxelles, Belgium, discussed leaching from lean concrete in a road subbase[400] with interesting conclusions. They used a tank test defined by NEN 7345 with concrete cubes of 10 × 10 × 10 cm after curing for 56 days. Tests were done after selected immersion periods of 6 hours, and 1, 3, 7, 14, and 36 days. The final sample was taken after immersion for 64 days after replacing 6 L of water with demineralised water. The concrete leachates corresponding to each immersion period were analysed by inductively coupled plasma (IPC) to quantify each of the heavy metals in the list of chemical parameters specified by the Drinking Water European Directive 98/83/EC (As, Cd, Cr, Cu, Hg, Mn, Ni, Pb, Sb, Se). Occasionally, the analysis was extended to include other heavy metals (Ba, Co, Mo, Sn, Ti, V, Zn). Tests were done on lean concrete with a compressive strength after 28 days of 15 MPa, made with calcareous aggregate with Belgian cement from three different sources and substitutions with two different mineral admixtures: fly ash and blast furnace slag. In the conclusion, the authors say:

> Lean concrete for road sub-bases, made from calcareous aggregates and type CEM II and III cements, containing fly ash and/or blast furnace slag within the limits defined by EN 197-1, demonstrate low levels of leaching capabilities in diffusion conditions; the release of heavy metals concentration are far lower than the maximum permissible values of the European Drinking Water Directive. Consequently, this type of concrete, despite its porous characteristics, can be considered not harmful to the environment.
>
> There is no systematic correlation between the total heavy metals concentration in the cement and the leachable fraction of these elements from concrete, whatever the mineral admixture type (fly ash and blast furnace slag) and proportion. Therefore the use of the total cement as a criteria of environmental acceptability is unfounded and unduly restrictive.
>
> The heavy metal concentration in leachates from lean concrete into demineralized water is generally significantly lower than in common drinking water.
>
> The total heavy metal concentration in lean concrete made with CAM II and III cements are much lower than allowable in Belgian soils. In view of the very limited capability of leaching of this kind of material, the risk of contamination of soil and ground water layer from lean concretes exposed on the site appears non-existent.[400]

5.5.3 Concrete to prevent leakage

In a paper from a RILEM conference in Jinan, China, in 2010, Li et al. reported on leaching tests in solidification of heavy metals and radioactive metal.[401] They tested solidifications with various types of geopolymer material combinations using silica fume, metakaolin, fly ash, cement, and ground granulated blast furnace slag. As alkaline activators they used sodium hydroxide, potassium hydroxide, and industrial water glass.

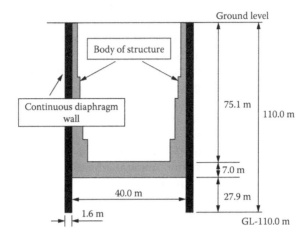

Figure 5.23 Illustration of structure.[402]

The reports tell little about the mixture design and qualities of the various concretes, but they claim that the various mixes gave concentrations of heavy metals in the leachate lower than the maximum allowable concentrations prescribed in the Chinese *Identification Standard for Hazardous Wastes*, and that the prepared solidification forms had no leaching toxicity. The mass percent of the heavy metals and radioactive metals was 0.2 to 0.8% of the geopolymer materials. Tests were performed at different temperatures and time intervals. The solidification rates were all above 99.8%. The solidification of heavy metals was from low to high for the various metals, as follows: cuprum, zincum, nickel, cadmium, and plumbum. Plumbum was solidified 100% by the various mortars. Results from the long-term leaching test proved that the leaching rate and the cumulative leaching fraction of strontium were remarkably lower than those of cesium at the same age. It is more difficult to solidify cesium than strontium.

In a paper in 1998, Yanai et al.[402] reported on a very interesting leakage challenge in the mixture design of the concrete for a 110 m deep diaphragm wall, at the time the largest structure of its kind in Japan (Figure 5.23).

The wall was built for a sewage disposal plant, with a diameter of 43.2 m and a wall thickness of 1.6 m. The concrete had a compressive strength requirement of 40 MPa and a slump of 21 cm. To avoid thermal stresses in the wall, three measures were taken:

- Use of low-heat blended cement containing 45% blast furnace slag and 20% fly ash
- Precooling of the concrete using liquid nitrogen during mixing
- Placement of the slurry-like concrete carefully through tremie pipes, and restricting the cement content

Tests showed that reducing the initial temperature of the concrete from 30°C to 10°C could improve the crack index (tensile strength/thermal stress) from 1 to 3.2. Two different types of cement were tested.

From the conclusions we note:

- The use of low-heat cement was effective in inhibiting thermal cracking.
- The inhibiting effect of precooling to thermal cracking was verified in the actual construction work.
- The use of low-heat Portland cement was the most effective method of the three measures taken to avoid cracking.

Solidification of hazardous waste and control of leakage in this respect are also mentioned in Section 5.10. Something worth mentioning is that heavy metals speciation in concrete made from such clinker can be different from that in cement-based solidification,[403] as the speciation of heavy metals in wastes would be changed in the process of calcinations. This would thus result in the release behaviours of heavy metals in concrete being different from those in cement-based solidification.

5.6 NOISE POLLUTION

Noise pollution is excessive, displeasing human, animal, or machine-created environmental noise that disrupts the activity or balance of human or animal life. The word *noise* comes from the Latin word *nauseas*, meaning "seasickness."

Noise health effects are both health related and behavioural in nature. The unwanted sound is called noise. This unwanted sound may damage physiological and psychological health. Noise pollution can cause annoyance and aggression, hypertension, high stress level, tinnitus, hearing loss, sleep disturbances, and other harmful effects.[404]

Norwegian authorities claim that noise pollution is the single environmental problem that affects most people in the country. In 2007 the environmental authorities put into force a plan with the aim to:

1. Reduce the noise problem by 10% before 2020, compared to 1999
2. Reduce by 30% the number of people exposed to sound increasing 38 dB indoors before 2020, compared to 2005[405]

In 1998, the Swedish cement company CEMENTA presented a brochure/report[406] called *Cement and the Environment* (translated from Swedish). In a chapter about healthy buildings it gives information about sound pollution. It compares interviews with inhabitants in a "silent" house in Lund

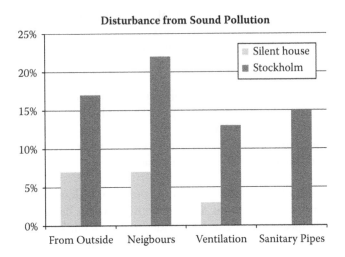

Figure 5.24 Swedish investigation about disturbance from noise pollution.

with an investigation with interviews with 10,000 people in Stockholm from 1991 to 1992 (Figure 5.24).

The investigation is interesting in that it shows how much it is possible to improve the environment by rather simple means, where good concrete design is an important tool.

Swedish technologists have for many years been among the "leading ladies" in the noise pollution efforts. In another Swedish publication from the same period[407] they try to give simple advice regarding good design with respect to sound insulation. This is not uncomplicated, as there are so many factors that influence the result: the load-bearing system, the structure of the partition walls, the structure of the floor design, the floor covering, the windows and the façade design, the dimensions of the rooms in height, width, and length, the total depth of the building, possible bad workmanship, and technical installations like the water and sewage pipes. However, some simple general recommendations are given; the floor partitions should have a minimum weight of 620 kg/m², and the weight of the exterior wall should be at least 480 kg/m². Looking at the partition walls, they should be at least 17 cm in concrete to give an air sound reduction of 52 dB, and 29 cm to give a reduction of 60 dB. Many regulations give 52 dB reduction as the minimum requirement in houses.

Sound is measured in decibels (dB). This is a logarithmic scale where 0 dB is equal to the lowest sound we can hear, and the top point (140 dB) is the highest level for a sound we can hear. When the sound level increases by 10 dB, the intensity increases 10 times. An increase of 3 dB means a doubling of the sound energy. If you sum up to different sound sources, each 30 dB, the total will be 33 dB.

If you have two sound sources of 60 and 70 dB, for example, the sum of these will be 70.4 dB. In practice, this means that only the strongest source is significant.

The human experience of sound is so that we will feel an increase of 10 dB as a doubling of the noise. Changes of less than 3 dB are difficult to distinguish, but a variation of 5 to 6 dB will definitely be observed.

Typical sound levels are:

- Leaves that moves in quiet weather: 20–40 dB
- Noise from a refrigerator in a quiet kitchen: 30–50 dB
- Normal voice at 1 m distance: 55–65 dB
- At the sidewalk of a busy street: 80–90 dB
- Drilling in the street at 40 m distance: 90–100 dB
- Climbing aircraft at 200 m distance: 110 dB

The noise pollution challenge affects the production and design of concrete in several ways, for example:

- Work to reduce noise from production of concrete
- Concrete structures to reduce noise pollution from traffic
- Heavy structures like concrete in walls and floors to reduce air-borne sound
- Correct design of concrete stairways to considerably reduce step sound impact to neighbour rooms

5.6.1 Noise reduction in concrete production

There are several sources for noise in concrete production, where noise levels from drilling, concrete saws, and vibrators are probably the most irritating. A concrete saw used for cutting hollow core ends in concrete precast component production traditionally had noise levels close to 100 dB. Today modern sawing equipment is often encased to reduce the noise level. In addition, water cooling in high-speed saws will reduce the noise level. However, ear protection is a must to avoid hearing damage from long-term exposure.

5.6.2 Noise reduction from traffic

Soft surfaces like a cornfield or snow give a certain damping of the sound, while hard surfaces like asphalt, concrete, or water give no damping. When

Figure 5.25 Architectonic noise reduction wall, Ring Road, Oslo, Norway. Architect: Arne Hensiksens Arkitektkontor. Double-concrete brick wall, with partly open joints and 20 mm air layer and 40 mm sound-absorbing mineral wool.[408] (Photo by Trond Opstad.)

sound goes through 100 m of forest, we might get 5 to 10 dB noise reduction. The effect of a sound barrier will vary considerably with the local conditions, the height, shape, and location in comparison to the receiver. However, sound reduction of more than 5 to 10 dB is rare, but even this reduction might be very important for people (Figure 5.25).

The height, shape, and other design factors of the sound barrier are important both to reduce the noise level behind the barrier and to minimise the rebounding effect on the other side.

5.6.3 Reduction of noise pollution in buildings

In general terms, the designers are faced with two different challenges:

- Reduction of noise transmission from one room to another
- Balancing and reducing the noise echo effect in the room with the noise pollution source, and thereby also reducing irritating noise

Heavy materials like concrete and masonry are perfect tools in reducing sound transfer from one room to another. Typically, a 17 cm thick wall will

Table 5.3 Absorption coefficient of common building materials[409] (referring to http://www.sengpielaudio.com/calculator-RT60Coeff.htm)

Material	Frequency		
	125 Hz	*500 Hz*	*4000 Hz*
Concrete	0.01	0.15	0.02
Carpet on concrete	0.02	0.03	0.02
Carpet on foam	0.08	0.57	0.73
Wood flooring on joists	0.15	0.10	0.07
Unglazed bricks	0.03	0.03	0.07
Painted brick	0.01	0.02	0.03
Glass window	0.35	0.18	0.04
Smooth plaster			
Coarse concrete block	0.36	0.31	0.25
Painted concrete block	0.10	0.06	0.08
Acoustic ceiling tiles	0.70	0.72	0.75

Source: http://www.sengpielaudio.com/calculator-RT60Coeff.htm.

reduce the sound transmission in the order of 52 dB, provided that there is no flank transmission.

In reducing the noise level inside a room, concrete is far from the most efficient surface material, as it is a highly sound-reflective material. If we like to expose concrete surfaces to utilise the advantages of its thermal mass, some absorbing materials might be useful, and will in particular be useful if we place the absorbents just above, around, or near the noise pollution source (Table 5.3).

5.6.4 Step sound reduction in stairways

A condominium developer described the problem with this story: "5 o'clock in the morning the newspaper boy comes to deliver the newspapers. He takes the lift to the top floor. Then he tries to break his own record, running down the stairs while throwing out the papers. The possible noise effect and irritation effect on sleeping people is obvious."

Many countries have made regulations for the maximum impact of sound or step sound. According to Cliff Billington,[410] examples of allowable impact sound in various European countries are:

- Norway, Finland: 53 dB
- UK: 63 dB
- Germany: 53 dB

Norwegian regulations differ between different types of buildings. Here are some typical maximum values:

- Apartments: 53 dB
- Schools: 58 dB
- Hospitals: 58 dB
- Hotels: 58 dB
- Meeting rooms: 58 dB
- Offices: 63 dB

A typical *in situ* cast concrete stairway with rigid connection between the walls and the stairway will transmit a sound of 55 to 66 dB. Consequently, some action is necessary to meet the regulations.

By applying a carpet or other covers, the transmission might be reduced for higher frequencies, but not the lower ones.

By breaking the rigid connection between the wall and the stairway, and using an elastic support, the transmission might be reduced on all frequencies (Figure 5.26).

Effective solutions to break the step sound transmission have been utilised for several decades for precast concrete stairways (Figures 5.27 to 5.29).

There are several systems for such noise reduction available. Probably the most well-known one internationally is the Norwegian Invisible Connections® system, with the stairway connectors RVK and TSS in various models and load classes (Figure 5.30, www.invisibleconnections.no).

The step sound/impact sound reduction from the use of connection details like the ones shown (Figures 5.31 and 5.32) might give a reduction from 15 to 31 dB, depending on design of the rubber damping.

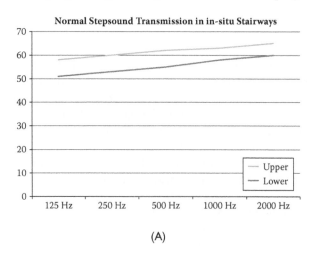

(A)

Figure 5.26 Step sound transmission in a stairway.

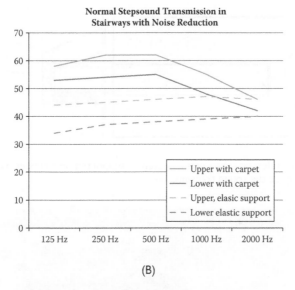

Normal Stepsound Transmission in Stairways with Noise Reduction

Legend:
- Upper with carpet
- Lower with carpet
- Upper, elasic support
- Lower elastic support

(B)

Figure 5.26 (Continued)

Figure 5.27 Recesses in the stairway walls are made for installing connections for precast concrete stairway landing components or stairway components.

Figure 5.28 A telescopic connection system transfers the force from the stairways to the wall.

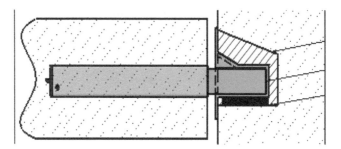

Figure 5.29 The telescopic connection system is supported on a rubber pad to further reduce the sound transmission.

Figure 5.30 Air traffic control tower, United States, with RVK type Invisible connection. (Photo by JVI, Inc.)

Figure 5.31 Air traffic control tower, United States, with RVK-type Invisible connection. (Photo by JVI, Inc.)

Figure 5.32 TSS-type connection in a precast concrete landing component ready for erection.

5.7 RADIATION

Not far from the devastating earthquake on March 11, 2011, with its epi-centre approximately 72 km east of the Oshika Peninsula of Tōhoku, Japan, a disastrous tsunami was triggered, as well as the following explosion of nuclear power plants and radioactive leakage. The aftermath of the disaster, according to the National Police Agency, Japan, resulted in 15,867 confirmed deaths.[411] Estimates of the cost of the damage range well into the tens of billions of U.S. dollars. Radioactive cesium, iodine, and strontium were detected in the tap water, soil in some places in Fukushima, food products, etc.[412, 413] "Radiation fears have since become part of daily life in Japan after cases of contaminated water, beef, vegetables, tea and seafood were discovered." Agence France-Presse reported on Monday, January 16, 2012, "Japan's government was investigating how an apartment had been built with radioactive concrete in the latest scare from the country's ongoing nuclear crisis."

NHK (the Japan Broadcasting Company) reported that same day (Figure 5.33):

> The Japanese government is investigating the distribution of crushed stones that may contain radioactivity from the accident at the Fukushima Daiichi nuclear power plant. It has found that concrete made of the stones has been sold to more than 200 firms. The probe comes after radioactive cesium (Cs-137) was detected in a new apartment building in Nihonmatsu, Fukushima Prefecture where the concrete was used. Readings of radiation levels up to 1.24 microsieverts per hour have been recorded inside the building which is higher than outside.

The catastrophic radiation and radioactivity again shocked the world and leads to a rethinking and change of plan for utilising nuclear energy for the present and future.

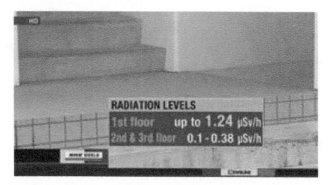

Figure 5.33 NHK screenshot.[411] (From National Police Agency of Japan, Damage Situation and Police Countermeasures (from "deaths" template), July 11, 2012, http://www.npa.go.jp/ accessed July 15, 2012.)

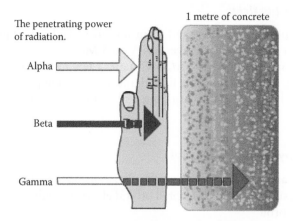

Figure 5.34 The penetrating power of radiation.[412] (From Shimbun Y., Japan, April 18, 2011, http://www.yomiuri.co.jp/science/news/20110319-OYT1T00743.htm (accessed March 19, 2011).)

5.7.1 Effects of radioactive radiation on the human body

As we know, human beings are exposed to different kinds of naturally occurring radiation, which includes radiation from outer space as well as radiation from natural sources on earth. Outer space-originated radiation is mainly absorbed by the atmosphere. The exception includes air travel and space travel that can expose people inside the crafts to increased radiation from space, compared to sea level, including cosmic rays and from solar flare events.[414] An example of a measured dose (not a simulated dose) is 6 µSv per hour from London Heathrow to Tokyo Narita on a high-latitude polar route.

There are various kinds of radiation that can be classified in electromagnetic radiation (EM) and particle radiation (p). The x-rays and gamma rays are part of the electromagnetic spectrum; both have a wavelength range between 10^{-4} and 10^1 nm; they differ only in their origin (Figure 5.34).

Particulate radiations are subatomic particles with mass (e.g., alpha and beta particles, electrons, neutrons). Electromagnetic radiations (x-rays and gamma rays) have no mass or charge.

5.7.1.1 Alpha particles (or alpha radiation)

- Helium nucleus (2 neutrons and 2 protons); +2 charge; heavy (4 AMU); typical energy = 4 to 8 MeV
- Limited range (<10 cm in air; 60 µm in tissue)
- High linear energy transfer (LET) (quality factor (QF) = 20) causing heavy damage (4000 to 9000 ion pairs/µm in tissue)
- Easily shielded (e.g., paper, skin) so an internal radiation hazard

5.7.1.2 Beta particles

- High-speed electron ejected from nucleus; –1 charge; light 0.00055 AMU; typical energy = several KeV to 5 MeV
- Range approximately 12 ft/MeV in air, a few mm in tissue
- Low LET (QF = 1) causing light damage (6 to 8 ion pairs/μm in tissue)
- Primarily an internal hazard, but high beta can be an external hazard to skin

5.7.1.3 X-rays and gamma rays

- X-rays are photons (electromagnetic radiations) emitted from electron orbits, such as when an excited orbital electron "falls" back to a lower energy orbit
- Gamma rays are photons emitted from the nucleus, often as part of radioactive decay
- Gamma rays typically have higher energy (MeVs) than x-rays (KeV), but both are unlimited
- No mass; charge = 0; speed = C; long range (km in air, m in body); light damage (QF = 1)
- An external hazard (>70 KeV penetrates tissue); usually shielded with lead or concrete

The most important naturally occurring radionuclides present in soil and rocks are ^{226}radium (^{226}Ra), ^{232}thorium (^{232}Th), and ^{40}potassium (^{40}K). Since these radionuclides are not uniformly distributed, the knowledge of their distribution in soils and rocks plays an important role in radiation protection and measurement.

The worldwide average specific activities of ^{226}Ra, ^{232}Th, and ^{40}K in the earth's crust are estimated as 50, 50, and 500 Bq kg^{-1}, respectively.[415] European Commission Radiation Report 112[416] gives somewhat smaller figures of 40, 40, and 400 Bq kg^{-1}, respectively, for this world average concentration.

Based on the report of the National Council on Radiation Protection (NCRP), in 1987 and 2006, the annual estimated average effective dose equivalents from natural radiation have remained almost unchanged during those 20 years, i.e., 300 mrem (3000 μSv) and 310 mrem (3100 μSv) per adult, respectively. The total dose from natural and artificial radiation, however, almost doubled, due mostly to the increased utilisation of computed tomography (CT) and nuclear medicine, which is 5 times higher than 20 years ago. Radon and decay products take about 70% of natural background radiation (Table 5.4).

For Years people have been utilising different types of radiation for industrial and R&D purposes as well as medical ones. The commonly used x-ray for a number of applications, such as crystalline materials phase

Table 5.4 Annual estimated average effective dose equivalent received by a member of the population of the United States

| Source | Average annual effective dose equivalent | | | |
| | NCRP 95[a] | | NCRP 160[a] | |
	(μSv)	(mrem)	(μSv)	(mrem)
Inhaled (radon and decay products)	2000	200	2290	229
Other internally deposited radionuclides	390	39	310	31
Terrestrial radiation	280	28	190	19
Cosmic radiation	270	27	270	27
Rounded total from natural source	3000	300	3100	310
Rounded total from artificial sources (medical, industrial, etc.)	600	60	3100	310
Total	3600	360	6200	620

[a] NCRP 95 was published in 1987,[417] and NCRP 160 was published in 2006.[418]

identification, and Co-60 gamma rays, used to determine the surface density of materials, are examples of such application. Besides the use of radiation for medical diagnosis, external radiation treatment by ionising rays to damage fast-growing cells has been widely used. Modern medical technology now can make use of protons and heavy ions to treat brain tumours because these can damage deep-seated tumours.

Radiation, however, can be very risky and present a health hazard if used improperly. Exposure to ionising radiation can cause damage to living tissues, and can result in mutation, radiation sickness, cancer, and death. It can cause not only immediate effects (radiation sickness), but also long-term effects, which may occur many years (cancer) or several generations (genetic effects) later. The potential biological effects and damages caused by radiation depend on the conditions of the radiation exposure. The risk assessment depends on the age of the exposed person; different organs have a different response to radiation, and therefore the risk of cancer differs considerably. The body consists of cells of different radiation sensitivity; a large dose of radiation delivered acutely does larger damage than the same dose delivered over a long period of time.

A presentation by Daniel Demecz et al.[419] explains the effects of radioactive radiation on the human body as follows:

- 0.3 mSv/a: Limit for an effective dose of radioactive discharge to the population through water and air.
- 2.0 mSv/a: Medical radiation.
- <3 mSv/a: Natural radiation when living in a cement or granite building.
- 250 mSv: First clinical measurable effects of radiation when a patient has one session of radiation to his or her whole body.

- Ca. 1000 mSv: Temporary radiation sickness, one-off radiation of whole body.
- Ca. 4000 mSv: Serious radiation sickness, 50% mortality rate if victim remains untreated, one-off radiation of the whole body.
- Ca. 7000 mSv: Fatal dose, no medical treatment, one-off radiation of whole body.

The Nuclear Regulatory Commission of the United States specifies annual radiation dose limits for the general public as well as occupational limits:

- General public limit: 100 mrem (1 mSv).
- Occupational limit: 5000 mrem (50 mSv).
- Remember: We get approximately 300 mrem (3 mSv) per year from natural background exposure.

The National Council on Radiation Protection and Measurements (NCRP) Report 116,[420] *Limitation of Exposure to Ionizing Radiation,* and *Recommendations of the International Commission on Radiological Protection (ICRP),* Publication 60,[421] established the basic exposure limits for occupations and the public (see Table 5.5).

The average exposure for Americans is about 360 mrem (3.6 mSv) per year, 81% of which comes from natural sources of radiation. The remaining 19% results from exposure to human-made radiation sources such as medical x-rays, most of which are deposited in people having had CT scans. However, in some areas, the average background dose can be over 1000 mrem (10 mSv) per year. An important source of natural radiation is

Table 5.5 Basic exposure limits from NCRP 116 and ICRP publication 60

	NCRP-116	ICRP-60
Occupational exposure		
Effective dose		
Annual	50 mSv	50 mSv
Cumulative	10 mSv × age (y)	100 mSv in 5 y
Equivalent dose		
Annual	150 mSv lens of eye; 500 mSv skin, hands, feet	150 mSv lens of eye; 500 mSv skin, hands, feet
Exposure of public		
Effective dose		
Annual	1 mSv if continuous 5 mSv if infrequent	1 mSv; higher if needed, provided 5 y annual average ≤ 1 mSv
Equivalent dose		
Annual	15 mSv lens of eye; 50 mSv skin, hands, feet	15 mSv lens of eye; 50 mSv skin, hands, feet

radon gas, which seeps continuously from bedrock but can, because of its high density, accumulate in poorly ventilated houses.

The background rate for radiation varies considerably with location, being as low as 1.5 mSv/a (1.5 mSv per year) in some areas and over 100 mSv/a in others. People in some parts of Ramsar, a city in northern Iran, receive an annual absorbed dose from background radiation that is up to 260 mSv/a. Despite having lived for many generations in these high-background areas, inhabitants of Ramsar show no significant cytogenetic differences compared to people in normal background areas.[422] This has led to the suggestion that high but steady levels of radiation are easier for humans to sustain than sudden radiation bursts.

In addition to following the guidelines for the limits of radiation exposure, we also need to understand the way to minimise the radiation exposure:

- Time: Minimise the amount of time spent near sources of radiation for people who are exposed to radiation in addition to natural background radiation.
- Distance: Maximise distance, as radiation intensity decreases sharply with distance, according to an inverse-square law (in an absolute vacuum).[423]
- Shielding: Air or skin can be sufficient to substantially attenuate low-energy alpha and beta radiation. Barriers of lead, concrete, or water give effective protection from more energetic particles such as gamma rays and neutrons. Some generally accepted thicknesses of attenuating material are 5 mm of aluminium for most beta particles, and 3 in. of lead for gamma radiation.

5.7.2 Natural radioactivity in building materials

The world in fact is radioactive, and has been since it was created. Over 60 radionuclides (radioactive elements) can be found in nature, and they can be:

- Primordial: From before the creation of the earth.
- Cosmogenic: Formed as a result of cosmic ray interactions.
- Human produced: enhanced or formed due to human actions (minor amounts compared to natural).

Every day we ingest and inhale radionuclides in our air and food and water. Natural radioactivity is common in the rocks and soil that make up our planet, in water and oceans, and in our building materials and homes.

All building raw materials and products derived from rock and soil contain various amounts of mainly natural radionuclides of the uranium (^{238}U) and thorium (^{232}Th) series, and the radioactive isotope of potassium (^{40}K). In the ^{238}U series, the decay chain segment starting from radium (^{226}Ra) is radiologically the most important, and therefore reference is often made to ^{226}Ra instead of ^{238}U. These radionuclides are sources of external and internal radiation exposures in dwellings.

The external exposure is caused by direct gamma radiation. For example, an inhabitant living in an apartment building made of concrete with average activity concentrations (40 Bq kg^{-1}, 30 Bq kg^{-1} and 400 Bq kg^{-1} for radium, thorium and potassium, respectively) receives an annual effective dose of about 0.25 mSv (excess to the dose received outdoors). The inhalation of radioactive inert gases radon (^{222}Rn, a daughter product of ^{226}Ra) and thoron (^{220}Rn, a daughter product of ^{224}Ra), and their short-lived secondary products leads to internal exposure of the respiratory tract to alpha particles. Typical excess indoor radon concentration due to building materials is about 10 to 20 Bq/m^3.[424]

The specific activities of ^{226}Ra, ^{232}Th, and ^{40}K in the building raw materials and products mainly depend on geological and geographical conditions as well as geochemical characteristics of those materials.

Typical and maximum activity concentrations in common building materials and industrial by-products used for building materials in the EU are shown in Table 5.6.[425] Typical concentrations are population-weighted national means of different member states.

NCRP 94 (1987)[426] also listed a few common building materials and their estimated levels of uranium, thorium, and potassium (Table 5.7).

Table 5.6 Typical and maximum activity concentrations in common building materials and industrial by-products used for building materials in the EU[416]

Material	Typical activity concentration (Bq kg^{-1})			Maximum activity concentration (Bq kg^{-1})		
	^{226}Ra	^{232}Th	^{40}K	^{226}Ra	^{232}Th	^{40}K
Most common building materials (may include by-products)						
Concrete	40	30	400	240	190	1600
Aerated and lightweight concrete	60	40	430	2600	190	1600
Clay (red) bricks	50	50	670	200	200	2000
Sand-lime bricks	10	10	330	25	30	700
Natural building stones	60	60	640	500	310	4000
Natural gypsum	10	10	80	70	100	200
Most common industrial by-products used in building materials						
By-product gypsum (phosphogypsum)	390	20	60	1100	160	300
Blast furnace slag	270	70	240	2100	340	1000
Coal fly ash	180	100	650	1100	300	1500

Source: European Commission, Radiological Protection Principles Concerning the Natural Radioactivity of Building Materials, Radiation Report 112, 1999.

Table 5.7 Estimates of concentrations of uranium, thorium, and potassium in building materials (NCRP 94, 1987, except where noted)

Material	Uranium		Thorium		Potassium	
	ppm	mBq/g (pCi/g)	ppm	mBq/g (pCi/g)	ppm	mBq/g (pCi/g)
Granite	4.7	63 (1.7)	2	8 (0.22)	4.0	1184 (32)
Sandstone	0.45	6 (0.2)	1.7	7 (0.19)	1.4	414 (11.2)
Cement	3.4	46 (1.2)	5.1	21 (0.57)	0.8	237 (6.4)
Limestone concrete	2.3	31 (0.8)	2.1	8.5 (0.23)	0.3	89 (2.4)
Sandstone concrete	0.8	11 (0.3)	2.1	8.5 (0.23)	1.3	385 (10.4)
Dry wallboard	1.0	14 (0.4)	3	12 (0.32)	0.3	89 (2.4)
By-product gypsum	13.7	186 (5.0)	16.1	66 (1.78)	0.02	5.9 (0.2)
Natural gypsum	1.1	15 (0.4)	1.8	7.4 (0.2)	0.5	148 (4)
Wood	—	—	—	—	11.3	3330 (90)
Clay brick	8.2	111 (3)	10.8	44 (1.2)	2.3	666 (18)

The European Commission (EC) set guidelines on the radiological protection principles concerning the natural radioactivity of building materials (RP-112 document) for the member states.[427]

The following activity concentration index (I) is derived for identifying whether a maximum dose criterion of 1 mSv/y due to the excess external gamma radiation in a building is met:

$$I = \frac{C_{Ra}}{300 \text{ Bq kg}^{-1}} + \frac{C_{Th}}{200 \text{ Bq kg}^{-1}} + \frac{C_{K}}{3000 \text{ Bq kg}^{-1}} \leq 1$$

where C_{Ra}, C_{Th}, C_{K} are the ^{226}Ra, ^{232}Th, and ^{40}K activity concentrations (Bq/kg) in the building material, and a background dose rate of 50 nGy h-1 corresponding to an average value outdoors in Europe has been used in deriving the activity concentration index.

For a dose criteria of 0.3 mSv/y, the same activity concentration index calculation as above can be used, yet its limit value is set at 0.5 instead of 1.

[Table 5.8] gives the values of the activity concentration index limit depending on the dose criterion and the way and the amount the material is used in a building as specified in EC RP-112.

General evaluation on the possibility of exceeding 0.3 mSv or 1 mSv because of the use of certain building materials is also given in [Table 5.9].

Table 5.8 Limits of the activity concentration index

Dose criterion	0.3 mSv/y	1 mSv/y
Materials used in bulk amount, e.g., concrete	I ≤ 0.5	I ≤ 1
Superficial and other materials with restricted use: tiles, boards, etc.	I ≤ 2	I ≤ 6

Table 5.9 Possibility of exposure exceeding limits due to the use of certain building materials

Building material	Exposure above 0.3 mSv/ circumstances or explanation	Exposure above 1 mSv/ circumstances or explanation
Concrete	Possible/almost anywhere where bulk amounts are used	Possible/if bulk amounts are used and the concrete contains large amounts of blast furnace slag, fly ash, or natural sand or rock rich in natural radionuclides
Aerated and lightweight concrete	Possible/if blast furnace slag, fly ash, or natural materials rich in natural radionuclides are used	Not likely/used only in walls
Clay bricks	Possible/if clay rich in natural radionuclides is used	Not likely/used only in walls
Sand-lime bricks	Not likely/low activity concentrations, limited use (only walls)	Not likely/low activity concentrations, used only in walls
Natural building stones	Not likely/superficial or other minor use	Possible/if used in bulk amounts
Not likely/superficial or other use	Possible/if used in bulk amounts	Gypsum boards or blocks
Not likely/natural gypsum	Possible/if radium rich by-product gypsum is used	Not likely/superficial use or used only in walls

As seen from Table 5.9, doses exceeding 1 mSv/y within the EU should be taken into account from the radiation protection point of view. Countries having low radioactivity background levels, such as Denmark, the Netherlands, and Israel, generally choose a stricter dose criterion. In Denmark and Israel an excess dose of 0.3 mSv/y was applied. Some EU states, as in Poland, have stricter activity concentration limits for ^{226}Ra than in the RP-112 recommendation (200 Bq/kg instead of 300 Bq/kg) in order to ensure indoor radon concentrations lower than 200 Bq/m^3.[428]

It should be noted that restricting the use of certain building materials might have significant economical, environmental, and social consequences at a local or national level. For example, some natural building materials that have been traditionally used for decades or centuries may contain natural radionuclides at levels such that the annual dose of 1 mSv might be exceeded. In these cases, the detriments and costs of giving up the use of such materials should be assessed, taking into account the financial and social costs and benefits, especially when establishing binding regulations.

5.7.3 Radiation from cement and concrete

The radiation question with cement and concrete is discussed in one direction: possible radiation from cement and concrete materials.

Cement is an important construction material for houses and buildings. The fact that people normally spend about 80% of their time inside offices and homes makes it important to understand the detailed information of the specific activities of ^{226}Ra, ^{232}Th, and ^{40}K in cement products and their raw materials in order to estimate the radiological hazards on human health. Unfortunately, these data are often unavailable in many countries.

As we know, the contents of ^{226}Ra, ^{232}Th, and ^{40}K in cement can vary considerably depending on their geological source and geochemical characteristics. Trace and radioactive elements are obtained from natural feedstock, fuels, and alternative fuels and raw materials (AFRs, normally wastes). Tables 5.10 and 5.11 give the comparison data of radioactivity behaviour of the raw materials and resultant cement in Pakistan and some other countries in the world.[429]

Large variations can been observed between the radioactivity content of Portland cement and the raw materials of which it is made, both in the present study and in the literature values for other countries of the world. These variations and the large spread in the data again are a reflection of the different geological origins of the raw materials. Such data are of value in determining the radioactivity content of the buildings and the possible radiological risks associated with these structures.

China conducted a systematic investigation of the radiation from cement:[430] 373 cement samples had been analysed covering 27 provinces and municipalities. The average specific activities of Ra, Th, and K reported

Table 5.10 Comparison of total specific activity (Bq kg/y) due to ^{40}K, ^{226}Ra, and ^{232}Th in the raw materials of cement of Pakistani origin with those of other countries of the world

Country	Limestone	Gypsum	Slate	Latrite	Reference
Australia	11.0	11.1	—	—	Beretka and Mathew (1985)
Brazil	236.3	160.3	—	—	Malanca et al. (1993)
Canada	—	60.0	1153	—	Zikovsky and Kennedy (1992)
Finland	—	33.5	—	560.0	Mustonen (1984)
Mexico	131.8	81.6	—	—	Espinosa et al. (1986)
Netherlands	296.4	244.0	—	—	Ackers et al. (1985)
Pakistan	104.2	193.2	890.4	545.6	Khan and Khan (2001)

Table 5.11 Comparison of specific gamma activities (Bq kg/y) of the Pakistani Portland cement samples with those of other countries of the world

Country	^{40}K	^{226}Ra	^{232}Th	Total	Reference
Australia	114.7	51.8	48.1	214.6	Beretka and Mathew (1985)
Austria	210.0	26.7	14.2	240.9	Sorantin and Steger (1984)
Bangladesh	505.7	120.2	132.4	758.3	Mollah et al. (1986)
Belgium	—	62.0	76.0	—	Proffijin et al. (1984)
Brazil	564.0	61.7	58.5	684.2	Malanca et al. (1993)
Finland	251.0	40.2	19.9	311.1	Mustonen (1984)
Germany	325.0	15.1	22.9	363.0	UNSCEAR (1977)
Italy	316.0	46.0	42.0	404.0	Sciocchetti et al. (1984)
Malaysia	203.5	81.4	59.2	354.1	Chong (1982)
Mexico	—	26.0	52.6	—	Espinosa et al. (1986)
Netherlands	230.0	27.0	19.0	276.0	Ackers et al. (1985)
Norway	259.0	29.6	18.5	307.1	Stranden and Bertag (1980)
Pakistan	272.9	26.1	28.7	327.7	Khan and Khan (2001)

are 53.6, 28.4, and163.6 Bq/kg, respectively. Five samples were identified to exceed the national limit in terms of the specific activity of Ra. The maximum activity can be as high as 341.5 Bq/kg due to the use of coal ash, with high activity ranging from 689 Bq/kg to 3133 Bq/kg. Fourteen clinker samples were also examined with the average specific activities of 52.6, 19.3, and 159.8 Bq/kg, respectively.

It has been well proved, as shown in Section 5.7.2, that attention has to be given on the increments of radiation exposure with the recycled industrial by-products containing the technologically enhanced naturally occurring radioactive materials (TENORMs) extensively used in the construction industry, specifically for cement and concrete production. The case of high radioactivity of cements is mainly found with the high activity of additives (fly ash, blast furnace slag) as clinker substitute. The AFRs normally containing relatively high levels of radioactivity are:

- Coal ash, produced as waste in the combustion of coal, is used as an additive to cement and concrete, and in some countries bricks are made from fly ash. Coal slag is used in floor structures as insulating filling material. A U.S. Geological Survey in 1997 showed the presence of U, Th, Ra, and Rn. During coal combustion, most of the radon is lost in stack emissions; the less volatile elements (U and Th) and the majority of their decay products are retained in the ash.[431] Australia (1984)[431a] indicated that "some sources of fly ash have radioactivity above a safety level, but not considered a hazard as the ash is used in small proportions in concrete; the ash is also concentrated in finer particles and can be removed prior to use." Australia exports coal containing 1.6 ppm uranium and 3.5 ppm thorium. The Dutch power companies became aware in 2000 of the presence of ^{210}Pb in the boilers, which originated from the decay of ^{238}U present in the coal. The measured levels of total radioactivity exceeded the limit of 100 Bq/g. Hungarian coal from the Mecsek mountains in the south is high in radioactivity due to U, K, and Th (reports dating back to 1948). Coal in the Calaf area, Barcelona, Spain, is characterised by its U content (1983).

- Blast furnace slag, a by-product of the iron and steel industry, is used as either a raw materials component or clinker replacement, or both. The low concentration of the radionuclides originates from iron ores, and processing leads to radionuclide enrichment in the slag, assimilating 98% of all the radionuclides from the beginning (Germany, 2002). High radon exhalation rates using blast furnace slag cement (Egypt, 1999) were banned from use in residential buildings.[432]

- Phosphogypsum, a by-product in the production of phosphorous fertilisers, is used as a setting time adjuster of cement and raw materials of other building materials. Phosphogypsum was found to contain ^{226}Ra, ^{232}Th, and ^{40}K" (Australia, 2003; Germany, 2002).

- Red mud, a waste from primary aluminium production, is used in bricks, ceramics, and tiles.[433] Bauxite ores, the main raw materials of the aluminium industry, can contain significant levels of ^{238}U and ^{232}Th and their respective decay products (Australia, 2003; Germany, 2002).

- In some cases, high radioactivity of clay, such as Albanian clays (2000) high in uranium (>10 ppm), thorium (20 ppm), and potassium (>5%), has been found.

Cementitious materials are certain to contain relatively low levels of heavy metals and radioactive elements. The radioactivity in raw materials and final products of cement varies from one country to another, and also within the same type of material from different locations. The manufacturing operation reduces the radiation hazard parameters. The levels are usually too low to affect the performance; cement products thus do not pose a significant radiological hazard when used for building construction. Yet they could affect the chemical composition of the solidified radioactive waste streams. The results also may be important from the point of view of selecting suitable materials for use in cement manufacture.

Now we come to the resultant product—concrete. The radon exhalation rate from concrete varies according to the age of the concrete, the water content, and the addition of fly ash. The exhalation increases almost linearly with the moisture content up to 50 to 60%, peaks at 70 to 80%, and decreases steeply for higher moisture levels.[434] The radon exhalation rate from concrete based on the research results of Sarah C. Taylor Lange[432a] et al., from the University of Texas at Austin, are dependent upon the ^{226}Ra specific activity of the constituent material and the material porosity. A reduction in material pore size could reduce the adsorption of emanated radon and reduce the interconnectedness of pores, in turn limiting the ^{222}Rn exhalation rate. Usage of uranium (^{238}U) containing aggregates dictated total concrete ^{226}Ra specific activities and ^{222}Rn exhalation rates, mainly because aggregates comprise most of the concrete mass.

The addition of fly ash to concrete generally increases the ^{226}Ra activity, while the radon exhalation rate slightly increases or even decreases.[435,436]

The reference radon concentration for this purpose could be the lower limit of the action level for radon in dwellings (200 Bq/m^3) recommended by the ICRP,[437] or some fraction of it. In Hong Kong a high-rise building action level of 200 Bq/m^3 was proposed for existing buildings and 150 Bq/m^3 for new buildings.[438]

The regulation standards commonly including an activity concentration index I are introduced specifically in Tables 5.8 and 5.9.

It is also worth mentioning that exposure of concrete structures to neutrons and gamma radiation in nuclear power plants and high-flux material testing reactors can induce radiation damages in their concrete structures. Paramagnetic defects and optical centres are easily formed, but very high fluxes are necessary to displace a sufficiently high number of atoms in the crystal lattice of minerals present in concrete before significant mechanical damages are observed.[439]

5.7.4 Radioactivity risk reduction with cement and concrete

The radiation with cement and concrete can also be discussed in another direction: use of cement and concrete to reduce radiation risk.

There two ways to achieve this purpose:

- Concrete as a shield of radiation (Section 5.7.4.1)
- Encapsulation of radioactive materials with cement and concrete (Section 5.7.4.2)

5.7.4.1 Concrete as a shield of radiation

Living isolated from radiation is impossible in the modern world, as the population is subjected to increased radiation and artificial radiation in particular over recent years. The construction of nuclear power plants has increased for many purposes, especially for energy supply all around the world. This results in the particular demand for protection from fatal rays such as neutron and gamma, that have the ability to penetrate objects.

Lead has normally been used for shielding structures such as nuclear power plants, yet for some cases, it has been insufficient to shield neutrons.

Concrete, a relatively inexpensive material that can be easily handled and cast into complex shapes, contains a mixture of many light and heavy elements, and therefore has good nuclear properties for the attenuation of photons and neutrons. It is therefore considered an excellent and versatile shielding material widely used for shielding in nuclear power plants, particle accelerators, research reactors, laboratory hot cells, and medical facilities, and also as a structural and shielding material for the storage and disposal of radioactive wastes.[440]

Generally normal concrete is specified for radiation shielding where space is available. However, heavyweight concrete comprising natural or synthetic heavyweight aggregates can be used to reduce the thickness of shielding concrete if space is limited. Heavyweight concrete not only has a higher density, 2700 kg/m^3 to 5500 kg/m^3, but also has more desirable attenuation properties than normal-weight concrete.

ACI 304R-34[441] specifies that raw materials such as cements, admixtures, and water used in heavyweight concrete should conform to the standards generally required for normal-weight concrete; only the aggregate is different and may require special consideration during handling, batching, mixing, transporting, and placing. However, moderate-heat or low-heat Portland cement is preferential for heavyweight concrete in the authors' view because heavyweight concrete is normally prepared with high cement content and low w/c ratio, which can exhibit a high-temperature rise at early ages and may cause undesirable localised cracking from the thermally induced stress. This can be easily understood with the excellent performance of low-heat belite-based Portland cement and concrete, introduced in Chapter 6.

Aggregates, which are the largest constituent (about 70 to 80% of the total weight of normal concrete), are important components for radiation protection properties of concretes, as mentioned above. ACI 304R-34[441] lists the typical radiation-shielding aggregates for heavyweight concrete, as shown in Table 5.12.

When heavyweight concrete is used to absorb gamma rays, the density is of prime importance. When the concrete is used to attenuate neutrons, material of light atomic weight containing hydrogen should be included in the concrete mixture.[442] Some aggregates are used because of their ability to retain chemically bound water at elevated temperatures (above 85°C), which ensures a source of hydrogen. Composition of aggregates for use in radiation-shielding concrete is described in ASTM C 638, and aggregates should meet requirements of ASTM C 637. The commonly used types of heavyweight concretes can be categorised based on the types of aggregates used: barites (density 2.5 to 3.5 g/cm^3), magnetite (3.5 to 4.0 g/cm^3), and hematite (4.0 to 4.5 g/cm^3). Occasionally, even denser aggregates such as iron are incorporated. Density of those concretes can reach up to 5.0 g/cm^3 to enhance the shielding properties.[443]

Table 5.12 Typical radiation-shielding aggregates[441]

Natural mineral					Synthetic mineral			
			Percent by weight				Percent by weight	
Aggregate	Source[b]	Specific Gravity[a]	Iron	Fixed water	Aggregate	Specific Gravity[b]	Iron	Fixed water
Hydrous ore					**Crushed aggregate**			
Bauxite	—	1.8–2.3	0	15 to 25	Heavy slags	5.0	0	0
Geothite	Utah, Michigan	3.4–3.8	0	8 to 12	Ferrophosphorous[c]	5.8–6.3	0	0
Limonite	Utah, Michigan	3.4–3.8	55	8 to 12	Ferrosilicon	6.5–7.0	70	0
Heavy ore					**Metallic iron products**			
Barite	Nevada, Tennessee	4.0–4.4	1–10	0	Sheared reinforcing bars	7.7–7.8	99	0
Magnetite	Nevada, Wyoming, Montana	4.2–4.8	60	1.0–2.5	Steel punchings	7.7–7.8	99	0
Ilmentite	Quebec	4.2–4.8	40	0	Iron and steel shot	7.5–7.6	99	0
Hematite	South America, Australia	4.2–4.8	70	—				
Boron additives					**Boron products**			
Boro calcite	Turkey	2.3–2.4	0	0	Boron frit	2.4–2.6	0	0
Colemanite	California	2.3–2.4	0	0	Ferroboron	5.0	85	0
					Borated diatomaceous earth	1.0	0	0
					Boron carbine	2.5–2.6	0	0

Source: Society of Automotive Engineers (1993), Davis (1967), and Anon (1955).

a   Material water saturated with its surface dry.
b   Other sources may be available.
c   Ferrophosphorous when used in Portland cement will generate flammable and possibly toxic gases that can develop high pressure if confined.

Regarding the performance requirement for the radiation-shielding concrete, it is important to mention that concrete used in nuclear applications must have adequate and satisfactory structural and engineering properties such as compressive strength, shrinkage, workability, tensile strength, and modulus of elasticity. The physical properties of the radiation-shielding concrete specified by ACI 304R-34(2007) are as follows:

> High modulus of elasticity, low coefficient of thermal expansion, and low elastic and creep deformation are ideal properties for heavyweight concrete. High compressive strengths may be required if heavyweight concrete is to be subjected to high stresses. Heavyweight concrete with a high cement content and a low w/cm can exhibit increased creep and shrinkage, and in a massive concrete placement could generate high temperatures at early ages, causing undesirable localized cracking from the thermally induced stresses. When structural considerations require this cracking potential to be eliminated, it is necessary to use appropriate temperature control measures, which could include precooling or postcooling the concrete, or both.

One common practice is to add boron, e.g., mineral colemanite, manufactured boron frits, or ferroboron, and boron carbide as neutron attenuators to concrete in order to enhance the thermal neutron attenuation properties and to suppress secondary gamma ray generation. Take the use of colemanite $(2CaO \cdot 3B_2O_3 \cdot 5H_2O)$, a calcium borate mineral with a hardness between 4 and 4.5 and a specific gravity of about 2.4 g/cm^3 in radiation-shielding concrete, as an example. Gencel et al.[444] reported that increasing the colemanite ratio in volume has reduced engineering properties of the resultant concrete. Colemanite replacement up to 30% of the natural river stone or crushed stone both in coarse and fine parts can be considered acceptable in terms of workability and strength. Caution should be exercised because of the possibility of retardation due to the presence of soluble borates.[445]

Another commonly used heavyweight aggregate is metallic fillers—usually ilemenite, magnetite, hematite, etc. Hematite, for example, a natural red rock that contains iron oxide, has a Mohs hardness between 5.5 and 6.5 and a specific gravity between 4.9 and 5.5 g/cm^3 when pure. However, physical properties of rocks with hematite as the main constituent may vary considerably. Gencel and his team[446] studied concretes containing hematite for use as shielding barriers. The results revealed that after 30 freeze-thaw cycles the plain concrete loses 21.3% of its compressive strength, while the composite containing 10% hematite loses only 7.8% of the strength. Concrete and hematite composites have lower drying shrinkage than plain concrete, thus lowering stresses resulting from the shrinkage.

El-Sayed et al[447] also conducted an investigation of radiation attenuation properties for barite concrete as a biological shield for nuclear power plants, particle accelerators, research reactors, laboratory hot cells, and different radiation sources. Barite concrete with a specific gravity of 3.49 g·cm^{-3} was prepared, and the transmitted fast neutron and gamma ray spectra through cylindrical samples were measured.

5.7.4.2 Encapsulation of radioactive materials with cement and concrete

The risk of radioactive wastes is mainly due to the long time they need to eliminate the radioactivity levels so as to become stable wastes. Stabilisation and solidification (S/S) processes have been applied to ensure the isolation of the hazardous radiation from the biosphere.

In the stabilisation/solidification systems, for the immobilisation and solidification of low (LRW) and medium (MRW) radioactive wastes, different kinds of cementitious materials have been employed; ordinary Portland cement (OPC) is the most frequently used.[448]

Concrete to reduce radiation from radioactive waste has been utilised in several countries. The immobilisation happens mostly by both physical absorption and chemical means. Heavy metal (e.g., Ba, Cd, Cr, Cu, Pb, Mn, Ni, Zn, etc.) salts or radioactive elements in the cement S/S system are converted to low-solubility forms under the high pH environment of cement pastes, i.e.:

Precipitation reaction:

$$M^{n+} + Ca(OH)_2 \rightarrow M(OH)n + Ca^{2+}$$

The main hydrate of Portland cement, C-S-H, is a colloidal and amorphous compound with a high surface area. C-S-H therefore has the ability to immobilise metal ions by addition and substitution reactions as follows:[449]

Addition reaction:

$$C\text{-}S\text{-}H + M \rightarrow M\text{-}C\text{-}S\text{-}H$$

Partial substitution reaction:

$$C\text{-}S\text{-}H + M \rightarrow M\text{-}C\text{-}S\text{-}H + Ca^{2+}$$

It should be noticed that the cement S/S system does not reduce the radioactivity of the material, but will immobilise the radionuclides within the treated material, and prevent release of these materials into the environment. Over time the level of radioactivity emitted from the immobilised radionudles reduces itself through a process of radioactive decay.

Another thing worth mentioning is that the design of cementitious materials for sealing a nuclear waste repository should address the question of long-term durability as well as the compatibility with other sealing materials or the host rock. Time for isolation of the repository system from the biosphere may vary from hundreds to thousands of years.

In Spain, the storage place for low- and medium-level radioactive wastes was designed to ensure their confinement for 300 years, which is the time calculated to exhaust their radioactive emission levels.[450]

El-Dakroury and Gasser[451] studied the effects of different concentrations on compressive strength and bulk density, the effect of the gamma ray dose on resistance of the concrete to radiation, and the leaching of radioactive wastes (Cs-137). Test results revealed that the compressive strength and shielding of the concrete increased with an increase of coarse ilemenite growth. The leaching of waste appears to decrease, and the early strength increases with 10 wt% of silica fume used as the mineral admixture. The 15% (mass fraction) silica fume addition did not significantly alter the leaching rates of these nuclides.

At a conference in Jinan, China, in 2010, Japanese researchers gave an interesting report about their efforts to find the best mixture design.[452] The purpose of the tests was to find the best possible concrete mortar for a low-level radioactive waste (LLRW) disposal facility, in an engineered barrier system for a subsurface disposal facility planned in Japan. The facility will contain a higher level of radioactive waste in LLRW and will require durability of several tens of thousands of years. The engineered barrier materials are also used with the combination of two kinds of cementitious materials and bentonite. A goal in the research was to produce a mortar with a minimum of internal cracks and to reduce the voids of 100 nm in pore diameter to improve low diffusion. They tested 33 different mixture designs with different materials, and ended up with a design with low-heat Portland cement, crushed limestone sand, fine limestone powder, fly ash, and expansive, superplasticising, and air-entraining admixtures as their final choice. The ratio of the low-heat Portland cement:fly ash:fine limestone powder was 338:153:307. Upon conclusion of the tests, the chosen mixture design was tried out in a full-scale field demonstration inside a test cave about 18 m wide and 16 m high excavated about 100 m under the ground.

At a conference in Trondheim, Norway, in 1989, Scheetz et al.[453] reported on a test with a specially designed concrete as a candidate for sealing a geological nuclear waste repository in a tuff host rock environment, as part of the Nevada Nuclear Waste Storage Investigations projects. One of the tested grouts is described in the paper, using ash and silica fume in order to achieve a higher SiO_2 and Al_2O_3 content more compatible with the tuff geochemistry than a plain Portland cement.

Effects of elevated temperatures of 150 to 300°C were also tested. Initial compressive strengths of materials cured at 38°C for 7 to 900 days ranged from 100 to 125 MPa. Samples heated to 150°C for extended periods (28 days), either dry or hydrothermally, maintained their strength and well-bonded microstructure, while the results of heating at 300°C were mixed, with some strengths remaining high (95 to 110 MPa) and others diminishing (42 to 51 MPa). The water permeability did not increase much at 150°C, but did decrease at 300°C.

In a paper at a symposium in Hyderabad, India, in 2005, the director of the Civil and Structural Engineering Division of the Atomic Energy Regulatory Board (AERB), Mumbai, India, Prabir C. Basu, described the present situation at the time, of use of supplementary cementing materials for improving concrete durability in India.[454]

He reported that design life of a nuclear power plant is generally 40 years, which could be extended further by 20 years. Some waste management facilities have been designed even for 300 years. He said that blended cements were not commonly used, but that use of supplementary cementing materials added at the batch plant at the site had increasing use. High-performance concrete by using silica fume as an admixture was first used during the reengineering of a delaminated containment dome at the Kaiga generation station in the 1990s.

He further reported on tests using fly ash, slag, silica fume, and metakaolin as possible tools to ensure proper durability.

We also draw attention to the following papers, mentioned in other sections:

- A paper by Li et al.,[401] mentioned in Section 5.5, solidification of heavy metals and radioactive metal was tested.
- In a paper at a conference in Trondheim, Norway, C.A. Langton reports on the stabilisation of approximately 400×10^6 L of low-level radioactive alkaline salt solution at the Savannah River Plant Defence Water Processing Facility in the United States[455] (Section 5.10).

The American PCA shows three interesting examples of solidification and stabilisation on its website:[456]

The Feed Material Production Center (Fernald Site) is located 18 miles northwest of Cincinnati, Ohio. From 1951 to 1989, the Fernald site was a uranium processing facility. Its primary mission was to produce high-purity uranium metal products in the form of ingots, derbies, billets, and fuel cores for other sites within the nuclear weapons complex. Department of Energy used Fernald products as fuel for nuclear reactors to produce plutonium. When operations closed, some of the production wastes remained at Fernald. In the spring of 2005, cement-based S/S will treat radioactive wastes stored in Silos 1 and 2 for safe disposal.... 8900 cubic yards of high-activity, low-level waste material will be removed from the two silos, treated with cement, and shipped off-site for disposal. Once the waste is removed from the silos, cement and other supplemental cementitious materials will be blended to create a grout. The mix design calls for a 20% loading of waste (80% cement/cementitious material). The very cement rich mix will not only produce a monolithic waste form but also provide shielding from radioactivity.

West Valley Demonstration Project is a radioactive waste-cleanup project located approximately 35 miles south of Buffalo, N.Y. The site is the location of the only commercial nuclear fuel reprocessing facility that ever has operated in the United States. Low-level radioactive waste (LLW) was treated with cement and cast into specially made drums for storage and disposal.

Also mentioned are the Weldon Springs facilities in Missouri. The facilities were used in the 1950s and 1960s to produce uranium metal. Contaminated sludge from the production has been solidified and stabilised. A cleanup was completed in 1999.

5.7.5 Clearance of radioactive concrete

Back to the beginning of Section 5.7, on the disaster of radioactively contaminated concrete in Japan, scientists at Argonne National Laboratory developed a technology that is introduced in Figure 5.35[457] for cleaning certain types of radioactive materials from concrete using a superabsorbent gel and engineered nanoparticles. In the event of a terrorist attack using a dirty bomb or other radioactive dispersion device, the gel would be used to clean a site so people could safely enter it.

To remove the radioactive substances, a wetting agent and a superabsorbent gel are applied on the contaminated surface from a remote location. The wetting agent causes the radioactive material to resuspend in the water in the pores. The superabsorbent polymer gel then draws the radioactive-laden water out of the pores, while the engineered nanoparticles irreversibly capture the radioactive molecules. The dried gel is then vacuumed and recycled, leaving only a small amount of radioactive waste.

The work has shown that a single application of the gel can remove up to 90% of the radioactive element under observation. Right now it is common practice to demolish the contaminated materials in hopes of getting rid of the radioactivity. The gel technique would allow surfaces to be preserved, which means that we wouldn't have to deface monuments or buildings just to remove the radiation.

Figure 5.35 Argonne researchers[457] designed a system to safely capture and dispose of radioactive elements on porous structures, such as buildings and monuments, using this spray-on, superabsorbent gel and engineered nanoparticles.

5.8 SAFETY

When the "world's greatest inventor," Thomas Alva Edison, with 1093 U.S. patents, at a dinner speech in New York in August 1906, introduced his new invention, a concrete home, cast in one operation, he argued about the safety: fire proof, insect-free, and clean (Figure 5.36).

When Captain Edv. Kolderup in 1893 introduced his first Norwegian book on concrete, *Handbook in Building Art,* he gave a number of recommendations why society should use the new material *reinforced concrete.* It is interesting to look at this list from a concrete enthusiast well over a century ago:

1. Long life
2. Fire safety
3. Great load-carrying ability
4. Great resistance against shock and vibrations
5. Space savings

Figure 5.36 Safety—there might be several meanings of the word (Tamil Nadu, 2005).

6. Savings in foundations and anchorage
7. Faster construction
8. Sanitary advantages
9. Low cost investment
10. No maintenance
11. Absolutely water- and airtight
12. Bulletproof
13. Dryness
14. Safe against burglars
15. Great shaping possibility
16. Aesthetically advantageous

Nearly half of these arguments are important factors looking at concrete from a sustainability focus more than a century later, and at a time where the cement consumption per annum has grown nearly 1000 times (the cement consumption in the world around 1900 had reached 10 million tonnes per year).

Several authors have also argued that the introduction of reinforced concrete is one of the more important factors in the increased life expectancy over the last century.

Safety has been defined in a number of ways. One of them is *the control of recognised hazards to achieve an acceptable level of risk.*

It is important to remember that safety is relative. It had hardly been a possible, optimum, and sustainable solution by eliminating all the risk completely. We therefore also often speak of safety margins when we have some indications of the limitations of a structure and some knowledge of the magnitude of the possible risk.

While sustainability often focuses on restoring balance, managing growth, and other such efforts to shift our future course, the fact is that disaster mitigation and management are central to a sustainable future. Concrete provides superior resistance to damage, reducing the overall loss of life and cost of repair. For large-scale events it helps ensure that critical services like roads, hospitals, communications, data transmission, and emergency services can remain in operation.

The authors here wish to again stress the evolution of the focuses along the direction below when viewing the development of world concrete and construction:

Safety → Durability → Serviceability/Functionality → Sustainability

It is important in this context to learn at least two things:

- All the focuses in the evolution process are closely linked to each other and function upon needs instead of occurring and existing independently or replacing one by another.
- Sustainability is not only evolved from the previous focuses but works as a function of them as well.

We therefore can believe that sustainability is a holistic thinking/approach that can be considered as the function of safety, durability, functionality and economical feasibility, environmental compatibility, and social responsibility. The level/magnitude of each factor to sustainability varies depending on the specific requirement of the target and local boundary conditions.

It may, for example, be that a structure that has adequate safety according to the existing building regulations might also be increased in:

- Dimensions/capacity due to the need for higher thermal capacity, or be a better noise reduction barrier
- Dimensions to optimise concrete toward a more emission-friendly mix design
- Load-bearing capacity and concrete quality to be optimised toward increased durability and longevity

- Dimensions, such as safety measures, as the structure has essential safety tasks for ensuring a local community from infrastructure collapse during possible extreme loads from catastrophic incidents, such as tsunamis, floods, etc.
- Capacity due to possible increased loads from a tougher weather regime due to global warming half a century into the future

We have chosen to look at concrete and safety in a sustainability context in two different directions:

- Concrete as a safety tool
- Concrete safety levels in a climate change perspective

5.8.1 Concrete as a safety tool

Concrete is a powerful tool in taking better care of our environment and in protection of the environment from natural disasters. Concrete can be and is used to control and minimise the impact and risk of damage from coastal and inland flooding and the effect of climatic changes. In addition, concrete is a powerful tool in reducing the effects of human errors like fires, collisions, and explosions.

We have already mentioned erosion protection and various water regulation products in Chapter 1. As a reminder of all the other uses, we mention two other examples:

- Our nearly daily observation of collision and partition barriers on the roads
- The fantastic flood control system in the Netherlands called Oosterscheldekering, built in 1986

An English proverb says: God created the world, but the Dutch created Holland.

The American Society of Civil Engineers has named the dam works in the Netherlands to be one of the modern wonders of the world.

The storm barrier Oosterscheldekering (*kering* = barrier) was officially opened by Queen Beatrix on October 4, 1964. The cost was 2.5 billion euros, and it is claimed that the project has reduced the flood possibility for the country to once every 4000 years (Figure 5.37).

The Netherlands was always fighting to tame the water forces, and the big spring flood in 1953 gave the starting signal to the work with a barrier and water regulation project beyond any other previous projects. The largest spring floods occur when extreme flooding (high tide) comes at the same time as storms.

Figure 5.37 Oosterschelde (east part of the river Schelde) is located in Zeeland in the southwest part of the Netherlands.

One end of the Oosterscheldekering starts at the artificial island Neeltje Jans. Here is found a placard that says: "Hier gaan over het tij, de wind, de maan en wij" [Here the tide is conquered by the wind, the moon and us].[458]

Originally it was the plan that the connection between Oosterschelde and the North Sea would be shut by a dam. The work started at the end of the 1960s by constructing several artificial islands, among others, Roggenplaat (1969), Neeltje Jans (1970), and Noordland (1971).

Toward the end of 1973, the dam was 5 km long, but environmental protests stopped the work in July 1974. In 1976, it was decided that the remaining 4 km should be equipped with locks. Normally these are not closed, but open, so that the tidal water can run. The reason is to keep the saltwater isotope inside intact.

The work with the most difficult and costly part of the delta project started again in April 1976. A 4 km² dry dock was constructed at Neeltje Jans, where the bottom was 15.2 m below sea level. Here they cast the gigantic concrete components that should form the pillars for the lock gates. Every pillar was between 30.25 and 38.75 m high, and had weights up to 18,000 tonnes. This means that 7000 m³ of concrete was used in each pillar. The construction time for each pillar was 1.5 years, and starting work on a new pillar took place every third week. The work went on 24 hours a day, and up to 30 pillars were being worked on at the same time. All together, 65 pillars were cast, plus 2 reserve pillars. When the casting work was finished, the water was let into the dock, and the pillars were

Figure 5.38 Oosterscheldekering.

transported out with specially designed ships. The two reserve pillars can today be found at Neeltje Jans.

The pillars are the foundations for sixty-two 42 m wide lock gates. The last gate was lifted in position on June 26, 1986, and on October 4 the opening took place (Figure 5.38).[458]

The road across the dam was opened by Princess Juliana on May 10, 1987.

The annual costs for running the gigantic project are estimated to be 17 million euros.[458]

The 62 gates are closed during storms (wind higher than 9 Beaufort), or with extreme high tide, so that the high tide does not reach Oosterschelde. If there are no people at command central at Neeltje Jans, the gates are closed automatically if the high tide exceeds 3 m.

Oosterscheldekering is the largest of the 13 dams and barriers in the delta project.

On May 8, 2002, Oosterschelde was declared part of the national parks in the Netherlands. With the very special underwater fauna it contains, it is a popular place to visit for both national and international divers.

5.8.2 Concrete safety levels in a climate change perspective

When we design a structure, safety factors are added from our best estimations regarding testing and approving the material strength, as safety margins on the specified strength, on the dead loads (i.e., the weight of the structure), the live loads, and the eventual environmental loads.

From a purely economical point of view, there is an attempt to treat all the various materials equally to avoid unfair and economically unoptimised competition.

From an environmental safety point of view, however, there are differences, and sometimes they are considerable. In a typical concrete structure, the dead load, or the weight of the structure, is typically about the same as that of the live load, and consequently about half of the total loading. In a lighter structure, the dead load might be a considerably much smaller portion than the live load or the environmental loads that the structure is designed for. Consequently, concrete is most often more robust to unexpected loads that might come that are higher than the design loads.

In a situation where we must expect "more weather," meaning warmer and dryer, and when it is warm, more and stronger precipitation in rainy periods and stronger winds in windy periods, this might be a considerable advantage to concrete as a material, even if we do not take this into consideration in an economical analysis between material options at the design stage. However, in a life cycle assessment (LCA) evaluation including a statistical evaluation of extreme load expectancy, this could be given an important and more realistic weight in evaluations between options.

Another aspect of the safety level challenge is that more unforeseeable weather, leading to more unforeseeable environmental loads, probably should lead to new thinking regarding requirements for the robustness of the structures we build. We have over the last few decades seen disasters with much more serious consequences than before. Whether due to more dense concentrations of people in places where these concentrations were once more limited, human structures and routines for expressing needs and activities that we did not have before, cyclic or other casualties, or climate change having some influence, is probably of secondary interest. The disasters and their disastrous consequences for thousands and thousands of people can possibly be reduced in the future with renewed thinking of the need for more robust design in some of the structures we are building.

The society has used concrete as:

- Protection from fires
- Protection from collisions
- Protection from explosions
- Protection from flooding
- Protection from mudslides
- Protection from avalanches

- Protection from quakes
- Protection from tsunamis

There will in the future be a need for rethinking and probably more robust protective concrete designs. Rethinking of safety philosophy must also include second and third barrier safety devices for society damage minimising. The robustness and high-mass properties of concrete must be more innovative and be utilised for protection against extreme winds, debris from tornados, mud, snow, and ice slides, flooding from rivers and extreme wave activity, etc.

5.9 WATER

Water is said to be the largest commodity in the world, but water is also much more than that; 71% of the globe is covered by water, and the seas have an average depth of 3.5 km.

Even small increases in the water temperature will, due to temperature expansion, increase the sea level, and this means that more water evaporates and increases the intensity and amount of rainfall.

The approximate total water volume on earth is 1338 million km^3 (Figure 5.39).[459] Of this,

$$0.3 \times 0.03 \times 0.2\% = 0.018\% \text{ is freshwater in rivers and brooks}$$

In many countries some of this freshwater is bound in ice and snow during winter months (Figure 5.40).

When we discuss water it might be useful to remember the different sources of water, and that water has three phases: as a liquid, as we normally think of water; as a solid; and as a gas. As seen from Figure 5.41, the solid state is particularly important with respect to freshwater on the planet. The fresh surface water as normal is regarded as our water, NS only makes out a very limited part of the global water resources.

Water to concrete is the same as it is to plants, and even human beings. In normal construction, the water demand is roughly 10 to 20% of the volume of concrete used, for concrete mixing to gain suitable workability and make the hydration reaction happen, and for curing to continue the hydration and avoid the cracking of concrete. Water therefore can be considered an indispensable ingredient for concrete, although too much water in concrete functions negatively to the quality of concrete. On the other hand, concrete has been intensively used in different types of water projects, such as dams, harbours, pipes, and even ships.

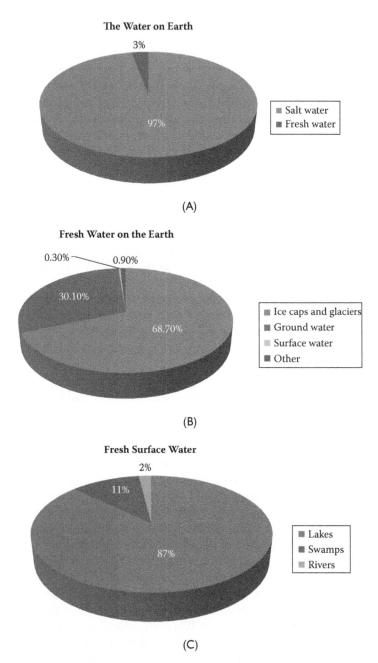

Figure 5.39 The water on earth.[459] (From Water, http://en.wikipedia.org/wiki/water (accessed February 23, 2012).)

Figure 5.40 A small early spring brook waterfall, Norway.

Climate changes with "more weather"; both more dry periods and more heavy rainfalls will produce several challenges to the concrete industry:

- Water shortage (Section 5.9.1) is a regional problem that also affects the use of water in concrete production in some areas.
- Increased attention and need for managing the increased precipitation (Section 5.9.2) will increase the demand for both concrete design solution and new thinking in the way we use concrete.

Figure 5.41 Life is coming back. The ice is melting. Early April. Local lake, Asker, Norway.

- We get 1% of our food supply (Section 5.9.3) from the waters (marine catch and fish farming). Theoretically, this could have been 20 times larger.

Another possible challenge for the concrete industry is the possible environmental effect of wash water from concrete production:

- On reuse of water (Section 5.9.4) in concrete production
- On escape of wash water (Section 5.9.5) to freshwater or the sea

Concrete has for many years also been an important tool in erosion protection (Section 5.9.6) from water.

5.9.1 Water shortage

Access to safe drinking water is essential to most life on earth, including humans. Over the last decades this has improved in most parts of the world; 1.6 billion people have gained access to safe water sources since 1990, and the proportion of people in developing countries with access to safe water is calculated to have improved nearly linearly from 30% in 1970 to 84% in 2004.[459] One of the Millennium Development Goals was to halve the number of people without access to safe water or basic sanitation by 2015. This goal is projected to be reached. A United Nations report in 2006 stated that "there is enough water for everyone," but that access to it is hampered by mismanagement and corruption.[459]

However, this is very challenging, looking at the fact that nearly 2 million children die each year due to unsafe water and inadequate sanitation, and more than 1 billion people lack access to clean water, and 2.6 billion do not have basic sanitation facilities. In China, the world's most populated country, the water resources are affected by both severe water quantity shortages and severe water quality pollution due to the growing population and rapid economic development, as well as lack of environmental oversight. Some observers have estimated that by 2025 more than half of the world population will be facing water-based vulnerability. Even most of the United States is expected to face water shortages in the next few years.

In 2001, P. Kumar Mehta wrote an interesting paper about concrete and the environment,[460] with a number of important thoughts. About water he says:

> So far, fresh water is abundantly available almost everywhere, and is being freely used for all purposes by the concrete industry. In fact, construction practice codes routinely recommend the use of potable water for concrete mixing and curing. But now, the situation has changed. Hawken et al. (Hawken P., Lovins E., and Levins H., *Natural Capitalism—Creating the Next Industrial Revolution*, Little Brown and Co., New York, 1999) report that fresh, clean water is getting more and more scarce every day. Although there is a lot of water on earth, less than 3% is fresh and most of that is either locked up in fast-melting glaciers and ice caps, or is too deep in the earth to retrieve. In recent press reports, the Indian government expresses deep concern over the future water shortage in the country because, due to global warming, the Himalayan glaciers, which are the primary source of water for Indian rivers, have receded by 30 m (100 ft) during the past 2 years alone.

Further, Mehta argues for reducing water in concrete mixtures by using water-reducing admixtures, which is definitely a very important solution to the reduction of water consumption as well as the improvement of concrete quality and durability.

Mehta's analysis and arguments are important. Even if using less water in the concrete mix design might be wise advice to increase durability of concrete, we believe that other actions in the concrete industry might also be effective in reducing the possible negative impact on the shortage of water, such as recirculation of wash water, more optimised use of water in curing, use of membrane curing to replace normal water curing, use of heat (steamed, electrical, or infrared radiation) curing, and use of other sources than tap water, recycled water use, for example. From different sources we have learned that the total annual consumption of wash water in the precast industry is nearly the same as that of water in the concrete produced annually—probably the wash water consumption is slightly higher. For example, Muszynski et al.,[461] from Florida, claim that a typical batch plant generates an average of 100 L of wash water per cubic meter of ready-mixed concrete. While efforts to reduce the water in the concrete only can reduce percentages, all the wash water can be recycled.

Taking care of our water resources is definitely a global challenge, while shortage of water is at most a regional, and mostly a local, problem. However, it is a local problem in too many local communities on earth, and is therefore a challenge that must be addressed globally.

During national seminars on concrete sustainability in four cities in India in 2011, the president of the Indian Concrete Industry, Vijay Kulkarni,[4] said:

> India is the second largest country in the world that has high precipitation. However, due to increasing population and pollution due to human activity, the supply of water is reducing. Further, rainfall is not uniformly distributed and erratic. Total annual precipitation is about 4000 km^3 and an average rainfall received is 1200 mm. There are rainfalls of 11 000 mm in Cherrapunji and the minimum average rainfall in West Rajasthan of about 250–300 mm. As per the World Watch Institute, India will be a highly water-stressed country from 2020 onwards. The meaning of water stress is that less than 1000 m^3 of water will be available per person per annum. India's population, is projected to go up to 1333 million by AD 2025 and further to 1640 million by AD 2050. It is projected that the per capita water availability in India may be reduced to about 1200 m^3/year by 2047.
>
> Many areas in India are perpetually drought-prone. A recent study, which identifies such areas, concludes that in most parts of India the probability of moderate drought ranges from 11% to 20%. Water is required for concrete production and curing in large quantity. The availability of water may therefore be one of the major constraints to growth of the concrete industry in perpetually drought-prone areas.

In a paper in 2009, V.M. Malhotra gave some alarming data regarding what can be expected:[462]

> The IPCC report on climate change issued in 2007 warns that global warming will affect very seriously the availability of water in the future. The Himalayan glaziers are melting fast. This could lead to water shortage for hundreds of millions of people. The glaciers regulate the water supply to the Ganges, Indus, Brahmaputra, Mekong, Thanklwin, Yangtze and Yellow Rivers and are believed to be retreating at a rate of about 10 to 15 meters every year. It is estimated that 500 million people on the planet live in countries critically short of water, and by 2025, the above number will leap to 3 billion.

... USA is the leader in water usage as shown below:

Continent	Water Consumption in litres per day
North America	600
Europe	300
Africa	30

In spite of the looming water crisis in the not too distant future, there is a huge wastage of water worldwide. For example, 9.5 billion liters of water are needed to support 4.76 billion people for their daily needs as set by the United Nations. On the other hand, currently 9.5 billion liters of water are being used to irrigate the world's golf courses.

... The intense irrigation has dramatic effect on the water tables. For example, the number of bore holes that pump irrigation water to India's farmland was 10,000 in 1960. And the number increased to 20,000,000 in 2007. This has caused declines of water tables from 100 to 150 meters in some places.

As an additional comment it should be noted that approximately 70% of freshwater used by humans goes to agriculture.[459]

In an interesting paper at a conference in Lillehammer, Norway, in 2007, representatives from the UK Precast Concrete Industry gave some good examples of sustainable handling of the water source challenge.[463] Among others they mention an example from the company Aggregate Industries, which at its Hulland Ward site annually consumes 110,000 m³ of water. Their mains water cost is approaching 1 euro/m³, which is considerable. Since 2002, three methods have been used to reduce the volume of mains water required for production:

- Approximately 60,000 m³ of water per annum is abstracted under license from a large natural pond located within the site boundary. In addition to providing a resource, the pond also provides a local amenity and a habitat for a wide variety of flora and fauna.
- Approximately 70% of the site rainfall is collected in storage lagoons for reuse in production; the area available for rainwater collection is over 200,000 m².
- All process water is collected and reused in the production process.

The report argues that the financial costs of the measures are minimal compared to the economic, social, and environmental benefits associated with reduced mains water consumption (Figure 5.42).

Figure 5.42 Rainwater harvesting 110,000 m³ of nonmains water, 100,000 euro savings.[463] (From Holton I., et al., Case Studies Demonstrating Reductions in the Consumption of Natural Resources and Energy by the UK Precast Concrete Industry. Presented at International Conference on Sustainability in the Cement and Concrete Industry, Lillehammer, Norway, September 16–19, 2007.)

Also taking into consideration the other factors mentioned in later sections of this chapter, it might be wise to consider the use of local buffer ponds for precipitation and excess process water, for production use and for possibly beautifying the environment as an interesting alternative when planning new concrete production facilities.

5.9.2 Managing the increased precipitation

From an engineering point of view it might sound simple. Most of the globe is oceans (70.9%), and with an average depth of 3.5 km, increased temperature must lead to at least two reactions. The water level must increase due to temperature expansion, and the moisture in the air will increase due to increased evaporation. The increased temperature in the air will increase energy in the moisture as well as producing larger raindrops. Consequently, we get more precipitation, and heavier rainfalls and wind will come with it, when it rains.

However, the experienced climate researcher will soon tell of a more complicated picture and considerably many more factors to take into consideration. We have therefore chosen to quote a few comments from a typical debate article in a climate journal:[464]

> "Observations show that global precipitation in our region has increased the last decade. However, it is uncertain how much is due to human created global warming, and how much that is due to natural variations."
> "Short stay for water in the atmosphere (typically 10 days) will quickly give global balance between precipitation and evaporation."

"Due to shortage of data (over the oceans) it has only recently by the help of satellite data been possible to give reasonable and reliable estimates of the precipitation development. Wentz et al. reported that in the period 1987–2006 the increased precipitation increased with 7.4% per degree of temperature increase."

The paper reports that various models indicated an increase in global precipitation of 17% over 100 years, but that the observations so far indicate higher values, and with very big regional and local variations.

Managing the increased precipitation gives us a number of challenges. Among them are more periods with heavy rainfall, combined with periods of drought.

Concrete pipes and channels are important tools in leading away water from rainfall. The wide spectre in precipitation volume that shall be taken away in a certain time interval creates new challenges. The pipe should have a large enough area, and the shear friction at the bottom of the pipe must be large enough to ensure cleaning of it to reduce precipitation of sludge, etc.

In ancient times, and for the first products in modern concrete history, the pipes were often egg shaped, where the smaller diameter in the bottom of the pipe gave a good design for low-water periods (Figure 5.43).

Modern production technology favoured circular pipes for practical and economic reasons. The egg-shaped pipes went out of fashion, but in later years, with more focus on and experience with variation in the weather, has seen the old shapes or similar solutions brought back (Figures 5.44 to 5.46).

The challenge to manage the effects of heavy rainfalls is not new, but the combination of climate change and increased urbanisation has increased and intensified the problems considerably. The change in the water volume to be handled, from a natural environment with pervious ground to a urban environment with impervious ground, is considerable, and might in extreme cases increase the water problem from five- to ten-fold (Figures 5.47 and 5.48).

Many of the flood catastrophes we have seen in later years have increased in consequence due to this challenge. Among the first principles in coping with this intensified water-handling challenge is to try to handle as much of the problems as possible locally, and in no way accelerate the water flow. This might mean reducing the water flow to centralised recipients, by increasing the permeability in the ground, increasing the use of green roofs, increasing the delaying capacity in the piping system, increasing effective tree planting in urban environments, using local ponds and small lakes to delay water flow (and make a more enjoyable environment), etc. The optimum solution seems to utilise as many of the mentioned tools as possible simultaneously. The increased speed for the water and the reduced filtration through soil also have a significant effect on the water quality and increase the concentration of harmful substances in the water.

Figure 5.43 Egg-shaped pipe in operation for over 90 years in Narvik, Norway.

To improve this challenge, concrete and concrete solutions might play an important role and are some of the most effective tools.

We mention a few examples:

- Pervious concrete
- Pervious ground with concrete paver systems
- Delaying systems in the pipework

5.9.2.1 Pervious concrete

Pervious concrete is a type of concrete where the cementitious paste forms a thick coating around the coarse aggregate particles, with a maximum amount of voids around the coated particles. It has also been called no-fine concrete. The technology challenge is to make a concrete with the highest

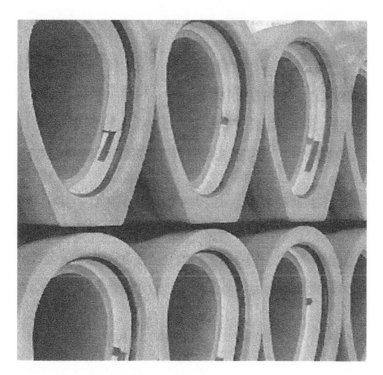

Figure 5.44 Modern egg-shaped pipes, Finland.

possible void percentage, and at the same time have a coating paste that has durable and acceptable strength properties. Typically, with a void percentage of 15 to 25% the concrete might allow water seepage through the concrete of 200 L/m²/minute. Pervious concrete might be used for a number of applications, where use in pavements and use in erosion protection probably are the most important, from a climate point of view (Figure 5.49).

The high flow rate of water through pervious concrete pavements allows rainfall to be captured and percolate into the ground, reducing storm water runoffs, and possibly improving the hydrological balance in the area. In addition to the advantages of reducing or eliminating surface water in pedestrian walks, parking areas, and roads with moderate traffic, and delaying the runoff of storm water, pervious concrete will reduce the loading on public piping systems, and it has been shown that the system has an important cleaning function with respect to heavy metals, hydrocarbons, and oil spillage (Figures 5.50 and 5.51).

In colder areas pervious concrete will speed up melting of snow and reduce surface ice. However, special care should be taken regarding freeze-thaw properties, as only pervious concrete with extreme high quality in the matrix will have the same freezing properties as normal concrete.

Figure 5.45 Alternative solution, Norway.[465] (From Reiersen T., *Fordrøyning i betngrør-dimensjonering, løsninger og erfaringer*, Report 3, Norwegian Concrete Society, Environmental Committee, October 2011.)

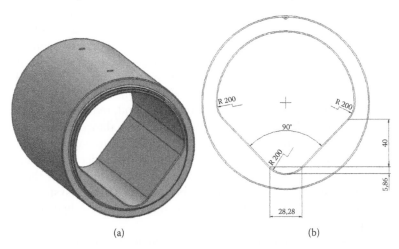

(a)   (b)

Figure 5.46 Alternative solution, Norway.[465] (From Reiersen T., *Fordrøyning i betngrør-dimensjonering, løsninger og erfaringer*, Report 3, Norwegian Concrete Society, Environmental Committee, October 2011.)

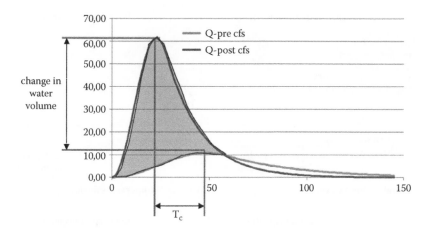

Figure 5.47 Changes in the maximum water volume and concentration time from natural to urban precipitation area. The low curve shows the natural area, and the high curve the situation after urbanisation.[466] (From Muthanna T.M., *Det koster å ikke gjøre noe*, Report 3, Norwegian Concrete Society, Environmental Committee, October 2011.)

In addition to the use of pervious concrete on road surfaces, this interesting "new" type of concrete has also found use as a possible growth medium—in "green" parking spots, as retaining walls, as green erosion protection, and even as floating fish aggregating devices (FADs) (Figures 5.52 and 5.53).

5.9.2.2 Pervious ground with concrete paver systems

An important alternative to pervious concrete is pervious ground made by concrete paver systems. For both of these concrete alternatives, the major advantage is the possibility to handle storm water as locally as possible, and not to send it out in an unnecessary and costly journey in the pipe systems, where the problems increase as you move downstream in the water handling—the accumulated volume increases. Even if not all the rainwater can be infiltrated and taken care of locally, the delaying function is also of major importance (Figures 5.54 to 5.56).

5.9.2.3 Delaying systems

When the community drainage system does not have sufficient capacity to take extreme rainfalls, the design of delaying systems might be a rational way to cope with the challenge. The normally easiest way, in nonurban areas, will probably be to design ponds or small lakes, which might be dry or partly dry in the dry season, but might be valuable delaying magazines in heavy rainfalls. In urban areas there might be a need to design such

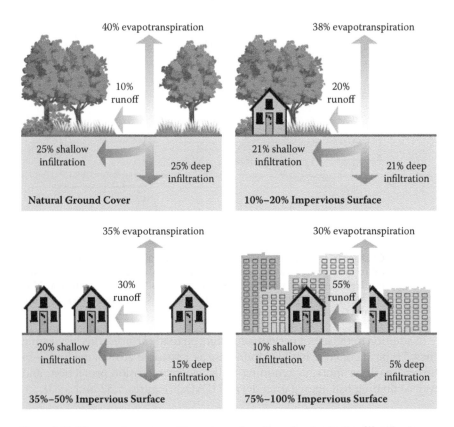

Figure 5.48 Changes in water volume as a function of urbanisation,[466] referring to Norwegian Concrete Society.[467] (From Norwegian Concrete Society, Environmental Committee, *Økt focus på Miljø og Klimaendringer-nye muligheter for betong*, Report 2. Oslo, Norway, 2010.)

delaying systems underground. In big cities large underground grottos have been established to cope with the challenge since ancient times. Another possibility is to design such delaying magazines by concrete pipes locally (Figure 5.57).

5.9.3 Reuse of wash water from concrete production

Reuse or recirculation of water from concrete production is normal in a number of countries in the concrete industry. The most typical example is probably in the spinning of concrete pipes, poles, and pillars. In this type of production a very flowable concrete is pumped into the concrete mould. During the spinning process, the water is run off from the product centre at the end of the mould, while dense concrete is compacted toward the mould in the perimeter.

Figure 5.49 Pervious concrete.

In a report from 2005, Justnes gives some recommendations and conclusions regarding the reuse of wash water.[469] He concludes that there is no problem in the reuse of wash water, in particular if sludge is removed through filtration or sedimentation, but warns that the water might have an effect on the early binding of new concrete, probably in the sense that the binding might come earlier. He also refers to Norwegian Standard NS-EN 1008, *Mixing Water for Concrete*, which also includes the reuse of wash water, with some particular boundary values for lead, zinc, phosphates, sugar, and sodium oxides. He concludes that the measured values from practice are far below the allowable values (Figure 5.58).

Recycling of wash water in the ready-mixed concrete industry has been done at various locations around the world for several decades, but still there is a long way to go before this becomes standard procedure all over the world. Several places report on waste concrete and solids from wash water in the order of 1% of the annual concrete production not being recycled. The wash water alone normally makes approximately the same quantity as used in production. In addition to the positive environmental effect of recycling, the economical savings can often pay for the equipment investment in a surprisingly short period of time. Most companies pay for the water they consume, and also for the water that later goes to the recipients.

Figure 5.50 During the ACI conference in Cincinnati, Ohio, in October 2011, a competition was arranged between American universities to make the best pervious concrete. Twenty-nine teams participated, where the proposals were judged on five items: cylinder splitting strength, perviousness measured by time for a defined quantity of water to run through, price, carbon footprint, and the report. The winning team came from the University of Florida.

Figure 5.51 Pervious concrete in the old Japanese capital, Nara, just after a heavy rainfall, November 2000.

Figure 5.52 Professor Tamai, Kinki University, Japan, showing an alternative use of permeable concrete (2001).

In addition to the savings in water cost, the recycling of aggregate might also show some interesting savings.

A report from a French author in *Concrete International* in 2002[470] claimed that recycling facilities for the ready-mixed concrete in France had been regarded as unproductive, but that opinions have changed. The case story tells about analysis made at an operation in the Lyon area, concluding with interesting economic gains in addition to the environmental advantages (Figures 5.59 and 5.60).

The recycling of wash water and other process water from concrete will require local testing based on the local use of cement, aggregate, mineral, and chemical admixtures, to avoid unforeseen changes in setting time, slump loss, or retardation. However, modification chemicals to aid the modified

Figure 5.53 Concrete blocks in pervious concrete in a retaining wall in a garden in Asker, Norway.

Figure 5.54 Permeable paving ground to the left, versus asphalt.[468] (From Myhr K., *Veileder i dimensjonering og bruk av permeable dekker av belegningsstein*, Report 3, Norwegian Concrete Society, Environmental Committee, October 2011.)

mix design are available from a number of admixture suppliers. The type of washing and recycling equipment to be used will also depend on the local types of concrete to be produced, the wash water volume to be handled, and how the recycled materials best can be utilised in the production.

In a paper at a conference in 2004, Nakamoto et al. from Wakayama National College of Technology, Japan,[471] stressed the importance of adding set retarders to the wash water sludge. They explained how Japanese Industrial Standard A 5308 permits usage of under 3% solids from sludge water of the cement content of the new concrete in Japan. Somewhat depending on the temperature, they claimed that this might be increased to

Figure 5.55 Use of pervious pavement, compared with asphalt pavement, in Beijing, China. (Provided by Rechsand Science & Technology Group, www.rechsand. com.)

Figure 5.56 Pervious pavement around the Bird Nest, 2008 Olympic Stadium, Beijing, China. (Provided by Rechsand Science & Technology Group, www.rechsand. com.)

Figure 5.57 Delaying system with concrete pipes, Norway.[465] (From Reiersen T., *Fordrøyning i betngrør-dimensjonering, løsninger og erfaringer*, Report 3, Norwegian Concrete Society, Environmental Committee, October 2011.)

6 to 9%, and that this will help in enabling the concrete industry to recycle all its wash water.

They also claimed that at the time, the annual quantity of sludge water in the Japanese ready-mixed industry was 0.7 million m³, and that too much of this was dried to a sludge cake and abandoned in a landfill. As background for their conclusions, they referred to two test series, one with 20°C and one with variable temperatures from 5 to 35°C. The sludge water tested had about 15% of solid content, where fine aggregate under 0.15 mm occupied 20% of this. The setting retarder used to stabilise the sludge was of alkyl aminophosphine acid. The sludge tested then consisted of 12% cement, 3% fine aggregate, 85% water, and 2% setting retarder. The sludge was used in new concrete 1 or 3 days after collection.

As a comparison, we mention that Heimdal, in a presentation in Norway in 2000,[472] claimed that experience from three ready-mixed operations from one of the larger ready-mixed concrete companies in Norway showed that

Figure 5.58 Wash water treatment plant.[463] (From Holton I., et al., Case Studies Demonstrating Reductions in the Consumption of Natural Resources and Energy by the UK Precast Concrete Industry. Presented at International Conference on Sustainability in the Cement and Concrete Industry, Lillehammer, Norway, September 16–19, 2007.)

with a solid content of 5% in the sludge water, all the water can be recirculated. If the solid content is 10%, only half of the sludge water can be recirculated. He also reported that 100% of sludge water was recirculated for concrete qualities up to C30. For higher qualities, the content was lower.

The Norwegian Ready Mixed Concrete Association–FABEKO in 1997 assigned a consulting company to prepare recommendations for sedimentations for ready-mixed operations.[473] In the report it claims that each concrete truck uses from 500 to 3000 L of water each day for washing. As an average in planning, it recommends using 1000 L. In addition, approximately 1500 L each day comes from the mixer. The association also claims that each car washing adds 200 to 400 L of sludge to the basin.

Finally, we also want to mention an interesting study by Meininger from 1973,[474] who published his results on recycling mixer wash water. Data were compared between concrete batches using tap water, clarified wash water (about 3000 ppm total dissolved solid), and slurry water of two different volumes, as shown in Figure 5.61 (about 100,000 ppm total solid, or 10% by weight). He demonstrated that cement-based wastewater did not have much effect (≤10% compared to tap water batch) on the concrete drying shrinkage when less than 49.5 L/m^3 was used.

(A)

Figure 5.59 Collection and storing of aggregate to be recycled. (Photo Thrane & Thrane Teknikk AS, Norway).

5.9.4 Escape of wash water from concrete production to freshwater and the sea

In the earlier mentioned report from Justnes,[469] he had studied the effect of high alkalinity and possible heavy metals from concrete wash water on nature. Justnes emphasised the importance of the use of sedimentation, as this will both reduce the amount of sludge and increase the carbonation, which will aid in neutralisation of the high pH in the fresh concrete.

Well-cured concrete might contain about 25% calcium hydroxide $(Ca(OH)_2)$. The calcium hydroxide together with the effect of the alkali oxides in the cement will create an environment with a relatively high pH value. This is somewhat reduced over time through the reaction with CO_2 in the air. Most surface water in the ground is normally acidic, and will quickly neutralise the basic concrete water. The seawater, however, is naturally basic (pH 8 to 9), and a considerable dilution is necessary to bring the pH down to that of natural water. Practical use of concrete in seawater, however, indicates that there should be no harm to the life in the sea. On the other hand,

(B)

Figure 5.59 (Continued)

as the dilution challenge is a fact, care should be taken to handle the problem properly, through moderate disposal, neutralisation processes, etc.

Regarding the possible effect of heavy metals, there is hardly any direct harm to be expected, but comparing the amount of heavy metals in concrete wash water to the natural level in seawater, dilution is necessary from 0 to nearly 5000 times to reach a natural level, depending on what metal is in question.

In a report from the Norwegian Agricultural Research Institute from 2008,[475] where it tested the effect of using concrete wash water sludge as a stabilizer for the soil, it concluded that of the levels of various heavy metals, only As and Ni show a conclusive increase in the soil, and all values are

Figure 5.60 Sludge collection tank, Jærbetong Norway. (Photo by Thrane & Thrane Teknikk AS, Norway).

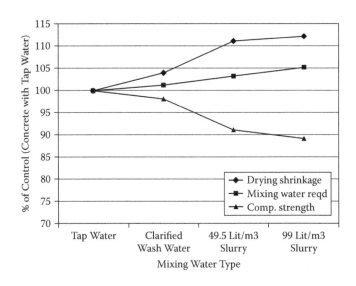

Figure 5.61 Study of concrete strength, mixing water requirement, and compressive strength.[474] (From Meininger R.C., *Recycling Mixed Wash Water—Its Effect on Ready-Mixed Concrete*, Technical Information Letter 298, National Ready-Mixed Concrete Association, Silver Spring, MD, March 1973.)

far below what can be regarded as harmful. The sludge, however, has a positive effect on increasing the pH, and 500 kg of concrete sludge per day every 5 years would be favourable.

5.9.5 Food supply—artificial fish reefs (AFRs)

Water is essential for our food supply on land. Irrigation takes up 90% of water withdrawn in some developing countries, and significant proportions in economically developed countries (in the United States, 30% of freshwater usage is for irrigation). Wikipedia claims that producing food for the 6.5 billion people or so who inhabit the planet today requires the water fill a canal 10 m deep, 100 m wide, and 7.1 km long—that's enough to circle the globe 180 times.[459] As mentioned in Chapter 2, concrete is an important tool to distribute this water through water channels, etc.

However, as mentioned previously, the oceans cover 71% of the earth's surface, but give us 1% of our food. The basic source for production of food is photosynthesis. From a theoretical point of view, taking into account both the fact that the light is deflected and less efficient in water, and the possible available area, there is a potential to increase the food production from the oceans 20 to 40 times. However, more than 50% of all our fish resources have been overfished, and as can be seen from Figure 5.62, the world marine catches are stagnating and partly going down.[476]

Of the four largest families of fish, with an annual capture of more than 10 million tonnes, all have reduction in annual capture from 1996 to 2002.[476] Artificial fish reefs might be an important tool in solving this challenge in the future.

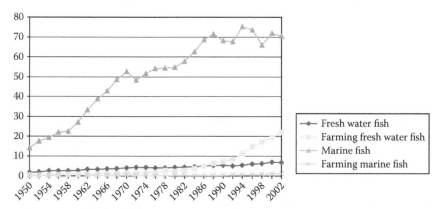

Figure 5.62 Development in fish harvest.[476] (From *FAO Yearbook Fishery Information: Capture Production*, vol. 94/I, Rome, 2004.)

In 1990 Seaman introduced in his book[477] that the French scientist Francois Simard made an investigation in Japan among 40 members of the Fisheries Cooperative Association (FCA) regarding the use of artificial fish reefs:

- 38 of 40 answered that they were active users of the reefs.
- 77% answered that the reefs increased global fish resources.
- 77% said that they believed that the reefs gathered fish.
- 55% meant that the reefs played a positive role in reproduction.
- 37% said that the reefs gave more regularity in capture.
- 32% said that the reefs increased fish size.
- 15% said that the reefs increased the quality of the capture.

5.9.5.1 History

Already more than 100 years ago, it was a tradition among farmers in the eastern part of Norway to leave the cutoff branches of the trees in piles on the frozen lakes, when they were cutting timber in the winter. The next summer they knew that this was the best place to fish pike. However, the oldest documented use of artificial fish reefs is probably from Japan in 1795. Fishermen from Awaji Island outside Kobe got unusually good catches of fish near a sunken ship. After 7 to 8 years, the sunken ship disentegrated and the ship disappeared. They then filled timber rafts with bags of sand and sunk them 40 m depth. Three months later they got more fish in this spot than they had experienced with the sunken ship.[478] A stone monument from the town of Uoshima in Japan tells a similar story.

More than 100 years ago it became common practice in many places in Japan to sink old boats to improve the fishing conditions.

In 1920, the Japanese expression *jinco gyosho* (fish reefs built by men) was used for the first time, and in 1923 came the first Japanese government subsidies for the building of artificial fish reefs.

The first artificial reefs in concrete were built in Japan in 1954.

In the years 1988–1994, the Japanese government investments had grown to more than 1 billion USD per year.[479]

5.9.5.2 Where have AFRs been used?

Seaman[477] refers to a bibliographical investigation in 1986 that refers to AFR projects from 39 countries. Jahren made a state-of-the-art report[478] in 1998, and could record artificial fish reef activity in 61 different countries. In August 1997, there were 40,060 documents on the Internet about artificial fish reefs. Artificial fish reefs have been used nearly all over the world, both in the sea and in inland lakes. A pioneering nation in this has been Japan.

The first artificial fish reef in Norway was built in 1994. These were concrete structures of three different types, placed in two locations at Vestvågøy in the Lofoten Islands in northern Norway.

5.9.5.3 Motivations for establishing AFRs

There are great variations in the motivation for establishing artificial fish reefs. We have recorded the following:

- Gathering of fish resources to promote more cost-efficient and resource-efficient fishing procedures
- Increasing the existing or creating new breeding structures, and thereby increasing the production of biomass
- Bringing new life to the sea bottom desert
- Creation by destruction of corals by temperature changes
- Fishing with explosives
- Unbalanced fishing leading to destruction of kelp forests, and silting of the sea bottom due to deposits from man-made wastes
- Purification of water, thereby creating bases for new life
- Utilisation of waste nutrition for increased biomass production, and improving the sea bottom under fish farms
- Refining the sea bottom for increasing the growth possibility for particularly valuable sea food
- Securing the fishing potential for the coastal population
- Establishing predictable fishing grounds for tourists (increase tourist season for hotels)
- Provision of recreational diving sites
- Positive utilisation of abandoned offshore oil platforms
- Prevention of trawling in defined areas (see Figure 5.63)
- Control of beach erosion
- Provision of breakwaters
- Creation of areas for scientific experiments

The most important thing to bear in mind with this list is that the reef design should take into consideration the motive for the establishment of the artificial fish reef. Unfortunately, we see an example of reefs established only to get rid of wastes, with too little focus on the optimum design.

In 1984, the United States approved the National Fishing Enhancement Act. This act gives a good general summary for the motivation of establishing the artificial fish reef:

Properly designed, constructed, and located artificial reefs ... can enhance the habitat and diversity of fishery resources; enhance United States recreational and commercial fishing opportunities; increase the production of fishery products in the United States; increase the energy efficiency of recreational and commercial fisheries; and contribute to the United States and coastal economics.

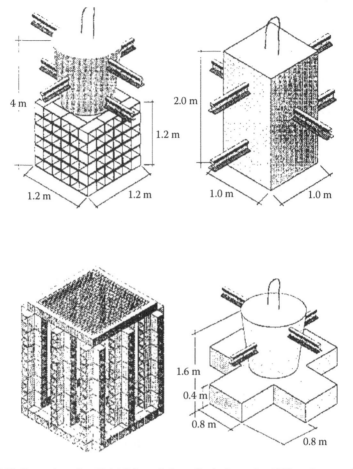

Figure 5.63 Examples of artificial fish reefs from Spain from the 1990s, where an important motive was to stop trawling in areas important for the coastal fishers. The weights of the components ranged from 3.5 to 5 tonnes. For the four types 88, 150, 150, and 265 units were used in each reef complex.[480] (From Rivenga, S. et al., European Artificial Reef Research, presented at Proceedings of the 1st EARRN Conference, Ancona, Italy, March 1996.)

5.9.5.4 Design factors

Design factors for artificial fish reefs will certainly depend on the motivations for building the reef system, but also other realities have to be considered:

- The sea bottom condition
- Current velocity and direction
- Wave height/depth
- Water depth

- Size, depth, and shape of the openings in the reef modules, depending on the present marine biology
- Suitability, durability, and porosity of building material
- Logistic factors, such as component size, jointing technique, transport, lifting capacity, and storage
- Distance between single items, structures, and size of the reef complex

Japanese researchers have observed that there are in principle three types of inhabitants in the reefs (Figure 5.64):[479]

1. Inhabitants that prefer physical contact with the reef, and that can be observed in holes, smaller openings, and narrow passages and along the sides of the reef. For these types, it is important that the reef construction has shapes and sizes in the same order as the individuals.
2. Inhabitants that live around the reef, without necessarily entering the reef. These individuals are attracted to the reef visually and by sound. For these species, it is important that the reef is big and variable enough to be sensed day and night.
3. Species that have a marked attraction to the reef, but who live in the water above the reef. Important for these types is the turbulence that is created by the reef. Observations show that such important turbulence is created in 80% of the water depth, if the height of the reef is more than 10% of the water depth. Research has also shown that there is an increased amount of plankton around the artificial fish reefs. This is probably due to the mentioned turbulence.

Different researchers have presented different models for the action and efficiency of artificial reefs. One of the better and instructive model is that shown by Prof. Hubert Jean Ceccaldi at Centre d'Etude des Resources Animals Marines (CERAM) at the University of Marseille in France (Figure 5.65).[481]

The size of the reef area and the number of modules present are also important factors in designing effective reefs. Researchers have argued that

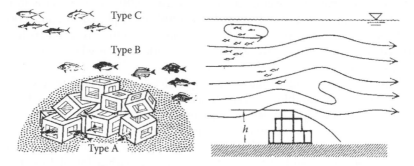

Figure 5.64 Types of inhabitants.

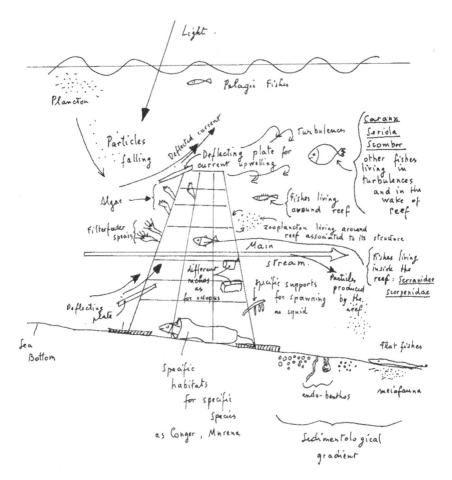

Figure 5.65 Theoretical model of the artificial fish reef.[481] (From Ceccaldi H.J., Figure in Private Letter, CERAM, Aix-Marceille, France, January 1998.)

effective reefs have to cover an area of at least 20,000 m² if a reef is to be constructed on new ground, or at least 5000 m² if placed in connection with existing natural coral reefs or rock structures. Other researchers have argued that these figures are somewhat dependent upon where you are in the world and the type of fish in the area.

With the complex challenge that proper design might lead to, concrete is often the most suitable material choice. Other materials, often combined with waste handling, have been used, and not always with the best results.

With its fantastic potential in shaping, concrete is a good choice if designed properly, but also there are material challenges for concrete. As an example, we would choose a dense concrete with a lot of limestone aggregate for restoration of coral reefs (slow growing with calcium base), while we would choose a porous concrete with a typical siliceous aggregate for restoring a kelp forest (better absorption regime for the roots, and a siliceous-rich medium).

5.9.5.5 Some examples

From what we have mentioned previously, it should be obvious that reef design covers a very wide variety of shapes and sizes. To illustrate this, we show some examples of larger reef constructions from Japan, and various reef modules used in France in the years 1986–1993 (Figures 5.66 and 5.67).[482]

5.9.5.6 Restoration of coral reefs

Unfortunately, many coral reefs have been destroyed. From out own diving experiences we have observed damaged or destroyed reefs by dynamite fishing in Tanzania, from earthquakes in Egypt, and from an intolerable temperature increase (whitening) in the South China Sea. It is definitely depressing to see this destruction of the most productive breeding grounds in the oceans.

Work has been done in several places in the world to restore damaged sea bottom and natural coral reefs. More than 20 years ago, Edwards and Clark[483] reported on work to repair coral reefs that had been destroyed by dynamite fishing in the Maldives. Different types of smaller concrete modules were tested. The philosophy behind the work was to let concrete modules calm the seabed, and thereby give the corals conditions to

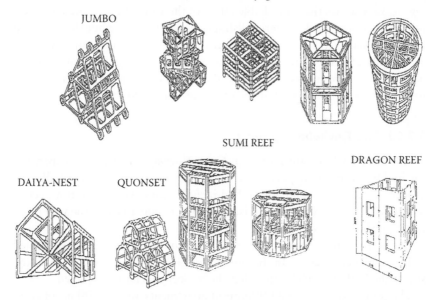

Figure 5.66 Examples of larger-size AFRs from Japan.

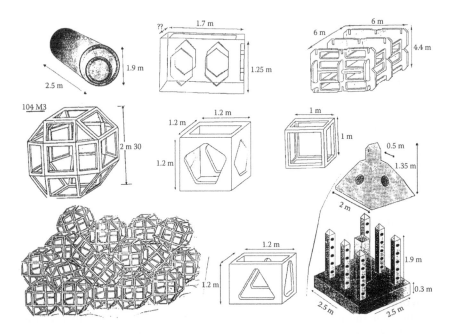

Figure 5.67 Examples of French reef modules from 1986 to 1993, with weights ranging from 260 kg to 27 tonnes.

grow again, and hopefully over time, grow through the concrete modules. Figure 5.68 shows the remarkable effect of the work. In nondamaged reefs nearby an average of 35 different fish species were found, while the damaged areas only had 5 to 10. From the figure it can be seen that in some repaired areas, only 250 days after emplacement, the same or higher number (39) of fish species was found as in the natural reefs. The least rewarding recording in the repaired area was 20 different species after 200 days.

5.9.5.7 The Tjuvholmen project

In 2002 the harbour authorities in Oslo, Norway, initiated a competition for building out a new part of the town, not far from the town hall. The group that won the competition had included an environmental aspect in its proposal, to install artificial fish reef components to clean the water and bring marine life back.

The construction work started in 2005, and the whole project is planned to be completed in 2014. In the summer of 2010, a concert was held on the site in the harbour, celebrating that the project was half completed.

The project consists of 84,000,m² of apartments and 33,000 m² of commercial buildings, a museum, etc. (Figure 5.69).

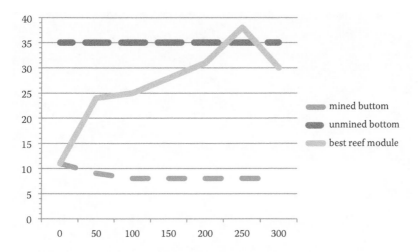

Figure 5.68 Number of fish species versus days after emplacement.

Figure 5.69 The Tjuvholmen project.

Before construction started, the sea bottom was muddy and had little or no marine life. Before the piling for the decks started, a 40 cm layer was placed on the bottom. All the building structures were placed on piles down to bedrock through the water and silt and clay. Under the foundation deck, there was 10 to 20 m water depth. Two 4-story parking structures

Figure 5.70 The construction design has never been used in Scandinavia.

were produced in a dock about 100 km down the Oslo fjord and towed into position and hung up on the pile heads (Figure 5.70).

The reef design had two purposes:

- To clean the water
- To bring marine life back

Five different components were used. Two components took care of the cleaning purpose:

- Under the 5200 m² of concrete deck were positioned 1500 mussel ropes that were each 8 m long. At 5 cm blue mussels (*Mytilus edulis*) filtrate 6 L of water per hour. Theoretically, with 10 mussels per 10 cm of rope, the mussels have the ability to filtrate the whole water table under the deck in 10 hours.
- On the sea bottom 400 cleaning modules were placed at 3 × 3 m intervals. The purpose of the cleaning modules was threefold:
  - To house smaller organisms that have a supplementary cleaning effect
  - To house organisms and fish that feed on dead mussels that fall to the bottom, to avoid an H_2S environment from old and dead mussels on the bottom
  - To contribute to a heterogeneous bottom/marine life by being a hard bottom component

Figure 5.71 Reef components, ready for setting out. The picture was taken during the Norwegian Concrete Day in October 2008.

Three types of components were placed in front of the decks with the purpose of attracting new marine life. All together, 20 groups of components, with 10 concrete components in each group, were set out (Figures 5.71 and 5.72):

- HD pack, for typical bottom fish and organisms
- JAFI 2000, for fish that live above the bottom
- LM 42, for lobster, etc., that prefer caves

The reef design has proven to be a success. NIVA, an independent government-founded Norwegian institute for water and marine biology, inspected the result in November 2008 and November 2010. In its report from 2008, it observed:[484]

- Dense growth of mussels down to 10 m
- Reefs in good condition
- Dense growth on the cleaning reefs
- Animal life starting to stabilise
- Improvement in water quality

In its report from 2010, it observed:

- The mussel ropes are densely populated with filtrating organisms.
- Organisms on the bottom clean up shells that fall down.
- The artificial reefs are grown with fixed organisms.
- Observations have been made of more than 10 types of fish plus lobster and crab.
- The design has been very successful for environmental improvement of polluted coasts and the harbour.

Figure 5.72 Artificial concrete reef components near the edge of a deck.

5.9.6 Erosion protection

Concrete has for years been a favourite material in protecting from physical and chemical erosion, as breakwaters along the ocean front, as soil protectors along rivers, as rock stabilisations along roads, as snow shelters along roads and railway lines, etc.

It is safe to assume that increased and intensified precipitations will increase the demands for such, and new development of solutions for this purpose, in the future.

In Chapter 1, we showed some examples. Figures 5.73 to 5.78 demonstrate more.

5.10 WASTES

All households and industries produce solid wastes. The wastes differ in characteristics and volume depending on culture and climate and have been changing over time. Some wastes, probably most wastes, have economical value, and the ideal solution for wastes is definitely recycling for

Figure 5.73 Examples of precast concrete snow and avalanche shelters, Japan.

Figure 5.74 Another Japanese example.

Figure 5.75 Simple precast concrete modules are protecting the church at the shore of Volga in the Rybinsk Reservoir, Uglich, Russia.

optimum utilisation of resources and reduction of landfills. However, what is waste for some people might be a valuable resource to others. Wastes for recycling and incineration are even exported from one country to another. Unfortunately, uncontrolled export of harmful waste also takes place.

According to the United Nations Statistical Division:

> Wastes are materials (that are not prime products that is products produced for the market) for which the generator has no further use in terms of his/her own purposes for production, transformation or consumption, and of which he/she wants to dispose. Wastes may be generated during the extraction of raw materials, the processing of raw materials into intermediate and final products, the consumption of final products, and other human activities. Residuals recycled or reused at the place of generation are excluded.[485]

According to the OECD, between 1990 and 1995, the amount of wastes generated in Europe increased by 10%. Most of what we throw away is either burnt in incinerators or dumped into landfill sites (67%). Both of these methods might create environmental damage. Landfills not only take

Figure 5.76 Combination of precast concrete components in normal concrete with pervious concrete for grass vegetation growth between the components. We have also seen summer idyllic examples of growth of wild strawberries in the erosion protection along canals and rivers.

up more and more valuable land space, but also cause air, water, and soil pollution, discharging CO_2 and methane into the atmosphere and chemicals and pesticides into the earth and groundwater. This is harmful to human health, as well as to plants and animals.[486]

In EU's Sixth Environmental Action Programme, waste prevention and management was identified as one of the four top priorities. The aim is to reverse the trend that is estimated by the OECD that we might generate 45% more waste in 2020 than we did in 1995 (Figure 5.79).[486]

Recycling of wastes in concrete and recycling of concrete were treated in Chapter 4. Waste as an energy source in cement production was mentioned in Chapter 3.

With respect to waste from concrete production, we are of the principal opinion that this should be recycled. There are, however, situations where the best environmental solution is local deposits for the concrete waste, such as leftovers from casting operations, washings of equipment, etc. Generally, from local testing around the world, the toxicity of this material

Figure 5.77 Erosion protection at ferry terminal in the harbour of Strømstad, Sweden, with concrete filled in textile blankets.

Figure 5.78 Erosion protection at ferry terminal in the harbour of Strømstad, Sweden, with concrete filled in textile blankets.

Figure 5.79 Sorting station, typical local community site for landfill and sorting of household waste, Heggedal, Asker, Norway. Typical sorting categories for waste to be recycled are metal, wood, paper and cardboard, glass, plastic, and electrical and electronic appliances. (Wood, paper, and cardboard are processed/shredded and incinerated in a cement kiln.)

is harmless, unless specific constituents are used, and will not harm drinking water. However, some organisms in nature might react negatively, and the deposition might have negative effects on neighbours in terms of filling sizes, etc. Local rules and regulations must therefore be observed.

We also draw attention to Section 5.5. For example, Li et al.[401] report on solidification of heavy metals and radioactive metal.

Unfortunately, not all waste can be recycled. In this chapter we have tried to concentrate on giving some examples of how concrete has been used to stabilise, solidify, and encapsulate waste that has to be deposited.

The use of concrete to protect against radiation was also discussed in Section 5.7.4.

Also, concrete containers of various shapes, sizes, and designs have been used in a number of countries for long-term storage of hazardous wastes. Such containers are normally designed with the possibility of inspection and control for minimisation of leakages.

Stabilisation and solidification (S/S) with cementitious-based solutions have taken place around the world for more than a century. What kind of concrete is produced in a S/S process is very dependent on the material to be solidified, the process, and the amount and type of cementitious materials utilised. The final compressive strength might range from 0.5 MPa to that of normal concrete. S/S treatments have been used for preventing or slowing down release of hazardous constituents from waste, contaminated soils, or sediments, and are regarded as an established treatment technology. There are no fixed mix designs for such treatments. The final recipe must be established through a test of the actual case, including leaching and extraction tests. Typical amounts of cementitious materials used might range from 5 to 30% of the total volume to be treated.

In a paper at a conference in Trondheim, Norway, Langton reports on the stabilisation of approximately 400×10^6 L of low-level radioactive alkaline salt solution at the Savannah River Plant Defense Water Processing Facility in the United States.[455] The stabilisation took place prior to disposal in concrete vaults. The treatment involved the removal of Cs^+ and Sr^{+2}, followed by solidification and stabilisation of potential contaminants in a salt stone, a hydrated ceramic waste form. The release of chromium, technetium, and nitrate from salt stone can be significantly reduced by substituting hydraulic blast furnace slag for Portland cement in the formation design. Slag-based mixes are also compatible with Class F fly ash used in sandstone as a functional extender to control heat of hydration and reduce permeability. She claims that chromium and technetium are less leachable from slag mixes than cement-based waste form because these species are chemically reduced to a lower valence state by ferrous iron or other ions, such as Mn, in the slag and are precipitated as relatively insoluble phases, such as $Cr(OH)_3$ and TcO_2.

The salt solution that was stabilised was a low-level hazardous waste with a pH higher than 12.5 and a metal toxicity with Cr above 100 ppm. The slag-based salt stone mix design contained 25% slag, 25% fly ash, 4% hydrated lime or Portland cement, and 46% salt solution.

In a paper at a conference in 2007,[487] Badanoiu et al. reported on solidification of toxic wastes with heavy metals in a concrete with blended cement with limestone filler and slag. They showed that industrial wastes in Romania, including those resulting from the mining and energy production industries, amounted to 372.3 million tonnes. Thirty-one percent of this was reused, eliminated by burning, and placed in landfills. In 2002, it was recorded that of 680 industrial waste deposits in Romania, with an estimated surface of 6400 ha, 147 of them contained hazardous wastes generated especially in the mining activities and metal production. Solidification and stabilisation are considered to be an effective technology for the treatment of a number of waste streams, including inorganic waste and contaminated soil. The main objective of the paper was to assess the capacity of blended cements to solidify and stabilise the hazardous waste, with heavy metal contents of Cr, Zn, Cd, Ni, and so on, and the influence on the main properties of the binding matrices, based on binary and ternary blended cements.

In their conclusion, they claim:

- Hazardous waste with heavy metals content can be solidified/stabilised in blended cements with limestone filler and slag content. The leaching characteristics of the solidified matrices comply with the European Community and Romanian environmental legislation. (In their tests they used 20% of supplementary materials and 80% Portland cements.)
- The presence of the hazardous waste, with heavy metal content, in the binding systems increases the water demand and reduces the workability of fresh pastes and mortars.
- The mechanical strength of the blended cements with hazardous waste content decreases with the increase of the heavy metal concentration in correlation with the increase of the setting time.
- The investigated properties, for example, water demand for normal consistency, setting time, and mechanical strength, are better for the blended cements with hazardous waste content compared with the Portland cement with the same amount of waste.

At the same conference in 2007, Ana Krol from Opole University of Technology in Poland gave an interesting report about some similar testing.[488] She reported on immobilisation of hazardous wastes in concrete with various combinations in the use of mineral admixtures like fly ash, blast furnace slag, and silica fume in substitution rates from 0 to 85%. The compressive strengths of the various mixtures ranged from 0 to about 30 MPa after 7 days, and from about 46 to 63 MPa after 28 days. Optimisation tests were not carried out. She tested heavy metals added as 1% of the binder mass in the form of salts: Pb^{2+}, Cu^{2+}, Cr^{6+}, Cd^{2+}, and Mn^{2+}. She also tested industrial wastes as dust extracted from metallurgical furnaces, as 20% of the binder mass. She reported that leaching of heavy metals from cement, as well as from the mineral admixtures, was on a very low level. The content of heavy metals in water extracts was on a level lower than allowed in regulations concerning drinking water. Water extracts from metallurgical dust had increased amounts of Pb, Cr, and Cd ions, which in practice eliminates its safe deposits in the natural environment without a disposal process. EN 12457-1–4:2002, *Characterization of Waste—Leaching—Compliance Test for Leaching of Granular Waste Materials and Sludge*, was used in order to evaluate the leaching level of heavy metals. A significant influence of decreased pH of extraction liquid on the immobilisation level of heavy metals in analysed composites was not observed. The immobilisation level of lead, zinc, and copper ions after just 7 days of hydration exceeded 99.7%. Along the development of hydration, the level of binding of these heavy metals increased and remained stable. Chromium stabilisation was lower than that of the other metals. It is difficult to be conclusive regarding which of the various mixture design blends gave the highest level of immobilisation. Of the interesting comments in the conclusion of the report, we mention:

- The addition of heavy metal ions into the matrix composition changes its mechanical and physical properties. The majority of the added heavy metal salts affected the extension of setting time and decrease the early strength.
- The immobilisation level of the major parts of the heavy metal (except for chromium ions) overcomes 99%, and increases with the hardening of the mortar over time.

Immobilisation and destruction of industrial waste in cement kilns have been done for several decades. With its high temperature, the cement kiln can be the perfect process equipment to destroy waste, and in addition, this can also give other environmental benefits, such as reduction of energy consumption and reduction of CO_2 emission from gas, oil, or coal combustion, and the need to invest in expensive purpose-built incinerators.

In Chapter 3, we have mentioned "improvements and more efficient cement production" as tool 10. Examples are given on the use of wastes as alternative fuel.

However, the destruction process is not without its challenges. In a paper at the concrete sustainability conference in Lillehammer, Norway, in 2007, Karstensen and Justnes from SINTEF in Trondheim, Norway, gave interesting views on the challenges.[489] The Chinese State Environmental Protection Administration, SEPA, has initiated a project to investigate the possibilities of waste destruction in cement kilns. The organisation of the authors, SINTEF, has been employed to manage and supervise the project.

SEPA claims that more than 1 billion tonnes of industrial wastes are generated annually in China. Up to 30 million tonnes of these are hazardous wastes, approximately 150 million tonnes are municipal solid wastes, and more than 30 million tonnes are sewage sludge. In 2007 China had 6000 cement plants, but only a few plants had started to co-process waste as a fossil fuel and raw material substitute.

When it comes to clear guidelines for co-processing of hazardous waste in cement kilns, the authors refer to the requirements from the Stockholm and Basel Conventions.[490,491]

In the project the authors refer to, they explain that it is the objective to get a quantitative and qualitative overview of the strengths and weaknesses of the current hazardous waste management system and to identify cost-efficient improvement possibilities.

In the work they have identified the following is to be considered:

- How are industrial and hazardous waste defined in national and local regulations?
- Are the definitions understandable and unambiguous?
- The basis for the definitions (chemical, physical, concentration, source, etc.).
- What is the awareness of industrial and hazardous waste and their possible impact?
- What is the current generation in the different sectors (sources, types, and quantities)?
- How and by whom are the wastes collected and transported?
- What is the current treatment option used, its performance, and its subsequent cost?
- Is there a willingness to pay for waste treatment?
- How does mismanagement impact health, the environment, and resource use?
- What is the potential for improvements with regards to resource recovery and reduced environmental impacts?
- What are the weak link(s) in the current management system (from regulation and enforcement to generation, collection, and transport to final treatment)?
- What kind of competence is available to assist industry and regulators on industrial and hazardous waste issues?

Tributylene (TBT) has been used in bottom paints for ships, to prevent water-living organisms on the ship hulls. When it was recognised that this was a chemical that was poisonous to the environment, restrictions were imposed. It has been forbidden in Finland since 1991, and the International Maritime Organization (IMO) approved an agreement at its conference in October 2001, where TBT should be forbidden on all ships and seagoing vessels from 2003.[492] TBT has relative moderate risks for human beings, but it might store in food, and thereby be harmful. The World Health Organization (WHO) has defined the maximum daily dose for human beings as 0.25 μg per kilogram of body weight. In an area in a harbour area in the capital of Finland, Helsinki, considerable sediments containing moderate to high dosages were registered. After thorough investigation and development work, work started to immobilise the harmful sediments. Several safety barriers were established before stabilising the sediments and using the stabilised soil for an extension in the harbour. Mixing equipment with 3 to 5 m action diameter was used to stabilise a total of 300,000 m^3 of sediment in two basins.

In a memorandum to the Norwegian Parliament (*Storting*) in 2001–2002, a plan was initiated to clean up contamination in Norwegian harbours and fjords. To test various technologies, five trial projects with financial support from the Norwegian government were initiated in 2003–2005. However, Norwegian environmental authorities identified 56 harbours as contaminated, where 26 of these were identified as very contaminated.[493] The cleanup program is anticipated to cost 5 to 500 billion NOK (1 to 86 billion USD).

One of the pilot projects in the program was the harbour in Trondheim. Contaminated sediments from the sea bottom were stabilised behind barriers onshore with moderate amounts of binders (100 to 200 kg/m^3) to increase the strength of the masses to 120 kPa. The total stabilised quantity of contaminated sediments was about 30,000 m^3. Eriksson[493] claims that the cost according to the method used for the pilot project in Trondheim was 50 euros per tonne, and that this could come down to 20 euros per tonne in a large-scale project. He mentions comparable costs from other countries (in euros per tonne): Hammarby, Sweden = 43, Nordsjø Harbour, Helsinki, Finland = 15.

In a paper at a sustainability conference in Lillehammer, Norway, in 2007, Laugesen and Eriksson give interesting summaries of experience on contaminated soil stabilisation and solidification.[494] They said that the solidification technology aims at both increasing the physical (bearing capacity, permeability) and chemical (resistance to leaching) properties of the material. As awareness and active political decisions of cleaning up of former pollution have increased, knowledge and improved technology have increased in the Nordic countries. They claimed that as a rule of thumb, the compressive strength of the final material should not be below 100 kPa after 7 or 14 days, and often much higher. From experience they also claimed that up to 50% of the Portland cement used in the stabilisation can often be substituted by fly ash. From experience in the Nordic countries, as the projects often are less than 100,000 m^3, and sometimes less than 10,000 m^3, mobile equipment is favourable. They summarised the advantages by stabilisation and solidification:

- The properties and land in the area increase their value.
- The ban on fish consumption can be lifted.
- Residual materials that are suitable as binders become a resource instead of a waste (e.g., fly ash and slag).
- The water quality is improved.
- The area gets more attractive for recreation activities.

Generally there are three principally different methods to clean up contaminated sediments in the sea:

- To remove the contaminated sediments from the sea bottom and neutralise and solidify them in an acceptable deposit—and in addition often bring fresh sand to the bottom—and sometimes add artificial reefs to the area to help normal marine life to be restored faster
- To cover the contaminated sediments with a permanent unleachable carpet, and then establish an acceptable sea bottom on top of this
- To clean the contaminated sediments and stabilise them for recycling

The method that is most favourable might differ from one case to another. However, cement and concrete are important tools independent of which method is chosen.

From a Norwegian project report regarding solidification and stabilisation (Figures 5.80 and 5.81)[495] we note that the necessary geotechnical strengths of stabilised sediments might differ:

- Sea bottom: 40–300 kPa
- Reclaimed land: 100–600 kPa
- Depth stabilisation: 200–600 kPa
- Binder content: 100–200 kg/m^3

In a somewhat complementary report from the Norwegian Public Road Administration,[496] soil stabilisation by producing so-called lime cement pillars for road building on difficult ground is discussed. The pillars were 11 m long and go 13 m down into the ground, ending 2 m under final ground

Figure 5.80 Soil stabilisation project, Kalvøya, Sandvika, Norway.

Figure 5.81 Waterfront view of Kalvøya, Sandvika, Norway, after stabilisation project completion.

level. The pillar diameter was 60 cm. Some of the Portland cement was substituted with fly ash from paper production, giving a binder combination per 100 kg of:

- 10 kg fly ash
- 22.5 kg lime
- 67.5 kg Portland cement

This blend was compared to a binder with 25 kg lime and 75 kg cement. Three different binder dosages with corresponding strengths were tested: 100, 67, and 42 kg per m^3. The substitution resulted in 25% increased shear strength for the fly ash-substituted pillars.

A very interesting stabilisation challenge was reported by researchers from Beijing, China, at a conference in 2011.[497] China has a long coastline, and in large areas the soil is highly saline, and also has saline sludge produced by the effect of tides over a long time. Currently a large-scale project "encircling the sea to make land" is underway. Since 2005 over 100 km^2 of land has annually been reclaimed. The major part of this land was reclaimed with seafloor sludge using hydraulic fill; however, this presents a serious problem regarding the salinity, created by chloride salts. The authors claim to have developed a new type of stabiliser that is better than Portland cement, in the sense that it gives both better neutralisation and considerably higher strength. The new stabiliser includes cement clinker and calcareous compound as the main components.

When it comes to solidification and stabilisation of soil with increasing amounts of toxic materials, it is impossible to generalise. Testing is needed for the various cases. As an example of the work done in this area, we refer to a paper from Dawadi et al. from 2004,[498] reporting on tests regarding arsenic-contaminated soil. They claimed that disposing of the solidified/stabilised contaminated soil in sanitary or secure landfills has been uneconomical in the past, but landfill space in secure landfills is at a premium, and incineration processes are becoming cost-prohibitive and socially unacceptable. Therefore, the incorporation of contaminated soils into usable structural concrete may be a valid disposal solution. Properly designed concrete could be used at contaminated sites to build the structures needed for parking lots, wash-down areas, control structures, and pads for retention ponds.

They claim that inorganic forms of arsenic are much more toxic than organic forms. Arsenical pesticides, natural geothermal sources, and mine tailings increase arsenic concentrations in soils, and the adsorption of arsenicals in soils depends on soil pH, texture, Fe, Al, and organic matters. The toxic amounts of arsenic in soils will limit the germination of seed and reduce the viability of seedlings having a concentration that is greater than 10 ppm. Organic arsenic is used in catalysts, glass manufacturing, alloys, electronics, and weed killer. Inorganic forms of arsenic are used to kill insects or rodents, to preserve wood, and as a medicine for asthma and psoriasis. Arsenic levels in municipal sewage are variable from 1 to 18 ppm. An upper limit of 0.2 ppm is recommended for arsenic in livestock drinking water, and an upper limit of 0.05 ppm for water intended for human consumption. In soils, the total arsenic concentration normally ranges from 1 to 40 ppm. The authors claimed that the U.S. Environment Protection Agency (USEPA) was considering an amendment to the Safe Drinking Water Act to set the drinking water equivalent level at 5 ppb.

The tests reported varied five parameters:

- The amount of contaminated soil was varied from 0 to 20% of the aggregate.
- The concentration of arsenic added to the soil as a contaminant was varied from 0 to –5, 50, 100, 200, 2000, and 20,000 mg/kg of soil.
- Replacement of cement in mixes was done with 0, 20, and 40% of fly ash.
- Replacement of cement in mixes was done with 0, 5, 10, and 15% of silica fume.
- Water/cement ratios of 0.485 and 0.52 were tested.

The results from the testing were very interesting, both for the possible leaching from arsenic stabilisation and as a principle for general evaluation of the complexity in the stabilisation and solidification of wastes. From the conclusions in the paper, we mention that the authors claim:

- The strength of the mortar mixtures decreased with the increase in the soil content, fly ash content, and w/c.
- The concentration of arsenic in the leachate and the compressive strength at 7, 28, and 90 days' testing were approximately the same for all concentrations up to 200 mg arsenic/kg soil.
- The testing of mortar encapsulation of arsenic-contaminated soil up to 200 mg of arsenic/kg soil indicates that the method was successful in reducing the arsenic concentration in leachate to approximately 5 ppb.

Finally, we mention that sludge from wash water in ready-mix operations has been used as a stabiliser in landfills of municipal waste.

Chapter 6

New possibilities and challenges

The increased awareness of environmental and climate change challenges in the society in general too often has a tendency to focus on problems and difficulties. However, they also create possibilities, new markets, and inspire technology development. One of the main topics in the Norwegian environmental database for the cement and concrete industry, www.miljobasen.no, is the key word *new possibilities*. Of 1165 references in the database in May 2011, 173 were about new possibilities or have information in this direction.

Many of the possibilities, we have covered in the previous chapters. Here is a reminder of some of them:

- CO_2 curing of precast concrete products (Section 3.3)
- Fish reefs (Section 5.9.5)
- NOx-absorbing concrete (Section 5.4)
- More flexible building design (Section 3.5.15)
- More efficient and new alternatives in ternary and quaternary binder blends (Section 3.4.9)
- Precipitation-delaying systems (Section 5.9.2)
- Use of recycled waste material, on an industrial scale, for new use and utilisation, for example (Section 4.2):
  - Mineral industry waste
  - Organic material
  - Plastic
- Used tires

In addition, we would also like to mention some items in particular that normally are not so much in focus when discussing sustainable development in the concrete environment:

- Demand for more renewable energy, for example:
  - Small hydroelectric power stations
  - Tidal water power stations
  - Windmills

- More energy-efficient buildings
- Restoration of polluted harbours, fjords, and sea areas
- Measures to improve protection against increased precipitation
- Traffic and air pollution improvements

Last, but not least, we would like to mention interesting future new developments of:

- New raw materials/low energy and low CO_2 cements, for example:
  - High-belite cement (HBC)
  - Sulphoaluminate cement (SAC)
- Advanced cement-based materials, for example:
  - U.S. advanced cement-based materials (ACBMs), densified systems containing ultra-fine powders (DSPs), macro-defect-free (MDF) cement, and chemically bonded ceramics (CBCs)
  - Ultra-high performance cement-based materials, e.g., Lafarge Ductal
- U.S. and Japanese engineered cementitious composites (ECCs)

We have below tried to give minor insights into some of these alternatives.

6.1 SMALL HYDROELECTRIC POWER STATIONS

The so-called hydropower stations are normally classified in:[499]

- Micro power stations: under 100 kW
- Mini power stations: 100 to 1000 kW
- Small power stations. 1000 to 10,000 kW

Analysis says that the technical potential annual hydropower in the world is 14,000 TWh (terrawatt-hours). Of this, 8000 TWh is economically possible to utilise. In 2005, the total hydropower production in the world was 2645 TWh/year. This was equal to 19% of the electricity production in the world, and about 90% of all renewable electric energy production. On a world basis only one-third of the possibilities have been utilised: 65% in Europe, 61% in North America, 7% in Africa.

Some countries are fortunate with respect to topography and renewable energy sources. On top of the list is Paraguay, with 100% hydroelectric power. Number two on the list is Nepal, with 99.8%. Another fortunate country is Norway, where hydroelectric power production is equal to 98% of the electricity consumption.

In 2006, the average annual production in Norway was about 120 TWh, and the installed effect in the power stations was about 28,000 MW, from

1100 power stations. Most of these power stations were built between 1950 and 1990. This automatically tells us about a considerable renewal potential.

The potential for upgrading of existing stations, increase of capacity, and unused potential for power stations larger than 10 MW is about 15 TWh.

The energy potential for so-called small power stations (less than 10 MW) is about 24 TWh. There is reason to believe that comparable numbers might also be found in other countries.

China, as another example, now is encouraging the development of small hydropower stations focusing on improving agricultural irrigation, remote rural area power supply, and CO_2 emission reduction. Based on the statistics of the China Small Hydropower Association, so far small hydropower covered one-third of China's remote counties with an annual output of 200 billion kWh; and around 300 million rural populations have benefited. Reservoir capacity has been updated to 250 billion m^3, which can ensure the irrigation of 370 million mu (equivalent to 247 thousand km^2). The small hydropower generated can be a substitute of 100 million tonnes of coal burning and a reduction of about 200 million tonnes of CO_2 emission.

Concrete might be a very interesting alternative for a number of purposes in utilising this potential.

6.2 WINDMILLS

In an article in a Norwegian newspaper in November 2001, called "Germany Will Be Run on Sun and Wind,"[500] the journalist refers to an interview with central German politicians that claims that the goal for Germany, the biggest economy and most populated country in Europe, is that 35% of its energy in 2020 shall come from windpower. The present situation is as shown in Figures 6.1 and 6.2.

The German case is only an example of Europe, where we see a considerable shift to renewable energy sources, with wind power as the most

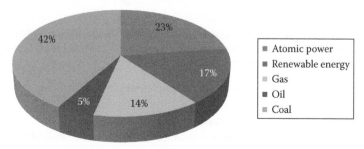

Figure 6.1 Energy production in Germany, 2010.[500] (From Brekke I., Tyskland skal drives av sol og vind, *Aftenposten*, Oslo, Norway, November 13, 2011.)

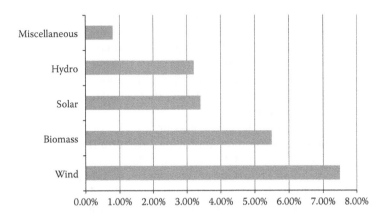

Figure 6.2 Renewable energy sources in Germany, 2010.[500] (From Brekke I., Tyskland skal drives av sol og vind, *Aftenposten*, Oslo, Norway, November 13, 2011.)

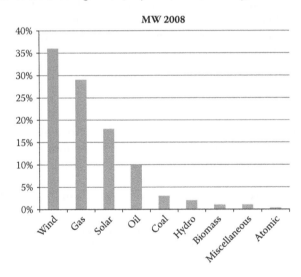

Figure 6.3 New energy capacity installed in EU in 2008. Total installed, 23.851 MW.[501] (From Markenes K.H., Norge griper ikke sjansen, *Teknisk Ukeblad*, Oslo, Norway, 43/2009, with reference to EWEA.)

interesting source. An article in the Norwegian technical periodical *Teknisk Ukeblad*[501] gives several interesting statistics. Figure 6.3 shows the effect, based on new energy installed in EU in 2008.

The article further shows the accumulated wind power installed in Europe by the end of 2008 (Figure 6.4).

The European Wind Energy Association (EWEA) believes that wind power in 2030 will represent more than 30% of the electricity production

Figure 6.4 Wind power installed in Europe, accumulated by 2008.[501] (From Markenes K.H., Norge griper ikke sjansen, *Teknisk Ukeblad*, Oslo, Norway, 43/2009, with reference to EWEA.)

in the EU, and that windmills in the sea will have a total production of 563 TWh. But, this will have a high cost. It believes that 20 billion euros will be invested in wind power in 2030 alone, and that the total investment up to that date will be close to eight times that sum.

An article in *China Daily* on December 15, 2011, relates that China had its first public auction of four offshore wind-power concessions for a generation capacity of 1000 MW in the Jiangsu Province in October 2010. A second round for offshore concessions was expected at the end of 2011, but had to be postponed. However, despite some difficulties, China has resolved to support renewable energy, including wind power, in the hope of reduction of both its carbon emissions and its reliance on fossil fuel. The National Energy Administration said that China aims to have the capacity to generate 5 GW from offshore wind by 2015. A promising development is that wind turbines have fallen in price in recent years. The increasing efforts to increase renewable energy sources in China are supported by the World Bank.

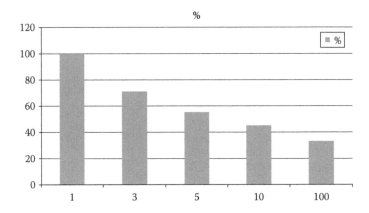

Figure 6.5 Typical costs for an offshore concrete windmill in relation to the number of units produced.

So far, windmills in concrete are relatively moderate in market share compared to steel, but this share might increase considerably for four reasons:

- Intensified product development efforts in the concrete industry
- The tendency against larger capacities and higher unit works in favour of concrete as a material
- Production in series increases the advantages for concrete, as the start-up cost and mould costs are higher for concrete than competing materials
- Increased offshore building of windmills is favourable to concrete

In 2010 we made an evaluation, based on information from several serious actors in the concrete industry, of four different construction principles for floating offshore windmills. As shown in Figure 6.5, the cost reduction with increased units in a series or order of concrete structures is considerable. This "learning curve" is considerably steeper than for steel.

Another important competition factor is the design life length for a structure. Increasing the life expectancy for the concrete structure from 20 years to 60 years will give a cost increase in the order of 2 to 5%. The similar figure for steel is considerably higher.

A third completion argument, as mentioned, is the height of the towers. The economy of wind power is extremely dependent on stable wind. This is increasing with higher towers and offshore structures. At the same time, technology progress works toward higher capacities per tower. In a paper in October 2010, Johansen[502] claimed that competitiveness of concrete towers increased considerably with tower heights above 110 to 120 m.

Spain is presently number two in Europe in wind power, after Germany. Travelling in the flat Spanish highlands, there are few places where windmills cannot be seen on the horizon. Some of them are in precast concrete.

Figure 6.6 Windmill in precast concrete from Inneo Torres, Spain.[503] (From Inneo Torres, Precast Concrete Wind Towers, http://www.inneo.es (accessed January 5, 2010).)

Inneo Torres shows that its precast concrete towers are for heights from 80 to 140 m, and with generators of 1.5 MW for 80 m and 2 to 4 MW for 100 m heights (Figures 6.6 and 6.7).

An email from another producing company in Spain (Hormifuste/Consolis) claims that nearly 100 concrete wind towers are produced in Spain with tower heights from 85 to 120 m (Figure 6.8).

A fascinating Belgian project was reported in *Concrete International* in 2008.[505] In July 2008 the first of 55 wind turbines was installed on a sand bank in the North Sea, 20 km offshore near the border between Belgium and the Netherlands. The 77 m high steel tower structure was set on the foundation of a 44 m high concrete structure and foundation. The bottom of the concrete structure is 27 m below sea level. When the project is finished, the windmill park will have a total capacity of 300 MW, with an energy output

Figure 6.7 Transport and erection of concrete towers.[503] (From Inneo Torres, Precast Concrete Wind Towers, http://www.inneo.es (accessed January 5, 2010).)

of 1000 GWh, or one-third of the Belgian 2010 renewable energy target, or enough electrical power for 600,000 Belgians. The intricate concrete substructure consists of a circular 23.5 m base plate with a thickness from 0.7 to 1.27 m, a conical-shaped base, and a cylindrical top. The structures were precast at a shore site yard in Ostende, Belgium, where six structures could be cast simultaneously. The concrete with compressive strengths from 45 to 55 MPa was utilising high dosages of blast furnace slag and fly ash.

Malhotra provided the statistics in Tables 6.1 and 6.2 for wind power capacity in 2009, at a conference in 2010.[506]

It is interesting to see the intensity in the interest in new wind power installations. For example, China doubled its capacity in 2009, the United States increased its capacity by nearly 50% in 1 year, etc.[507]

Figure 6.8 A 100 m high windmill tower, Finland, with the lower 46.5 m in concrete, 3 MW turbine.[504] (From de Jaime R., email to Per Jahren, June 20, 2011.)

6.3 NEW RAW MATERIALS/LOW ENERGY AND LOW CO_2 CEMENTS

Sustainability has become the global trend in this new century. The cement industry, as one of the key fundamental materials industries, on the one hand, plays a very important role in the social and economic development, and on the other, may also impose a great challenge in terms of its large consumption of natural resources and energy and the emissions of greenhouse gases. This issue is of great importance especially for China due to the huge volume of cement production, accounting for half of the world cement production, as well as to the high energy intensity resulting from the high burning temperature (1450°C) of alite (C_3S)-based Portland cement (PC)

Figure 6.8 (Continued)

and to the large CO_2 emission intensity (direct and indirect emission defined by the World Business Council for Sustainable Development (WBCSD) Cement Sustainability Initiative (CSI), an average 0.85 tonne of direct CO_2 emission per tonne of PC clinker mainly due to the decarbonation of limestone and combustion of fuel).

The latest progress on belite (C_2S)-based low energy and low CO_2 emission cements (low E cements) in China can be a good solution in dealing with the cement sustainability issue.[508,509] The low E cements refer to the belite-based clinker mineral composition design of the following two systems: belite-based Portland cement (C_2S-C_3S-C_3A-C_4AF) (high-belite cement (HBC)) and belite-calcium sulphoaluminate cement (C_2S-C_4A_3S-C_4AF) (BCSA).

6.3.1 Principle for clinker composition design

It is well known that the total embedded CO_2 content of the Portland cement clinker is equal to the sum of its raw materials-induced CO_2 and

Table 6.1 Addition of wind power in 2009

	Country	Capacity added (MW)
1	China	13,000
2	United States	9922
3	Spain	2459
4	Germany	1917
5	India	1271
6	Italy	1114
7	France	1088
8	Britain	1077
9	Canada	950
10	Portugal	673

Table 6.2 Total wind power capacity in 2009

	Country	Wind capacity (MW)
1	United States	35,159
2	Germany	25,777
3	China	25,104
4	Spain	19,149
5	India	10,926
6	Italy	4850
7	France	4492
8	Britain	4051
9	Portugal	3535
10	Denmark	3465

Table 6.3 Basic data of C_3S, C_2S, and C_4A_3S minerals

Mineral	Formation enthalpy, kg/kJ	Formation temperature, °C	CaO, %	Raw material-CO₂ per unit mass
C_3S	1848	1400	73.7	0.578
C_2S	1336	1200	65.1	0.511
C_4A_3S	~800	1300	36.8	0.216

fuel-derived CO_2 emissions during manufacture. The clinker mineral composition design therefore can be clearly seen in Table 6.3; the formation of C_3S requires more energy input and a high content of CaO or calcium carbonate, and accordingly emits more CO_2. That is why belite-based cement has been of intense interest to several generations of researchers in the world, particularly since 1970, after the global energy crisis.[510,511] It can also be seen that C_4A_3S embodies the lowest content of CO_2 due to its lower CaO content in mineral composition.

The HBC has been successfully put into commercial production and application in real construction projects such as the world's largest hydropower project of the Three Gorges Dam.[512] Industrial production proved that the production of low E cement can be up to 10 to 20% of energy savings, and thus giving a reduction in CO_2 emission of about ~10% in combination with the smaller amount of limestone needed in the raw meal proportion when compared with that of normal Portland cement.[513] Evaluation of the performance of the HBC cement and concrete also shows better performances in workability, heat liberation, later age strength, and durability than those of normal alite (C_3S)-based Portland cement and the resultant concrete. The C_2S-based low E cement therefore can be considered a good solution in dealing with cement and concrete sustainability not only for energy-efficient cement production, but also for high-performance concrete making.

6.3.2 Lower energy and low-emission clinker preparation

The typical mineral composition of an HBC clinker is shown in Table 6.4 in comparison with PC.

It has been demonstrated in lab research that the preparation of HBC clinker exhibits the advantages of a low burning temperature of 1350°C, which is 100°C lower than that of normal PC.

The preparation of BCSA cement clinker also reveals more promising results with a lower clinker temperature of 1350°C and more reduction of CO_2 emission, about 20% compared with those of PC.

6.3.3 Performance evaluation of HBC

HBC cement and concrete performances were evaluated compared with alite-based normal PC. The evaluation of HBC properties proves that it can better meet the requirement for high-performance concrete making. The following subsections are only some of the performance features of HBC.

Table 6.4 Typical clinker mineral composition of PC, HBC, and BCSA cements

Clinker minerals	PC, %	HBC, %	BCSA, %
C_3S	45–65	20–30	—
C_2S	15–30	40–60	40–60
C_3A	5–10	3–7	—
C_4AF	8–15	10–15	10–20
C_4A_3S	—	—	20–40

More comparison data for performance evaluation of HBC cement and concrete can be found.[514,515]

6.3.3.1 Strength

- Under standard curing temperature: HBC shows better strength gain when cured after 7 days, and much higher later age strength than PC. In addition, HBC exhibits lower heat evolution and adiabatic temperature rise. Therefore, cement with the performance of low heat and high strength can be achieved with HBC. A durability evaluation shows that HBC concrete has excellent properties on the resistance of chemical corrosion, freeze-thaw, carbonation, cracking, etc.

   Figure 6.9 gives the mortar strength development of the two cements, HBC and PC, under standard curing temperature.

   It can be seen that though the early strength of HBC prior to 7 days is lower than that of PC, the 28-day strength for HBC and PC is at the same level, and the long-term strength of HBC at 3 to 12 months of age is more than 10 MPa higher than that of PC. The strength development of concrete made of HBC and PC exhibits similar results.[515]

- Under elevated curing temperature: Strength development of cement and concrete tested in the laboratory is often different from that under actual working conditions. This is mainly due to the hydration heat accumulation inside the concrete, resulting in a certain temperature rise. Consequently, it is interesting to check the strength development of cement under elevated curing temperature.

   Figure 6.10 shows the comparison of HBC and PC under varying curing temperatures. It is shown that at a 3-day age HBC strength increases rapidly with the increase of curing temperature, while a

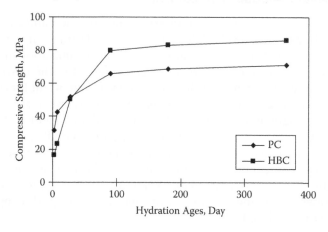

Figure 6.9 Motor strength development of HBC and PC under standing curing temperature.

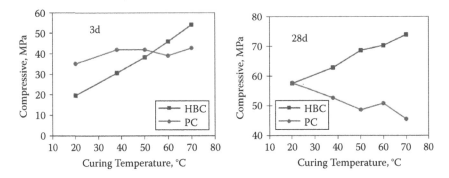

Figure 6.10 Comparison of HBC and PC under varying curing temperatures.[514] (From Sui T., et al., Strength and Pore Structure of High Belite Cement, in *Proceedings of the 5th International Symposium on Cement and Concrete*, Shanghai, October 2002, vol. I, pp. 261–265.)

very slight increase of PC strength is found. Interesting is the result for 28 days, which indicates a big difference between the strength development of both cements. The increase of strength for HBC with temperature elevation demonstrates the potential advantages of HBC over PC in the application of massive concrete, high-temperature concreting in summer or tropical areas, etc. One main reason for the difference between alite-based and belite-based cement is that the excessively fast hydration of alite under higher curing temperature encapsulates the mineral particles, and thus hinders the further hydration at the following hydration age.[516]

6.3.3.2 Heat evolution characteristics

Cement hydration is a process of an exothermic reaction and can be self-accelerated by the heat evolved. This may also be detrimental to the properties of cement and concrete because excess heat accumulation may result in the occurrence of thermal stress and cracking and in the retrogression of later compressive strength.[515] It is therefore very important for concrete, especially massive concrete, to control the temperature rise so as to improve long-term strength and volume stability and durability. Figure 6.11 shows the results for heat evolution of HBC in comparison with PC.

It can be seen that the heat evolved for HBC at corresponding ages decreases by 20% compared with that for PC. The adiabatic temperature rise of HBC massive concrete is 3 to 5°C lower than that of the concrete for Three Gorges Dam, and the time of the appearance of peak temperature for HBC is also delayed.[512]

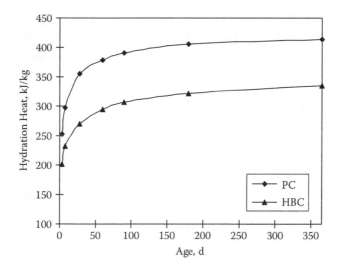

Figure 6.11 Hydration heat evolution of HBC and PC.

Table 6.5 Chemical corrosion resistance of HBC and PC

| | 3-month bending MPa/CRC coefficient % | | | |
Cement.	Freshwater	3 × seawater	3% Na$_2$SO$_4$	5% MgCl$_2$
HBC	9.68/100	7.87/81	10.44/108	8.87/91
PC	9.08/100	6.67/74	5.09/56	7.21/79

6.3.3.3 Chemical corrosion resistance

Comparison of chemical corrosion resistance (CRC) of HBC and PC was conducted using the corrosive media of artificial seawater with three times the concentration of natural seawater, 3% of NaSO$_4$, and 5% of MgCl$_2$ solution. It can be seen from Table 6.5[508] that under a 3-month age of corrosion, HBC exhibits better resistance to the chemical corrosion. In particular, excellent corrosion resistance to sodium sulphate for HBC is proved mainly due to the smaller amount of portlandite produced during the hydration of HBC compared with PC.

6.3.3.4 Drying shrinkage

Less drying shrinkage was also found for HBC in comparison with PC. As shown in Table 6.6,[508] the drying shrinkage of HBC only accounts for 50 to 60% of PC at a related age.

Table 6.6 Drying shrinkage of HBC and PC

Cement type	7-day	14-day	28-day	3-month	6-month
PC	0.060	0.083	0.103	0.115	0.096
HBC	0.030	0.042	0.057	0.058	0.057

Table 6.7 Existing standards for HBC

Standard	7-day compressive strength MPa	28-day compressive strength MPa
China, GB200-2003	≥13.0	≥42.5
United States, ASTM C 150	≥7.0	≥17.0
Japan, JIS R 521	≥7.5	≥22.5

6.3.3.5 Existing standards for HBC

Based on the results of experimental research, industrial application and field application of the HBC Chinese national standard was formulated in 2003. In Table 6.7 is the comparison of the current Chinese standard with international standards for HBC with a C_2S content over 40% in the clinker mineral composition.

6.3.3.6 Simplified explanation for the excellent performance of HBC

It has been proved through laboratory research and field application that HBC exhibits better performance than PC. Below we give a simplified hydration equation and calculation,[510] which is really a good explanation to these enhanced performances of HBC. For example, the hydration of C_2S needs less water for full hydration, and gives 30% more binding component as C-S-H gel and less calcium hydroxide (CH) formation than with C_3S hydration.

The simplified hydration formula and estimation are as follows:

$2C_3S + 7H \rightarrow C_3S_2H_4 + 3CH$
Theoretical $Q_{Alite} = -517$ kJ/kg
$2C_2S + 5H \rightarrow C_3S_2H_4 + CH$
Theoretical $Q_{Belite} = -262$ kJ/kg
100 g $C_3S \rightarrow$ 79 g C-S-H and 48 g CH
100 g $C_2S \rightarrow$ 105 g C-S-H and 21.5 g CH

6.3.4 Latest results on BCSA cement

As indicated above, the clinkering temperature for BCSA cement is 100 to 50°C lower than that of normal PC, which results in a more remarkable

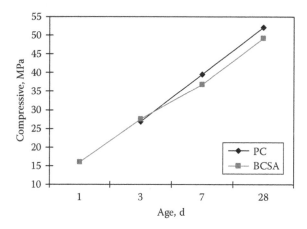

Figure 6.12 Strength development of PC and BCSA.

reduction of CO_2 emission of about 20%, not only due to the larger decrease in energy input but also due to the much lower demand of $CaCO_3$ in raw meal design. The strength development of BCSA and PC is shown in Figure 6.12.[517] It can be seen that the both BCSA and PC behave similarly in terms of strength development, which exhibits big potential in further development.

6.4 NEW CONCRETE PRODUCTS AND COMPONENTS

New and higher-strength concrete is a very interesting development aspect, with considerable sustainability possibilities. However, soberness must be shown regarding the impact this will have on the total volume. Jahren did an end-use evaluation of the impact of various concrete strengths in 1982[518] and claimed that the possible market impact of higher strength could be:

- High-strength concrete: 70 to 110 MPa—10% of the cement consumption
- Very high-strength concrete: 100 to 150 MPa—1% of the cement consumption

We see very few indications that this relationship has changed so far.

Chapter 7

The future

It is difficult to predict what will happen in the future. However, over the last decade, the attention for sustainable development has increased impressively in the concrete society nearly all over the world. National and international conferences, seminars, and workshops have been arranged in all major countries, and environmental or sustainability committees have been formed in most concrete societies, associations, and institutes. Intensified and interesting research has been initiated, and we are beginning to see promising results.

Looking at the history of concrete, the centre for the development:

- 10,000 years ago was somewhere east and south of the Mediterranean Ocean
- 2000 years ago was in Rome
- 1000 years ago was "nonexistent," as hardly anything happened because the secrets were hidden in religious archives
- 200 years ago was in England
- 100 years ago was probably somewhere in the Atlantic Ocean, midway between America and Europe

In the future, a very important part of the development will probably be in Asia. Japan has already, for many years, been one of the "leading ladies" in many fields in cement and concrete technology. China produced more than half of the cement and concrete in the world, and a few years ago, India took over the second place in the ranking from the United States. India exceeded the 200 million tonnes mark (216 tonnes) for cement production in 2010, and is expected to reach 860 million tonnes of cement production in 2030.

Statistics show that most mature countries today have a rather stable consumption of cement and concrete from year to year, with cement consumption per capita from about 250 to 600 kg per capita. Very few of these countries exceed 500 kg of cement per capita.

For the largest populations in the world, China and India, the situation is quite different. Analysts predict that the population in China will stabilise at

Table 7.1 The seven most populated countries in the world[519]

Country	Population (million)	Date	% of world population
China	1.347	November 10, 2011	19.3
India	1.204	March 2011	17
United States	312	November 10, 2011	4.48
Indonesia	238	May 2010	3.36
Brazil	195	November 2, 2011	2.8
Pakistan	177	November 10, 2011	2.55
Nigeria	158	2010	2.27

Source: http://en.wikipedia.org7wiki/World_population (accessed October 11, 2011).

its present size of in excess of 1.3 billion. Predictions say that the India will surpass China in population sometime between 2024 and 2032, and that it will stabilise at about 1.5 billion people sometime in the 2060s (Table 7.1).

The world population was estimated at 6.97 billion by the U.S. Census Bureau as of July 1, 2011.[519]

On November 1, 2011, the world media could report about the world citizen number, 7 billion, a newborn girl in the Philippines.

It took some 10 to 20,000 years for the population in the world to grow to 1 billion people around 1800, then a steady growth up to 2 billion people around 1930. Since then the world population has exploded, until we, in the last decade or so finally, have seen a tendency toward a possible stabilisation.

There are several projections for the future population in the world. Several predict that the population will stabilise on about 9 billion people sometime around 2060 (Figure 7.1).

One hundred years ago, 10% of the world population lived in cities. Today half the people live in cities or metropolises, many of them mega-cities of more than 10 million people. This tendency increases urbanisation habits in building and construction with, among others, and increase in cement consumption per capita. With an average cement world consumption of slightly less than 500 kg per capita, this leads to a possible total cement consumption of slightly over 4 billion tonnes of cement.

When we, for example, sets goals for emissions, etc., from production of concrete, it is this consumption we have to take into account, together with the reduction figures that are found necessary from recommendations from the Intergovernmental Panel on Climate Change (IPCC) and political negotiations and resolutions. In total, this will mean considerable additional improvements in the practice we experience today if we are also going to have considerable reduction in the emission from the concrete industry. Probably we will also need development of new technologies.

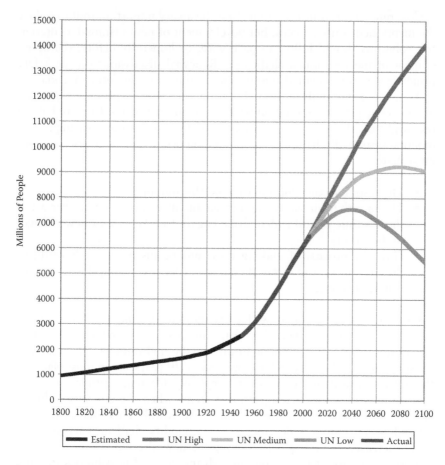

Figure 7.1 World population from 1800 to 2100, based on UN 2004 projections.[519] (Fromhttp://en.wikipedia.org7wiki/World_population(accessedOctober11,2011).)

With an increased part of the world population living in urban and more concentrated environments, new challenges are created with respect to safety. Safety is related to both probability and consequence. We do not know if natural disasters like volcanic eruptions, tsunamis, storms, and tornados are happening more often than before, but statistics show that the consequences in terms of economic losses and human tragedies have increased considerably over the last few decades. In addition, we will experience the consequences due to a gradual climate change, in terms of more precipitation in heavier and more intense rainfalls, etc. The higher population concentration, with a higher proportion of less absorbing surfaces and grounds, and less natural delaying systems, will further increase water concentration and flooding probability.

Concrete might, if utilised to its best, be a very important tool to reduce the mentioned consequences, but development of new sustainable alternative solutions should follow in pace with, or hopefully before, the development of the mentioned challenges. Renewed discussions on the safety principle we utilise seem also to be a natural item in the future agenda.

As mentioned initially in this publication, each year we produce a quantity of concrete equal to a mountain with a ground area of 1 × 1 km and about the height of Mt. Everest. Over a century, this makes quite an impressive mountain range. Probably, this is of less importance to the world environment than:

- The resource consumption to build this mountain
- The fact that an increasing percentage of what we build over a too short a time period is demolished again to take care of new changes in the society
- How the process of making these mountains might be harmful to the environment in terms of emission, energy consumption, runoff of water, dust, etc.

Over the last decade or so, we have witnessed a dramatic and very enjoyable change in the interest and priorities in the concrete industry regarding questions related to our common future and the environment. Sustainability questions were at best secondary topics some 20 years ago in most places in the concrete world. Today we see a completely different attitude.

Sustainability issues are top-priority items when the concrete industry and research environment meet around the world. This increased interest and attention to a wide range of sustainability challenges has also led to an increased brain capacity looking for new alternatives, possibilities, solutions, and not at least economic advantages and possibilities in more sustainable concrete design, production, use, and recycling. This evolution will create new scenarios that are difficult to foresee today. One of the main purposes behind this book has also been to try to be a tool in creating inspirations for such an evolution.

References

1. Bruntland G.H., et al. *Our Common Future. The World Commission on Environment and Development.* Oxford University Press, Oxford, 1987.
2. Malhotra V.M. Concrete Technology for Sustainable Development. Presented at Sustainable Development in Cement and Concrete Industries, Milano, Italy, October 17–18, 2003.
3. Cembureau Statistics. http://www.cembureau.eu/about-cement/key-facts-figures (accessed May 1, 2012).
4. Kulkarni V.R. Concrete Sustainability: Current Status in India and Crucial Issues for the Future. Presented at Concrete Sustainability through Innovative Materials and Techniques, Roving National Seminars, Bangalore, January 10–14, 2011.
5. Meyer C. The Economics of Recycling in the US Construction Industry. In *International Conference on Sustainable Construction Materials and Technologies*, June 11–12, 2007, pp. 509–513.
6. http://en.wikipedia.or.-7wiki/list-of countries by-GDP-(nominal) (accessed August 15, 2011).
7. Bygballe L.E. and Goldeng E. Den norske betongindustrien. Verdiskapning i perioden 2000–2009. Presented at Betong- samfunnsbygger og verdiskaper, Oslo, June 9, 2011. www.bi.no/bygg.
8. United Nations Environment Programme. *Buildings and Climate Change—Status, Challenges and Opportunities.* 2007.
9. Jahren P. *Concrete History and Accounts.* Tapir Academic Press, Trondheim, Norway, 2011.
10. Mehta P.K. Concrete Technology for Sustainable Development—An Overview of Essential Principles. Presented at CANMET/ACI, International Symposium on Sustainable Development of Cement and Concrete Industry, Ottawa, Canada, October 21–23, 1998.
11. Email from Nicolo Bini with information about Binishell. June 9, 2011.
12. Tongbo S. High Performance Low Heat Portland Cement—High Belite Cement, Preparation and Performance, Application and Perspective. Presented at International Workshop on Cement and Concrete Technology for Sustainable Development, Beijing, August 2007.

13. Jahren P. The Use of Artificial Concrete Fish Reefs—An Important Contribution to Increased Sustainability. Presented at Proceedings of the International Symposium on Ecological Environment and Technology of Concrete, Beijing, August 1–4, 2011.

14. Email from Rabago de Jaime, Homifuste, Spain. June 20, 2011.

15. Neimeyer T. Institute for Sustainable Infrastructure's "Envision" Infrastructure Sustainability Rating Tool. Presented at ACI Conference, Concrete Sustainability Forum IV, October 11, 2011.

16. Mahrer J.E., Kramer K.W. LEED-NC Version 2.2 Rating System Applications of Common Structural Materials. In *International Conference on Sustainable Construction Materials and Technologies*, Coventry, UK, June 11–13, 2007, pp. 549–562.

17. Bramslev K. BREEAM–nor kort innføring. Presented at Breeam-nor seminar, Lysaker, Norway, December 12, 2011.

18. Eldegard J. *BREEAM*. Report 3. Norwegian Concrete Society, Environmental Committee, October 2011.

19. Vold M. System for miljøvaredeklarasjon Type-III. Bakgrunn for deklarasjon av huldekkeelementer. Presented at Norwegian Concrete Day, Kristiansand, Norway, 2000.

20. Fib Commission 3, Task Group 3.1. Environmental Aspects of Design and Construction. Environmental Issues in Prefabrication. State of the Art Report by Task Group 3.1. In *Concise Encyclopedia of Science and Technology*. McGraw Hill, New York, 1982.

21. Fib Commission 3, Task Group 3.1. *Concise Encyclopedia of Science and Technology*. McGraw Hill, New York, 1982.

22. EUPAVE. *Concrete Roads: A Smart and Sustainable Choice*. Brochure-report. European Concrete Paving Association, Brussels, Belgium, September 2009.

23. Hasholt M.T., Berring A., Mathiesen D. *Anvisninger i grøn betong*. Center for grøn betong, Taastrup, Denmark, December 2002.

24. Horvath A. Life-Cycle Uses of Concrete for More Sustainable Construction. In *Third CNMET/ACI International Symposium on Sustainable Development of Cement and Concrete*, San Francisco, September 2001, pp. 419–430.

25. Sakai K., Douglas S. ACI St. Louis Workshop on Sustainability. *Concrete International*, February 2009.

26. Noguchi T., Fujimoto S. Evaluation and Minimization of Life Cycle Environmental Risk of Concrete Structures. In *International Conference on Sustainable Materials and Technologies*, Coventry, UK, June 11–13, 2007, pp. 769–777.

27. Kawai K., Fujiki A. A Case Study on Environmental Impact Assessment of Precast Concrete Products with Revegetation Function. In *Proceedings of Sustainable Construction Materials and Technologies*, Coventry, UK, June 11–13, 2007, pp. 163–176.

28. Sakai K. Contributions of the Concrete Industry towards Sustainable Development. In *International Conference on Sustainable Materials and Technologies*, Coventry, UK, June 11–13, 2007, pp. 1–10.

29. Schokker A.J. *The Sustainable Concrete Guide, Strategies and Examples*. U.S. Green Concrete Council, ACI 2010.

30. http://www.ivam.uva.nl/index.php?id=373&L=1 (accessed April 5, 2012).

31. http://www.misa.no/uk/sustainability-tools7eco_it-eco-edit/ (accessed April 5, 2012).
32. http://www.lisa.au.com/ (accessed April 6, 2012).
33. http://www.umberto.de/en/ (accessed April 6, 2012).
34. http://www.cenerg.ensmp.fr/english/logiciel/indexequer.html (accessed April 6, 2012).
35. http://www.eco-bat.ch (accessed April 6, 2012).
36. http://www.legen.de (accessed April 6, 2012).
37. http://www.ivam.uva.nl/index.php?id=373&L=1 (accessed April 5, 2012).
38. Day R.L., Thomas M., Nazir M. A Computer System for Sustainable Use of Supplementary Cementing Materials: SOS. In *International Conference on Sustainability in the Cement and Concrete Industry*, Lillehammer, Norway, September 16–19, 2007, pp. 467–475.
39. http://www.giacomo.lorenzoni.name/thermalwall/ (accessed April 6, 2012).
40. International Organization for Standardization. *Environmental Management Systems*. ISO 14000. Geneva Switzerland.
41. Smith M.R., Walls J. *Life-Cycle Cost Analysis in Pavement Design*. Report FHWA-SA-98-079. FHWA, Washington, D.C., 1998.
42. Sjunnesson J. Life Cycle Assessment of Concrete. Master's thesis, Department of Technology and Society, Environmental and Energy Systems Studies, Lund University, Sweden, 2005.
43. Weiland C.D. Life Cycle Assessment of Portland Cement Concrete Interstate Highway Rehabilitation and Replacement. Master's thesis, Science in Civil Engineering, University of Washington, 2008.
44. Athena Institute. *A Life Cycle Perspective on Concrete and Asphalt Roadways: Embodied Primary Energy and Global Warming Potential*. Athena Institute, Ottawa, Ontario, 2006.
45. ENDS. Report 262. November 1996.
46. Gajda J. Energy Use of Single-Family Houses with Various Exterior Walls. CD026. Portland Cement Association, Skokie, IL, 2001.
47. Guggemos A.A., Horvath A. Comparison of Environmental Effects of Steel- and Concrete-Framed Buildings. *ASCE Journal of Infrastructure Systems*, June 2005.
48. Mehta P.K. Concrete Technology for Sustainable Development. *Concrete International*, November 1999.
49. Pachauri, R. Speech to the Opening of the UN General Assembly. September 24, 2007. www.ipcc.ch.
50. Pileberg S. Himalaya Smelter (Himalaya Is Melting—From Norwegian). *KLIMA* (Norwegian journal for climate research), 2-2008.
51. Wilson K. Disasters Make It the Year of Living Dangerously. *China Daily*, December 16, 2011.
52. Editorial. The MBA's Climate Change Primer. *Stanford Business*, November 2005.
53. Cembureau Statistics. www.cembureau.be.
54. Sage R. Sustainable Development of the International Cement and Concrete Industry. Climate Change and Minerals and Metals. Presented at CANMET/ACI International Symposium on Sustainable Development of the Cement and Concrete Industry, Ottawa, Canada, October 21–23, 1998.

55. Glavind M., Damtoft J.S., Røtting S. Cleaner Technology Solutions in the Life Cycle of Concrete Products. In *Third CANMET/ACI International Symposium on Sustainable Development of Cement and Concrete*, San Francisco, September 2001, pp. 313–327.

56. Jahren P. *Greener Concrete—What Are the Options? The CO_2 Case*. Report STF22A03610. SINTEF, Trondheim, Norway, August 2003.

57. Sakai K. What Can We Do for Sustainability in Concrete Sector. Presented at 2008 International Expert Workshop on Cement and Concrete Technology for Sustainable Development, Beijing, October 7–12, 2008.

58. Punkki J., Hermstad A., Lindstrøm G. Active Utilization of Thermal Mass of Concrete Structures. Presented at International Conference on Sustainability in the Cement and Concrete Industry, Lillehammer, Norway, September 16–19, 2007.

59. Jahren P. How to Improve the CO_2-Challenge: A World Wide Review. In *International Conference on Sustainability in the Cement and Concrete Industry*, Lillehammer, Norway, September 16–19, 2007, pp. 136–148.

60. Humphreys K., Mahasenan M., Placet M., Fowler K. Towards a Sustainable Cement Industry—Sub-Study 8: Climate Change. World Business Council for Sustainable Development, March 2002. www.wbsdcement.org.

61. Sakai K. Standardization and Systems for Sustainability in Concrete and Concrete Structures. Presented at International Conference on Sustainability in the Cement and Concrete Industry, Lillehammer, Norway, September 16–19, 2007.

62. Damtoft J.S. The Sustainable Cement Production of the Future. In *International Conference on Sustainability in the Cement and Concrete Industry*, Lillehammer, Norway, September 16–19, 2007, pp. 75–84.

63. World Business Council for Sustainable Development, International Energy Agency. *Cement Technology Roadmap 2009*. Switzerland, 2009.

64. Gelli M. Industrial Ecology within the Cement Industry—The Holcim Case. Presented at International Seminar on Sustainable Development in the Cement and Concrete Industry, Plitectnico di Milano, Milan, Italy, October 17–18, 2003.

65. Gartner E., Quillin K. Low-CO_2 Cements Based on Calcium Sulfoaluminates. Presented at International Conference on Sustainability in the Cement and Concrete Industry, Lillehammer, Norway, September 16–19, 2007.

66. Ge Z., Wang K.J. Hydration of Sodium Silicate-Activated Binders Made with Cement Kiln Dust and Fly Ash. In *Proceedings of the First International Conference of Advances in Chemically-Activated Materials (CAM'2010)*, Ji'nan, China, May 9–12, 2010, pp. 118–130.

67. Sui T., Kong X. Status Report of China Cement Industry. Prsented at APP Cement Task Force Meeting, Charleston, SC, 2008.

68. WBCSD. *The Cement Sustainability Initiative (CSI) Climate Actions*. Switzerland, November 2008. www.wbcdcement.org.

69. WBCSD. http://wbcsdcement.org/GNR-2009/world/Graphs/GNR%20-%20 indic_ (accessed February 2, 2012).

70. Norcem AS. *Report on Sustainable Development 2007*. www.norcem.no.

71. Cornelissen A. Concrete Houses Reduce CO_2—Fact or Fiction? In *CANMET/ACI International Symposium on Sustainable Development of the Cement and Concrete Industry*, Ottawa, Canada, October 21–23, 1998, pp. 445–455.

72. Malhotra V.M. Role of Supplementary Cementing Materials in Reducing Green Gas Emission. Presented at CANMET/ACI International Symposium on Sustainable Development of the Cement and Concrete Industry, Ottawa, Canada, October 21–23, 1998.

73. Vanderborght B., Brodmann U. *The Cement CO_2 Protocol: CO_2 Emission Monitoring and Reporting for the Cement Industry, Guide to the Protocol.* Version 1.6. WBCSD Working Group Cement, October 19, 2001. http:/www. ghgprotocol.org.

74. Cusack D. Greenhouse Gas Abatement in the Australian Cement Industry. In *CANMET/ACI International Symposium on Sustainable Development of the Cement and Concrete Industry*, Ottawa, Canada, October 21–23, 1998, pp. 73–94.

75. Jacobsen S., Jahren P. Binding of CO2 by Carbonation of Norwegian OPC Concrete. Presented at International Seminar on Sustainability and Concrete Technology, Lyon, France, November 7–8, 2002.

76. Xu D. Developing Eco-Cement industry in China. Presented at 2009 International Workshop on Cement and Concrete Technology for Sustainable Development, Xi'an, China, May 7–10, 2009.

77. Klein M., Rose D. Development of CCME National Guidelines for Cement Kilns. In *CANMET/ACI International Symposium on Sustainable Development of the Cement and Concrete Industry*, Ottawa, Canada, October 21–23, 1998, pp. 16–29.

78. Sui T. Review of Rising Energy Efficiency and Emission Reduction of China Cement Industry. Presented at Russia-China Cement Summit, Moscow, Russia, November 30–December 1, 2011.

79. Sellevold E.J., Nilsen T. Condensed Silica Fume in Concrete: A World Review. In *Supplementary Cementing Materials for Concrete*, ed. V.M. Malhotra, 165–243. CANMET, Ottawa, Canada, 1987.

80. Maries A. Utilization of Carbon Dioxide in Concrete Construction. Presented at CANMET/ACI International Symposium on Sustainable Development of the Cement and Concrete Industry, Ottawa, Canada, October 21–23, 1998.

81. Neville A.M. *Properties of Concrete.* 4th ed. Longman Group Ltd., Harlow, England, 1995.

82. Djabarov N.B. Environmentally-Friendly Technology for Lightweight Concrete Units Based on CO_2 Waste Gases Treatment. Presented at CANMET/ACI International Symposium on Sustainable Development of the Cement and Concrete Industry, Ottawa, Canada, October 21–23, 1998.

83. Jønsson Å., Tillman A.M. *Livscykelanalys (LCA) av beton*, 49–83. Beton ock miljø–Fakta fron Miljøforum–AB Svensk Byggtjanst, Stockholm, Sweden, 1999.

84. Norcem. *Sementforbruk på segmenter.* Norcem, Heidelberg Cement Group, Oslo, Norway, February 2003.

85. Pade C., Guimaras M., Kjellsen K., Nilsson Å. The CO_2 Uptake of Concrete in the Perspective of Life Cycle Inventory. Presented at International Conference on Sustainability in the Cement and Concrete Industry, Lillehammer, Norway, September 16–19, 2007.

86. Jahren P. *Bærekraftig utvikling for sement og betongindustrien.* Betongindustrien, Oslo, Norway, April 1988.

87. Baetzner S., Pierkes R. *Release and Uptake of Carbon Dioxide in the Life Cycle of Cement.* Technical Report TR-ECRA 0004/2008. European Cement Research Academy, Duesseldorf, Germany, 2008.

88. Collepardi S., Ogoumah Olagot J.J., Troli R., Simonelli F. Carbonation of Concretes: The Role of Slag, Fly Ash, and Ground Limestone in Blended Cements. Presented at International Seminar on Sustainable Development in Cement and Concrete Industries, Milan, Italy, October 17–18, 2003.

89. Shi C., Wu. Y. CO_2 Curing of Concrete Blocks. *Concrete International,* February 2009.

90. Shao Y., Lin X. Early-Age Carbonation Curing of Concrete Using Recovered CO_2. *Concrete International,* September 2011.

91. Shao Y., Zhou X. CO2 Uptake by Concrete by Concrete Hardened in a Simulated Flue Gas. In *International Conference on Sustainable Materials and Technologies,* Coventry, UK, June 11–13, 2007, pp. 381–387.

92. Owens K.J., Basheer P.A.M., Sen Gupta B. Utilization of Carbon Dioxide from Flue Gases to Improve Physical and Microstructural Properties of Cement and Concrete Systems. In *International Conference on Sustainable Materials and Technologies,* Coventry, UK, June 11–13, 2007, pp. 59–69.

93. Ye Z., Wan L., Liu M., Chang J. Study on Effect of Carbonation Curing for Cement Minerals and Clinker. In *Proceedings of the International Symposium on Ecological Environment and Technology of Concrete,* Beijing, August 1–4, 2011, pp. 79–84.

94. Cheng X. Manufacture Building Materials by Using Industrial Waste Residue. Presented at 2008 International Expert Workshop on Cement and Concrete Technology for Sustainable Development, Beijing, October 7–12, 2008.

95. Zhao H.L., Wu H.Z., Chang J., Cheng X. Manufacturing Building Materials by Carbonating Steel Slag Mixed with Portland Cement. In *Proceedings of the 11th International Conference on Advances in Concrete Technology and Sustainable Development,* Jinan, China, May 9–12, 2010, pp. 800–804.

96. Philleo R.E. Slag or Other Supplementary Materials? In *Third CANMET/ACI International Conference on Fly Ash, Silica Fume, Slag and Natural Pozzolans in Concrete,* Trondheim, Norway, 1989, vol. 2, pp. 1197–1207.

97. Malhotra V.M. Sustainability and Global Warming Issues, and Cement and Concrete Technology. Presented at 2008 International Expert Workshop on Cement and Concrete Technology for Sustainable Development, Beijing, October 7–12, 2008.

98. Pocket World in Figures. *The Economist,* 2007.

99. Yeginbali A. Potential Uses for Oil Shale and Oil Shale Ash in Manufacturing of Portland Cement. Presented at CANMET/ACI Symposium on Sustainable Development of the Cement and Concrete Industry, Ottawa, Canada, October 21–23, 1998.

100. Malhotra V.M., Ramezanianpour A.A. *Fly Ash in Concrete.* 2nd ed. CANMET, Ottawa, Canada 1994.

101. Owens P. From Lednock to the Channel Tunnel. *Concrete,* May/June 1992, pp. 46–52.

102. Manz O.E., Faber J.H., Takagi H. Worldwide Production of Fly Ash and Utilization in Concrete. Presented at Third CANMET/ACI International Conference on the Use of Fly Ash, Silica Fume, Slag and Natural Pozzolans in Concrete, Trondheim, Norway, June 1989.

103. Aglave A., Feuerborn H.-J. Utilization of CCPs in Europe—Fly Ash in Concrete. Presented at International Seminar on Sustainability and Concrete Technology, Lyon, France, November 7–8, 2003.

104. Malhotra V.M. Reducing CO_2 Emissions. *Concrete International*, September 2006.

105. Jain A.K. Fly Ash Utilization in Indian Cement Industry: Current Status and Future Prospect. In *Concrete Sustainability through Innovative Materials and Techniques*, Roving National Seminars, Bangalore, January 10–14, 2011, pp. 46–51.

106. Mehta P.K., Manmohan D. High-Performance Green Concrete for the Building Industry. Presented at Proceedings of U.S. Green Building Council International Conference on Green Buildings, Austin, Texas, November 13–15, 2002.

107. Desai J.P. High Volume Fly Ash Concrete Technology and Its Application—Indian Scenario. In *International Conference on Sustainable Materials and Technologies*, Coventry, UK, June 11–13, 2007, pp. 54–59.

108. Malhotra V.M. High-Performance High-Volume Fly Ash Concrete. *Concrete International*, July 2002.

109. Ai H., Wang L., Wei J., Bai J., Lu P. Theoretical and Experimental Research on Compressive Strength of High Fly-Ash Content Concrete. In *Proceedings of the International Symposium on Ecological Environment and Technology of Concrete*, Beijing, August 1–4, 2011, pp. 257–262.

110. Niu Q.-L., Li C.-Z., Zhao S.-Q. Properties of a Low-Carbon Cement with 90% of Industrial Refuse. In *Proceedings of the International Symposium on Ecological Environment and Technology of Concrete*, Beijing, August 1–4, 2011, pp. 91–94.

111. Kokkamhaeng S. Utilization of Mae Moh Lignite Fly Ash in Roller Compacted Concrete Dam at Pak Mum. In *Sixth CANMET/ACI International Conference on Fly Ash, Silica Fume, Slag and Natural Pozzolans*, Bangkok, Thailand, 1998, pp. 17–31.

112. Ramme B.W., Fisher B.C., Naik T.R. Fly Ash Quality and Beneficiation Technologies at Wisconsin Electric Power Company (WE Energies). In *Second International Symposium on Concrete Technology for Sustainable Development*, Hyderabad, India, February 27–March 3, 2005, pp. 139–150.

113. van den Berg J.W., Moret J.B.M. Fly Ash Processing Plant. In *Sixth CANMET/ACI International Conference on Fly Ash, Silica Fume, Slag and Natural Pozzolans*, Bangkok, Thailand, 1998, pp. 213–220.

114. Lu Z. A New Technique of Manufacturing High Performance Supplementary Materials and Its Contribution to Cement and Concrete Technology for Sustainable Development. In *2009 International Workshop on Cement and Concrete Technology for Sustainable Development*, Beijing, May 6–12, 2009, pp. 167–178.

115. Justnes H. Concrete with High Volume of Supplementary Cementing Materials and Admixtures for Sustainable and Productive Construction. Presented at 2009 International Workshop on Cement and Concrete Technology for Sustainable Development, Xi'an, China, May 7–10, 2009.

116. Aitcin P.-C. *Binders for Durable and Sustainable Concrete*. Modern Concrete Technology 16. Taylor & Francis, London, 2008.

117. Hooton R.D. The Reactivity and Hydration Products of Blast-Furnace Slag. In *Supplementary Cementing Materials for Concrete*, ed. V.M. Malhotra, 245–288. CANMET, Ottawa, Canada, 1987.

118. World Steel Association. *Steel Statistical Yearbook 2010*. Accessed December 19, 2011. http://www.worldsteel.org/statistics/crude-steel-production.htm

119. Mehta P.K. Concrete Technology for Sustainable Development—An Overview of Essential Principles. In *CANMET/ACI International Symposium on Sustainable Development of the Cement and Concrete Industry*, Ottawa, Canada, October 21–23, 1998, pp. 1–14.

120. Mehta P.K. Role of Pozzlanic and Cementitious Materials in Sustainable Development of the Concrete Industry. In *Sixth CANMET/ACI International Conference on Fly Ash, Silica Fume, Slag and Natural Pozzolans*, Bangkok, Thailand, 1998, vol. 1, pp. 1–20.

121. Fidjestøl P. High Performance Concrete with SCM's. Presented at 2008 International Expert Workshop on Cement and Concrete Technology for Sustainable Development, Beijing, October 7–12, 2008.

122. Hardtl R. Utilization of Granulated Blast Furnace Slag. A Contribution to Sustainable Development in the Construction Industry. Presented at FIB Symposium, Concrete and the Environment, Berlin, Germany, October 2001.

123. Bernhardt C.I. SiO_2-støv som sementtilsetning. *Betongen I Dag*, 17, 152.

124. Markestad A. Tilsetning av SiO_2 støv til betong. Presented at Nordic Research Congress, Gothenburg, Sweden, August 1969.

125. Jahren P. Long-Term Experience with Silica Fume in Concrete. Presented at Ninth CANMET/ACI International Conference on Fly Ash, Silica Fume Slag and Natural Pozzolans in Concrete, Warsaw, Poland, May 20–25, 2007.

126. Nagataki S. Effects of Activated Silica on the Mechanical Properties of Concrete Submitted to High Temperature Curing (in Japanese). Presented at Cement and Concrete, Tokyo, May 1979.

127. Jahren P. Use of Silica Fume in Concrete. Presented at CANMET/ACI First International Conference on the Use of Fly Ash, Silica Fume, Slag and Mineral By-Products in Concrete, Montebello, Canada, July–August 1983.

128. Khayat K.H., Aitcin P.C. Silica Fume in Concrete—An Overview. In *CANMET/ACI International Workshop on Silica Fume in Concrete*, Washington D.C., April 7–9, 1991, pp. 3–46.

129. Pera J., Momtazi A.S. Pozzolanic Activity of Calcined Red Mud. Presented at Fourth CANMET/ACI International Conference on the Use of Fly Ash, Silica Fume, Slag and Natural Pozzolans in Concrete, Istanbul, Turkey, May 1992.

130. Sabir B.B. The Effects of Curing Temperature and Water/Binder Ratio on the Strength of Metakaolin Concrete. In *Sixth CANMET/ACI International Conference on Fly Ash, Silica Fume, Slag and Natural Pozzolans*, Bangkok, Thailand, 1998, pp. 493–506.

131. Marvan T., Pera J. Ambroise J. The Action of Some Aggressive Solutions on Portland and Calcined Laterite Blended Cement Concretes. In *Fourth CANMET/ACI International Conference on the Use of Fly Ash, Silica Fume, Slag and Natural Pozzolans in Concrete*, Istanbul, Turkey, May 1992, vol. 1, pp. 763–779.

132. Rossouw E., Kruger J. Review of Specifications for Additions for the Use in Concrete. In *First CABMET/ACI International Conference for the Use of Fly Ash, Silica Fume, Slag and Other Mineral By-Products in Concrete*, Montebello, Canada, July 31–August 5, 1983, pp. 143–172.

133. Kumar P., Sen A., Randey V.G. Performance Enhancement of High Strength Concrete through Metakaolin Addition. In *Proceedings of the Second International Symposium on Concrete Technology for Sustainable Development*, Hyderabad, India, February 27–March 3, 2005, pp. 95–101.

134. Mahoutian M., Bakhshi M., Bonakdar A., Shekarchi M. Effect of High Reactivity Metakaolin on the Gas Permeability of High Performance Concrete Mixture. In *Ninth CANMET/ACI International Conference on Fly Ash, Silica Fume Slag and Natural Pozzolans in Concrete*, Warsaw, Poland, May 20–25, 2007, pp. 139–147.

135. Zhang M.H., Malhotra V.M. High-Performance Concrete Incorporating Rice Husk Ash as a Supplementary Cementing Material. In *Fifth CANMET/ACI International Conference on Fly Ash, Silica Fume, Slag and Natural Pozzolans in Concrete*, Milwaukee, WI, 1995.

136. Mehta P.K. Rice Husk Ash—A Unique Supplementary Material. In *Advances in Concrete Technology*, ed. V.M. Malhotra, 2nd ed., 419–443. Natural Resources Canada, Ottawa, Canada, 1994.

137. Ashraf M., Srinivasa Rao P., Karthik S. Concrete Processing Technologies for Developing Countries. In *Proceedings of the Second International Symposium on Concrete Technology for Sustainable Development*, Hyderabad, India, February 27–March 3, 2005, pp. 615–622.

138. Azevedo A.A., Martins M.L.C., dal Molins D.C. A Study of the Penetration of Cloride Ions in Rice Husk Ash Concrete. In *Third CANMET/ACI International Symposium on Sustainable Development of Cement and Concrete*, San Francisco, September 2001, pp. 379–396.

139. Bui D.D., Stroeven P. Gap Graded Concrete Blended with Rice Hush Ash and Fly Ash. Presented at Seminar on Sustainability and Concrete Technology, Lyon, France, November 7–8, 2002.

140. Jain A.K. Sustainable Development and the Concrete Industry: Status Report from India. In *Second International Symposium on Concrete Technology for Sustainable Development, with Emphasis on Infrastructure*, Hyderabad, India February 27–March 3, 2005, pp. 33–37.

141. Joseph S., Baweja D., Crookham G.D., Cook D.J. Production and Utilization of Rice Husk Ash—Preliminary Investigation. Presented at Third CANMET/ACI International Conference on the Use of Fly Ash, Silica Fume, Slag and Natural Pozzolans, Trondheim, Norway, June 1989.

142. Ramezanianpour A.A., Mahdikhani M., Zarrabi K., Ahmadibeni G. The Production and the Optimization of Rice Husk Ash Quality on Sustainable Concrete. Presented at Proceedings of the International Conference on the Sustainability in the Cement and Concrete Industry, Lillehammer, Norway, September 16–19, 2007.

143. Horiguchi T., Sugita S., Saeki N. Effects of Rice-Husk Ash and Silica Fume on the Abrasion Resistance of Concrete. In *Sixth CANMET/ACI International Conference on Fly Ash, Silica Fume, Slag and Natural Pozzolans*, Bangkok, Thailand, 1998, pp. 517–526.

144. Tuan N.V., Ye G., van Breugel K., Dai B.D. Influence of Rice Husk Ash Fineness on the Hydration and Microstructure of Blended Cement Paste. In *11th International Conference on Advances in Concrete Technology and Sustainable Development*, Jinan, China, May 9–12, 2010, vol. 1, pp. 325–332.

145. Alhassan M., Mustapha A.M. Effect of Rice Husk Ash on Cement Stabilized Laterite. *Leonardo Electronic Journal of Practices and Technologies*, 11, 47–58, 2007.

146. Mehta P.K. Natural Pozzolans. In *Supplementary Cementing Materials for Concrete*, ed. V.M. Malhotra, 3–31. CANMET, Ottawa, Canada, 1987.

147. Jahren P. *Concrete—History and Accounts*. Tapir Academic Press, Trondheim, Norway, 2011.

148. Uzal B., Turanli L. High Volume Natural Pozzolans Blended Cements: Physical Properties and Compressive Strength of Mortar. In *Third CANMET/ACI International Symposium on Sustainable Development of Cement and Concrete*, San Francisco, September 2001, pp. 103–202.

149. Akman M.S., Mazlum F., Esenli F. A Comparative Study of Natural Pozzlans Used in Blended Cement Production. In *Fourth CANMET/ACI International Conference on Fly Ash, Silica Fume, Slag and Natural Pozzolans in Concrete*, Istanbul, Turkey, May 1992, vol. 1, pp. 471–494.

150. Mazlum F., Aköz F. Pozzolanic Properties of Diatomite and Durability of Cement-Diatomite Mortar in Sulfate Solutions. In *Fourth CANMET/ACI International Conference on Fly Ash, Silica Fume, Slag and Natural Pozzolans in Concrete*, Istanbul, Turkey, May 1992, vol. 1, pp. 917–933.

151. WBCSD/IEA. *Cement Technology Roadmap 2009*. Report 2010. World Business Council for Sustainable Development and International Energy Agency. www.wbcssg.org and www.iea.org.

152. Salazar A. Development of Early Age Strength in Concrete with Addition of Natural Pozzolan or Limestone. In *Sixth CANMET/ACI International Conference on Fly Ash, Silica Fume, Slag and Natural Pozzolans*, Bangkok, Thailand, 1998, pp. 479–491.

153. Hossain K.M.A., Lachemi M. Development of Volcanic Ash Concrete: Strength, Durability, and Microstructural Investigations. *ACI Materials Journal*, January–February 2006.

154. Long G.C.H., Li Y.S., Xie Y.J. Utilization of Natural Pozzolan from Yunan Province of China as a Mineral Admixture Used in Concrete. In *11th International Conference on Advances in Concrete Technology and Sustainable Development*, Jinan, China, May 9–12, 2010, vol. 1, pp. 695–700.

155. Bondar D., Lynsdale C.J., Ramezanianpour A.A., Milestone N.B. Alkali-Activation of Natural Pozzolan for Geopolymer Cement Production. In *International Conference on Sustainable Materials and Technologies*, Coventry, UK, June 11–13, 2007, pp. 313–317.

156. Erdem T.K., Meral C., Tokyay M., Erdogan T.Y. Effect of Ground Perlite Incorporation on the Performance of Blended Cements. In *International Conference on Sustainable Materials and Technologies*, Coventry, UK, June 11–13, 2007, pp. 279–286.

157. Douglas E., Malhotra V.M. A Review of the Properties and Strength Development of Non-Ferrous Slag-Portland Cement Binders. In *Supplementary Cementing Materials for Concrete*, ed. V.M. Malhotra, 371–427. CANMET, Ottawa, Canada, 1987.

158. Madej J., Ohama Y., Demura K. Assessment of Non Ferrous and Ferro Alloy Slags for Use in High-Strength Mortars. In *Fifth CANMET/ACI International Conference on Fly Ash, Silica Fume, Slag and Natural Pozzolans in Concrete*, Milwaukee, WI, 1995, pp. 413–431.

159. Damtoft J.S., Glavind M., Munch-Petersen C. Danish Centre for Green Concrete. Presented at CANMET/ACI International Symposium on Sustainable Development of the Cement and Concrete Industry, Ottawa, Canada, October 21–23, 1998.

160. Pera J., Ambroise J. The Use of Slag from the Ferro-Alloy Industry In Concrete. Presented at CANMET/ACI International Symposium on Sustainable Development of the Cement and Concrete Industry, Ottawa, Canada, October 21–23, 1998.

161. Pera J., Ambroise J. Properties of Concrete Incorporating Slag from the Ferro Alloy Industries. Presented at Seminar on Sustainability and Concrete Technology, Lyon, France, November 7–8, 2002.

162. Duricic R.L., Krizan D.M., Miletic S.R., Munitlak R., Drobnjak M. Researching of Possibilities for Utilization of Granulated Slag from Metal Magnesia Production According to ENV 197–1:1992. In *Sixth CANMET/ACI International Conference on Fly Ash, Silica Fume, Slag and Natural Pozzolans*, Bangkok, Thailand, 1998, pp. 425–439.

163. Cabrillac R., Coutial M., Estoup J.M. Improvement of Properties of Hydrated Magnesium Slag in Order to Recycle Them in Construction Block Form. In *Fifth CANMET/ACI International Conference on Fly Ash, Silica Fume, Slag and Natural Pozzolans in Concrete*, Milwaukee, WI, 1995, pp. 547–566.

164. Collepardi M., Fava G., Monosi S., Ruello M.L. Contribution to a Sustainable Progress by Using Non-Ferrous Slags in Cement and Concrete Industries. Presented at International Seminar on Sustainable Development in Cement and Concrete Industries, Milan, Italy, October 17–18, 2003.

165. Roper H., Kam F., Auld G.J. Characterization of a Copper Slag Used in Mine Fill Operation. In *First CANMET/ACI International Conference for the Use of Fly Ash, Silica Fume, Slag and Other Mineral By-Products in Concrete*, Montebello, Canada, July 31–August 5, 1983, pp. 1091–1109.

166. Hwang C.-L., Laiw J.-C. Properties of Concrete Using Copper Slag as a Substitute for Fine Aggregate. In *Third CANMET/ACI International Symposium on Sustainable Development of Cement and Concrete*, San Francisco, September 2001, vol. 2, pp. 1677–1695.

167. Qi G.-H., Peng X.-Q. Analysis on the Pozzolanic Effects of Phosphorous Slag Powder in Concrete. In *Proceedings of the International Symposium on Ecological Environment and Technology of Concrete*, Beijing, August 1–4, 2011, pp. 112–117.

168. Douglas E., Mainwarning P.R., Hemmings R.T. Pozzolanic Properties of Canadian Non-Ferrous Slags. In *Second CANMET/ACI International Conference on the Use of Fly Ash, Silica Fume, Slag and Natural Pozzolans in Concrete*, Madrid, Spain, June 1986, vol. 2, pp. 1525–1550.

169. Douglas E., Malhotra V.M. A Review of the Properties and Strength Development of Non-Ferrous Slags—Portland Cement Binders. In *Supplementary Cementing Materials for Concrete*, ed. V.M. Malhotra, 371–428. CANMET, Ottawa, Canada, 1987.

170. Warid Hussin M.-W., Abdul Awal A.-S.M. Influence of Palm Oil Ash on Sulfate Resistance of Mortar and Concrete. In *Sixth CANMET/ACI International Conference on Fly Ash, Silica Fume, Slag and Natural Pozzolans*, Bangkok, Thailand, 1998, vol. 1, pp. 417–429.

171. Van den Berg J., Visser J.L.J., Hohberg I., Wiens U. Fly Ash Obtained from Co-combustion—State of the Art in Europe. Presented at FIB Symposium on Concrete and the Environment, Berlin, Germany, October 2001.

172. Guerro A., Goni S., Macias A., Penandez E., Lorenzo M.P. Influence of Synthesis Temperature on the Hydration of New Cements from Fly Ash and Municipal Solid Waste Incineration. In *Third CANMET/ACI International Symposium on Sustainable Development of Cement and Concrete*, San Francisco, September 2001, pp. 267–284.

173. Zhou S.G. Research the Effect of Burned Ox Dung Ash on the Compressive Strength in Autoclaved Concrete Block. In *11th International Conference on Advances in Concrete Technology and Sustainable Development*, Jinan, China, May 9–12, 2010, vol. 1, pp. 243–247.

174. Naik T.R. Greener Concrete for Sustainable Concrete Construction. Presented at CANMET/ACI/UPC International Seminar on Sustainable Developments in Cement and Concrete, Barcelona, Spain, November 2002.

175. Brandstetr J., Drottner J. Composites Based on Solid Residues of Fluidized bed Coal Combustion and Other By-Products. In *Fifth CANMET/ACI International Conference on Fly Ash, Silica Fume, Slag and Natural Pozzolans in Concrete*, Milwaukee, WI, 1995, pp. 389–411.

176. Moir G.K., Kelharn S. Development in the Manufacture and Use of Portland Limestone Cement. In *ACI International Conference on High Performance Concrete*, Kuala Lumpur, Malaysia, 1997, pp. 797–819.

177. Lieberum K.-H. The Concrete in the Future Will Flow—Special Cement Makes Concreting Easier. *Concrete Plant International*, 5, 38–40, 2002.

178. Irassar E.F., Bonavetti V.L., Menendez G., Donza H., Cabrera O. Mechanical Properties and Durability of Concrete Limestone Cement. In *Third CANMET/ACI International Symposium on Sustainable Development of Cement and Concrete*, San Francisco, September 2001, pp. 431–450.

179. Brunatti C., Luco L.F. Blended Cements: 30 Years of Evolution in Argentina. In *Third CANMET/ACI International Symposium on Sustainable Development of Cement and Concrete*, San Francisco, September 2001, Supplementary papers, pp. 497–509.

180. Montgomery D.G., van B.K., Hinczak I., Turner K. Limestone Modified Cement for High-Performance Concrete. Presented at Sixth CANMET/ACI International Conference on Fly Ash, Silica Fume, Slag and Natural Pozzolans, Bangkok, Thailand, June 1998.

181. Bentz D.P., Irassar E.F., Bucher B.E., Weiss W.J. Limestone Fillers Conserve Cement. *Concrete International*, November 2009.

182. Kong X.Z., Ji G.J., Chen G.X. Effect of High-Volume Limestone Powder on Cement Hydration. In *11th International Conference on Advances in Concrete Technology and Sustainable Development*, Jinan, China, May 9–12, 2010, vol. 1, pp. 689–694.

183. Sakai E., Masuda L., Kikimana Y., Aikawa Y., Daimon M. Limestone Portland Cement Designed with the Packing Fraction and the Shape of the Particles. In *11th International Conference on Advances in Concrete Technology and Sustainable Development*, Jinan, China, May 9–12, 2010, vol. 1, pp. 107–113.

184. Wang J., Yang W., Sun K.P., Wu X. Effect of Limestone Powder on the Hydration Behavior of Cement. In *11th International Conference on Advances in Concrete Technology and Sustainable Development*, Jinan, China, May 9–12, 2010, vol. 1, pp. 279–284.

185. Dong Y., Xiao K.T., Yang H.Q. Influence of Limestone Powder on the Performance of Cementitious Materials. In *11th International Conference on Advances in Concrete Technology and Sustainable Development*, Jinan, China, May 9–12, 2010, vol. 1, pp. 794–799.

186. Toledo Filho R.D., Americano B.B., Fairbairn E.M.R., Filho J.-F. Potential of Crushed Waste Calcinated Clay Brick as a Partial Replacement for Portland Cement. In *Third CANMET/ACI International Symposium on Sustainable Development of Cement and Concrete*, San Francisco, September 2001, pp. 147–160.

187. Shao L.T., Rodriguez D. Waste Glass: A Possible Pozzolanic Material for Concrete. In *CANMET/CI International Symposium on Sustainable Development of the Cement and Concrete Industry*, Ottawa, Canada, October 21–23, 1998, pp. 317–326.

188. Aladine F., Laldji S., Tagnit-Hamou A. Glass Powder as an Alternative Cementitious Material in Concrete. In *Tenth ACI International Conference on Recent Advances in Concrete Technology and Sustainability Issues*, Seville, Spain, October 2009, pp. 683–687.

189. Høidalen Ø., Stoltenberg-Hansson E. Sementframstilling: Klinker, sement, energi-og miljøforholg. *Kjemi*, 6, 15–19, 1992.

190. Justnes H. Sement + Marl: Et kinderegg for godt til å være sant? Presented at Norwegian Concrete Day, Stavanger, Norway, 2010.

191. *Encyclopedia of Science and Technology.* McGraw-Hill, NJ, 1984.

192. Jahren P., Lindbak P. A New Pozzolan. In *Sixth CANMET/ACI International Conference on Fly Ash, Silica Fume, Slag and Natural Pozzolans*, Bangkok, Thailand, 1998, pp. 507–515.

193. Chandra S. Early High Strength Mortars and Concretes with Colloidal Silica Mixing. In *Sixth CANMET/ACI International Conference on Fly Ash, Silica Fume, Slag and Natural Pozzolans*, Bangkok, Thailand, 1998, pp. 323–334.

194. Laldji S., Tagnit-Hamou A. Properties of Ternary and Quaternary Concrete Incorporating New Alternative Cementitious Material. *ACI Materials Journal*, March–April 2006.

195. Thomas M., Hopkins D., Perreault M., Cail K. The Development of Ternary Blended Cement in Canada. In *Eight CANMET/ACI International Conference on Fly Ash, Silica Fume, Slag and Natural Pozzolans*, Las Vegas, May 23–29, 2004, pp. 9–26.

196. Burge T.A. Highly Reactive Ternary Blend with Portland Cement. In *Fifth CANMET/ACI International Conference on Fly Ash, Silica Fume, Slag and Natural Pozzolans in Concrete*, Milwaukee, WI, 1995, pp. 99–108.

197. Tiwari A.K., Jha D.N. High-Performance HVFA Concrete with Ternary Blends. In *Second International Symposium on Concrete Technology for Sustainable Development*, Hyderabad, India, February 27–March 3, 2005, pp. 265–270.

198. Sui T. Technological Innovation for Cement Sustainability—China's Effort. *ALITINFORM International Analytical Review Cement, Concrete and Dry Mixtures*, 3(20), 4–12, 2011.

199. Galitsky C. Benchmarking Tools for Emission Reductions and Energy Efficiency Improvement. Presented at First Stakeholder Meeting of Asia-Pacific Partnership on Clean Development and Climate Change, Xi'an, China, September 21, 2006.

200. IPCC. Integrated Pollution Prevention and Control: Reference Document on Best Available Techniques in the Cement and Lime Manufacturing Industries. European Commission, Directorate General, Joint Research Centre, March 2002.

201. IPCC. Intergovernmental Panel on Climate Change: Good Practice Guidance and Uncertainty Management in National Greenhouse Gas Inventories. Japan, 2000.

202. Noguchi T. Sustainable Development and the Concrete Industry: Status Report from Japan. In *Proceedings of the Second International Symposium on Concrete Technology for Sustainable Development*, Hyderabad, India, February 27–March 3, 2005, pp. 41–53.

203. Tokheim L.-A., Brevik P. Carbon Dioxide Emission Reduction by Increased Utilization of Waste-Derived Fuels in the Cement Industry. In *International Conference on Sustainability in the Cement and Concrete Industry*, Lillehammer, Norway, September 16–19, 2007, pp. 263–273.

204. Vanderborght B., Brodmann U. *The Cement CO_2 Protocol: CO_2 Emission Monitoring and Reporting for the Cement Industry, Guide to the Protocol*. Version 1.6. WBCSD Working Group Cement, October 19, 2001. http://www.ghgprotocol.org.

205. Brevik P. The CO_2 Challenge in Cement Production: Can CCS Be Part of the Solution? Presented at 2nd International Workshop on CO2: CCS and CCU in Germany, Norway, The Netherlands, Poland and Scotland—Challenges and Chances, Dusseldorf, November 10, 2011.

206. Lloyd F.K. Wood Firing in Calciner. Presented at 2008 International Expert Workshop on Cement and Concrete Technology for Sustainable Development, Beijing, October 7–12, 2008.

207. Gelli M. Industrial Ecology within the Cement Industry—The Holcim Case. Presented at International Seminar on Sustainable Development in the Cement and Concrete Industry, Plitectnico di Milano, Milan, Italy, October 17–18, 2003.

208. Duchesne J., Duong L., Bostrom T. Early Microstructure Development of Fly-Ash Based Polymer. In *First International Conference on Advances in Chemically-Activated Materials (CAM'2010)*, Jinan, China, May 9–12, 2010, pp. 94–99.

209. van Deventer J.S.J., Duxson P., Brice D.G., Provis J.L. Commercial Progress in Geopolymer Concretes: Linking Research to Applications. In *First International Conference on Advances in Chemically-Activated Materials (CAM'2010)*, Jinan, China, May 9–12, 2010, pp. 22–35.

210. Ma Y., Ye G., van Bruegel K. Effect of Alkaline Activating Solutions on the Mechanical Properties and Microstructure of Fly Ash Ash-Based Geopolymer. In *First International Conference on Advances in Chemically-Activated Materials (CAM'2010)*, Jinan, China, May 9–12, 2010, pp. 162–170.

211. Poulin R., Zmigrodzki S. Sulfur Polymer—An Environmentally-Friendly Concrete. In *CANMET/ACI International Symposium on Sustainable Development of the Cement and Concrete Industry*, Ottawa, Canada, October 21–23, 1998, pp. 531–542.

212. Shi C., Day R.L. A Caloriometric on Early Hydration of Alkali-Slag Cement. *Cement and Concrete Research*, 25(6), 1333–1346, 1995.

213. Tallin B., Brandstetr J. Present State and Future of Alkali-Activated Slag Concretes. In *Third CANMET/ACI International Conference on the Use of Fly Ash, Silica Fume, Slag and Natural Pozzolans in Concrete*, Trondheim, Norway, June 1989, vol. 2, pp. 1519–1545.

214. Bilec V., Opravil T., Soukai F. Searching for Practically Applicable Alkali-Activated Concretes. In *First International Conference on Advances in Chemically-Activated Materials (CAM'2010)*, Jinan, China, May 9–12, 2010, pp. 28–35.

215. Wenquan L., Zhen H., Beixing L., Yangbin O. Study on Properties and Applications of High Performance Mixes for Concrete (CRM). In *CANMET/ACI International Symposium on Sustainable Development of the Cement and Concrete Industry*, Ottawa, Canada, October 21–23, 1998, pp. 327–335.

216. Ding Z., Li Z. High-Early-Strength Magnesium Phosphate Cement with Fly Ash. *ACI Materials Journal*, 375–381, 2005.

217. Zhu H.J., Yao X., Zhang Z.H. Study on Non-Cement Based Alkali Activated Material for Oil and Gas Well Cementing at Low and Moderate Temperature. In *First International Conference on Advances in Chemically-Activated Materials (CAM'2010)*, Jinan, China, May 9–12, 2010, pp. 100–106.

218. Zhang Y.J., Li S., Wasng Y.C., Xu D.L. Synthesis of Geopolymer Composite and Its Mechanical Performance. In *First International Conference on Advances in Chemically-Activated Materials (CAM'2010)*, Jinan, China, May 9–12, 2010, pp. 181–186.

219. Lu D., Zhu Q., Xu Z., Effect of Dolomite on the Behavior of Geopolymers with Mixed Fly Ash and Metakaolin as Al-Si Sources. In *First International Conference on Advances in Chemically-Activated Materials (CAM'2010)*, Jinan, China, May 9–12, 2010, pp. 187–197.

220. Wang Q., Tu X., Li L., Ding Z.Y., Sui Z.T. Study on Shrinkage Performance of Slag-Based Geopolymer Concrete. In *First International Conference on Advances in Chemically-Activated Materials (CAM'2010)*, Jinan, China, May 9–12, 2010, pp. 198–205.

221. He J., Yang C.-H., Lv C.-F. Resistance of alkali-activated slag mortars to carbonation. In *First International Conference on Advances in Chemically-Activated Materials (CAM'2010)*, Jinan, China, May 9–12, 2010, pp. 216–224.

222. Shi C., He F. On the State and Roles of Anion or Anion Group of Activators during Activation of Slag. In *First International Conference on Advances in Chemically-Activated Materials (CAM'2010)*, Jinan, China, May 9–12, 2010, pp. 3–14.

223. Xiao L.G., Luo F. The Study of Alkaline-Activated Magnesium Slag Cementitious Material. In *First International Conference on Advances in Chemically-Activated Materials (CAM'2010)*, Jinan, China, May 9–12, 2010, pp. 66–71.

224. Pan Z., Yang N. Updated Review on AAM Research in China. In *First International Conference on Advances in Chemically-Activated Materials (CAM'2010)*, Jinan, China, May 9–12, 2010, pp. 45–55.

225. Xiao L., Hu H. The Influence of Activators on the Performance of Steam Curing Brick Prepared by Extracted Aluminum Fly Ash and Mechanics Studies. In *11th International Conference on Advances in Concrete Technology and Sustainable Development*, Jinan, China, May 9–12, 2010, vol. 1, pp. 519–529.

226. Liu R.J., Chen P., Ma S., Hu S.G. Mechanical Performance and Erosion-Resistance of Geopolymer Based on Manganese Slag. In *11th International Conference on Advances in Concrete Technology and Sustainable Development*, Jinan, China, May 9–12, 2010, vol. 1, pp. 570–576.

227. Wang Q., Ding Z.Y., Zhang J., Qiu L.G., Sui Z.T. Study on Slag-Based Geopolymer Hydration Process. In *Proceedings of the International Symposium on Ecological Environment and Technology of Concrete*, Beijing, August 1–4, 2011, pp. 67–71.

228. Byfors K., Klingstedt G., Lehtonen V., Pyy H., Romben L. Durability of Concrete Made with Alkali Activated Slag. In *Third CANMET/ACI International Conference on Fly Ash, Silica Fume, Slag and Natural Pozzolans in Concrete*, Trondheim, Norway, 1989, vol. 2, pp. 1429–1455.

229. Beauvent G., Holard E. Industrial Approach of Manufacturing Sulfo and Ferro-Aluminate Cement, including Use of Industrial By-Products. Presented at Seminar on Sustainability and Concrete Technology, Lyon, France, November 7–8, 2002.

230. Jirasit F., Ruscher C.H., Lohaus L. Property Development of the Slag- and Fly Ash-Based Geopoymeric Cement. In *International Conference on Sustainability in the Cement and Concrete Industry*, Lillehammer, Norway, September 16–19, 2007, pp. 149–162.

231. Bhanumatidas N., Kalidas N. FaL-G Brick and Cementitious Material: An Effective Sink for CO_2. In *CANMET/ACI International Symposium on Sustainable Development of the Cement and Concrete Industry*, Ottawa, Canada, October 21–23, 1998, pp. 563–575.

232. Bhanumatidas N., Kalidas N. The FaL-G Concrete for Housing and Infrastructure. In *Second International Symposium on Concrete Technology for Sustainable Development*, Hyderabad, India, February 27–March 3, 2005, pp. 289–299.

233. Rai M. Fly Ash Sand Lime Bricks in India. In *Fourth CANMET/ACI Conference on the Use of Fly Ash, Silica Fume, Slag, and Natural Pozzlans*, Istanbul, Turkey, May 1992, pp. 267–274.

234. Uomoto T., Kobayashi K. Strength and Durability of Slag-Gypsum Cement Concrete. In *First CANMET/ACI International Conference on the Use of Fly Ash, Silica Fume, Slag and Other Mineral By-products in Concrete*, Montebello, Canada, July 31–August 5, 1983, pp. 1013–1038.

235. Hashimoto S., Hashimoto C., Watababe T., Mizuguchi H. Experimental Study on Concrete Using a Binder Consisting of Three Industrial By-Products as Substitute for Cement. In *Eight CANMET/ACI International Conference on Fly Ash, Silica Fume, Slag and Natural Pozzolans*, Las Vegas, May 23–29, 2004, pp. 213–225.

236. Noguchi T. Sustainable Recycling of Concrete Structure. In *Concrete Sustainability through Innovative Materials and Techniques*, Roving National Seminars, Bangalore, January 10–14, 2011, pp. 86–97.

237. Malhotra V.M. *Supplementary Cementing Materials for Concrete*. CANMET, Ottawa, Canada, 1987.

238. Dubin J. Binder with Low Portland Cement Content for High Performance Concrete. In *ACI International Conference on High-Performance Concrete*, Singapore, 1994, pp. 283–295.

239. Bijen J., Waltje H. Alkali Activated Slag-Fly Ash Cements. In *Third CANMET/ ACI International Conference on Fly Ash, Silica Fume, Slag and Natural Pozzolans in Concrete*, Trondheim, Norway, 1989, vol. 2, pp. 1565–1578.

240. Buchwald A., Wierex J. Ascem Cement Technology—Alkali-Activated Cement Based on Synthetic Slag Made from Fly Ash. In *First International Conference on Advances in Chemically-Activated Materials (CAM'2010)*, Jinan, China, May 9–12, 2010, pp. 15–21.

241. Gjørv O.E. Alkali Activation of a Norwegian Granulated Blast Furnace Slag. In *Third CANMET/ACI International Conference on Fly Ash, Silica Fume, Slag and Natural Pozzolans in Concrete*, Trondheim, Norway, 1989, vol. 2, pp. 1501–1517.

242. Deja J., Malolepszy J. Resistance of Alkali-Activated Slag Mortars to Chloride Solutions. In *Third CANMET/ACI International Conference on Fly Ash, Silica Fume, Slag and Natural Pozzolans in Concrete*, Trondheim, Norway, 1989, vol. 2, pp. 1547–1563.

243. Ionescu I., Ispas T. Properties and Durability of Some Concretes Containing Binders Based on Slag and Activated Ashes. In *Second CANMET/ACI International Conference Proceedings*, Ottawa, Canada, 1986, vol. 2, pp. 1475–1493.

244. Georgescu M., Voinitchi C.D., Puri A., Voicu G. Durability Aspects of Concretes Containing Alkali Activated Slag. In *Ninth CANMET/ACI International Conference on Fly Ash, Silica Fume Slag and Natural Pozzolans in Concrete*, Warsaw, Poland, May 20–25, 2007, pp. 210–229.

245. Pavlenko S., et al. New Composite Cementless Binder from Industrial By-Products with the Use of Mechanochemistry. In *Third CANMET/ACI Internal Symposium for Sustainable Development of Cement and Concrete*, San Francisco, September 2001, pp. 181–192.

246. Pavlenko S.I. Fine-Grained Cementless Concrete Made with High-Calcium Fly Ash and Slag from Thermal Power Plants. In *Fourth CANMET/ACI International Conference on the Use of Fly Ash, Silica Fume, Slag and Natural Pozzolans in Concrete*, Istanbul, Turkey, May 1992, vol. 1, pp. 749–763.

247. Pavlenko S.I., et al. High Calcium Fly Ash to Silica Fume to Slag Sand Ratio Versus Compressive Strength and Density of Cementless Concrete. In *Sixth CANMET/ACI International Conference on Fly Ash, Silica Fume, Slag and Natural Pozzolans*, Bangkok, Thailand, 1998, vol. 2, pp. 1117–1126.

248. Komljenovic M., Bradic V., Bascarevic Z., Jovanovic N., Rosic A. The Nature of Industrial By-Products and Processes of Alkali-Activation. In *Tenth ACI International Conference on Recent Advances in Concrete Technology and Sustainability Issues*, Seville, Spain, October 2009, pp. 648–659.

249. Lee S.T., Lee J.H., Koh K.T., Ryu G.S., Jung H.S., Kim D.S., Hwang J.N., Kim H.J. Microstructural Observations of Fly-Ash Geopolymer Composites with Different Curing Conditions. In *First International Conference on Advances in Chemically-Activated Materials (CAM'2010)*, Jinan, China, May 9–12, 2010, pp. 155–161.

250. Criado M., Jimenez A.F., Palomo A. Corrosion Behavior of Steel Embedded in Activated Fly Ash Mortar. In *First International Conference on Advances in Chemically-Activated Materials (CAM'2010)*, Jinan, China, May 9–12, 2010, pp. 36–44.

251. Guerrero A., Goni S. Behavior of Class C Fly Ash–Belite Cement Mortar in a Mixed Sulfate Solution. In *Ninth CANMET/ACI International Conference on Fly Ash, Silica Fume Slag and Natural Pozzolans in Concrete*, Warsaw, Poland, May 20–25, 2007, pp. 85–98.

252. Krivenko P.V., Kovalchuk G.Y. A Heat-Resistant Gas Concrete with Geocement and Fly Ash for the Thermal Insulation of High-Temperature Industrial Equipment. In *Third CANMET/ACI International Symposium on Sustainable Development of Cement and Concrete*, San Francisco, September 2001, pp. 269–282.

253. Krivenko P.V., Petropavlovskii O.N., Vozniuk G.V., Pushkar V.I. Constructive Properties of the Concretes Made with Alkali-Activated Cements of New Generation. In *First International Conference on Advances in Chemically-Activated Materials (CAM'2010)*, Jinan, China, May 9–12, 2010, pp. 136–146.

254. Kavalerova E.S., Hrivenko P.V., Rostovskaya G.S. New Ukrainian Standard "Alkali-Activated Cements." In *First International Conference on Advances in Chemically-Activated Materials (CAM'2010)*, Jinan, China, May 9–12, 2010, pp. 242–247.

255. Vlasopoulos N., Cheeseman C.R. Use of Magnesium Oxide-Cement Binders for the Production of Blocks with Lightweight Aggregates. In *International Conference on Sustainable Materials and Technologies*, Coventry, UK, June 11–13, 2007, pp. 287–294.

256. Vandeperre L.J., Liska M., Al-Tabbaa A. Reactive Magnesium Oxide Cements: Properties and Application. In *International Conference on Sustainable Materials and Technologies*, Coventry, UK, June 11–13, 2007, pp. 397–410.

257. Neville A. High-Alumina Cements—Its Properties, Application and Limitations. Presented at Progress in Concrete Technology: Energy, Mines and Resources, Ottawa, Canada, June 1980, pp. 293–331.

258. Sui T. Innovation of Cement Based Materials for Sustainable Development. Presented at International Workshop on Cement and Concrete Technology for Sustainable Development, Beijing, August 2007.

259. Malhotra V.M. Sulphur Concrete and Sulphur Infiltrated Concrete: Properties, Applications and Limitations. Presented at Progress in Concrete Technology: Energy, Mines and Resources, Ottawa, Canada, June 1980, pp. 581–637.

260. Biasioli F., Øberg M. Concrete for Energy Efficient and Comfortable Buildings. In *International Conference on Sustainability in the Cement and Concrete Industry*, Lillehammer, Norway, September 16–19, 2007, pp. 593–599.

261. Ramachandran A.S. Recent Progress in the Development of Chemical Admixtures. In *Advances in Concrete Technology*, ed. V.M. Malhotra, 785–838. 2nd ed. Natural Resources Canada, Ottawa, Canada, 1994.

262. Jolicoeur V., Mikanivic N., Simard M.A., Sharman J. Chemical Admixture: Essential Components of Quality Concrete. Presented at Seminar on Sustainability and Concrete Technology, Lyon, France, November 7–8, 2002.

263. Marceau M.J., Vangeem M.G. Solar Reflectance Values for Concrete. *Concrete International*, August 2008.

264. Fidjestøl P., Grønvold T., Gulbrandsen K.O., Reierstølmoen L.J., Thorstensen R.T., Svennevig P. High Performance Concrete for More Sustainable Concrete Construction. In *Concrete Sustainability through Innovative Materials and Techniques*, Roving National Seminars, Bangalore, January 10–14, 2011, pp. 52–57.

265. Bellona. *Carbon Dioxide Storage: Geological Security and Environmental Issues—Case Study on the Sleipner Gas Field in Norway.* Bellona Report. Oslo, Norway, May 2007.

266. Malhotra V.M. Sustainability Issues and Concrete Technology. In *Concrete Sustainability through Innovative Materials and Techniques*, Roving National Seminars, Bangalore, January 10–14, 2011, pp. 1–10.

267. Markenes Hovland K. *Vil bygge CO2-fri sementfabrikk.* Teknisk Ukeblad, Oslo, Norway, 2009. www.tu.no.

268. http://calera.com/index.php/technology/ (accessed December 29, 2011).

269. Gullberg A.T. Skjerpet kvotesystem i EU etter 2012. *KLIMA* (Norwegian journal for climate research), 1-2008.

270. Swamy R.N. Designing Concrete and Concrete Structures for Sustainable Development. Presented at CANMET/ACI International Symposium on Sustainable Development of the Cement and Concrete Industry, Ottawa, Canada, October 21–23, 1998.

271. Betongelementer J.P. *Tekstmappe til lysbildeserie.* Precast Concrete Association, Oslo, Norway, 1972.

272. Tangen D.A. *Utradisjonelle gjenbrukstiltak. Eksempelsamling.* Technology Report 2377. Norwegian Public Road Administration, Oslo, Norway, September 2006.

273. Myren S.A., Mehus J. *Material Declaration of Recirculated Aggregate.* Reuse of Concrete Project Report 13. Norwegian Road Authorities, November 2007.

274. Noguchi T., et al. Resource-Flow Simulation in Concrete Related Industries by Using "ecoMA." Presented at International Conference on Sustainability in the Cement and Concrete Industry, Lillehammer, Norway, September 16–19, 2007.

275. Sakai K. Mail Correspondence. Numbers from Japan Crushed Stone Association. August 17, 2007.

276. Dosho Y. Development of a Sustainable Concrete Waste Recycling System: Application of Recycled Aggregate Concrete Produced by Aggregate Replacing Method. In *International Conference on Sustainable Materials and Technologies*, Coventry, UK, June 11–13, 2007, pp. 142–162.

277. Pietersen S.P., Fraay A.L.A., Hendriks C.F. Application of Recycled Aggregates in Concrete—Experiences from the Netherlands. Presented at CANMET/ACI International Symposium on Sustainable Development of the Cement and Concrete Industry, Ottawa, Canada, October 21–23, 1998.

278. Fraaij A.L.A., Pietersen H.S., de Vries J. Performance of Concrete with Recycled Aggregates. In *Third CANMET/ACI International Symposium on Sustainable Development of Cement and Concrete*, San Francisco, September 2001, pp. 463–479.

279. Poon C.S., Chan D. A Review on the Use of Recycled Aggregate in Concrete in Hong Kong. In *International Conference on Sustainable Materials and Technologies*, Coventry, UK, June 11–13, 2007, pp. 144–155.

280. WBCSD. *Recycling Concrete*. Executive Summary, Brochure 2009. World Business Council for Sustainable Development. www.wbcsd.org.

281. Corinaldesi V., Moriconi G. Recycled Demolition Wastes for Concrete and Mortar. Presented at International Seminar on Sustainable Development in Cement and Concrete Industries, Milan, Italy, October 17–18, 2003.

282. Smith E.D. Concrete Sustainability Opportunities. Presented at International Conference on Sustainability in the Cement and Concrete Industry, Lillehammer, Norway, September 16–19, 2007.

283. Chen H.-J., Liau C.-J. Use of Building Rubbles as Recycled Aggregate. In *Third CANMET/ACI International Symposium on Sustainable Development of Cement and Concrete*, San Francisco, September 2001, pp. 87–101.

284. Noguchi T. Optimum Concrete Recycling in Construction Industries—Japanese Experiences. Presented at 2008 International Expert Workshop on Cement and Concrete Technology for Sustainable Development, Beijing, October 7–12, 2008.

285. Mulder E., Feenstra L., de Jong T.P.R. Closed Cycle Construction—A Process for Separation and Reuse of the Total C&D Waste Stream. In *International Conference on Sustainable Materials and Technologies*, Coventry, UK, June 11–13, 2007, pp. 27–34.

286. Tangen D.A. *Gjenbruk av betong*. Report 2479. Norwegian Public Road Administration Technology, Oslo, Norway, December 2007.

287. Aurstad J. *Finstoff i gjenbruksbetong*. Report 2437. Norwegian Public Road Administration Technology, Oslo, Norway, December 2006.

288. Hauer B., Klein H. Recycling of Concrete Crusher Sand in Cement Clinker Production. Presented at International Conference on Sustainability in the Cement and Concrete Industry, Lillehammer, Norway, September 16–19, 2007.

289. Peng X., Huang T., Wang S. Research on Activation and Application of Fine Powders from Waste Concrete. In *Proceedings of the International Symposium on Ecological Environment and Technology of Concrete*, Beijing, August 1–4, 2011, pp. 3–9.

290. Fathfazl G., Razaqpur A.G., Isgor O.B., Abbas A., Fournier B., Foo S. Proportioning Concrete Mixtures with Recycled Concrete Aggregate. *Concrete International*, March 2010.

291. Bjegovic D., Miculic D., Rukavina M.J. Development of Sustainable Recycled Aggregate Concrete in Croatia. In *11th International Conference on Advances in Concrete Technology and Sustainable Development*, Jinan, China, May 9–12, 2010, vol. 1, pp. 423–431.

292. Yang K.-H., Chung H.-S., Ashour A.F. Influence of Type and Replacement Level of Recycled Aggregates on Concrete Properties. *ACI Materials Journal*, May–June 2008.

293. Dale O.H. Resikulerer betongstøvet. *Byggeindustrien*. Report 4-2007. Oslo, Norway, 2007.

294. Zhang H.B., Guan X.M. Influence of Rubber Crumb Addition to Portland Cement Concrete. Presented at Proceedings of the 7th International Symposium on Cement and Concrete (ISCC 2010), Jinan, China, May 9–12, 2010.

295. Liu W., Wang B.M. Research Progress of Cement Concrete Containing Scrap Rubber in Abroad. Presented at Proceedings of the 7th International Symposium on Cement and Concrete (ISCC 2010), Jinan, China, May 9–12, 2010.

296. Skripkiunas G., Grinys A., Dauksys M. Using Rubber Waste for Modification of Concrete Properties. In *International Conference on Sustainable Materials and Technologies*, Coventry, UK, June 11–13, 2007, pp. 85–99.

297. Khorami K., Ganjian E., Vafaii A. Mechanical Properties of Concrete with Waste Tire Rubber as Coarse Aggregate. In *International Conference on Sustainable Materials and Technologies*, Coventry, UK, June 11–13, 2007, pp. 85–90.

298. Yu L., Yu Q., Liu L. Mechanical Properties of Cement Mortar with Hybrid Modified Rubber Powder. Presented at Proceedings of the 7th International Symposium on Cement and Concrete (ISCC 2010), Jinan, China, May 9–12, 2010.

299. Wang B., Liu W., Zhao L. Experimental Research on Basic Mechanical Properties of Concrete with Used Rubber Powder. In *Proceedings of the International Symposium on Ecological Environment and Technology of Concrete*, Beijing, August 1–4, 2011, pp. 239–244.

300. Libo B., Shaomin S. Research on Mechanical Properties of Crumb Rubber Concrete. In *Proceedings of the International Symposium on Ecological Environment and Technology of Concrete*, Beijing, August 1–4, 2011, pp. 290–295.

301. Wong S.-F., Ting S.-K. Use of Recycled Rubber Tires in Normal- and High-Strength Concretes. *ACI Materials Journal*, July–August 2009.

302. Lupo R., Tyrer M., Cheeseman C.R., Donatello S. Manufactured Aggregate from Waste Materials. In *International Conference on Sustainable Materials and Technologies*, Coventry, UK, June 11–13, 2007, pp. 763–767.

303. Thawornsak V., Samrankrank S., Norrat P. Characteristics and Performance of Bio-Lightweight Aggregate Developed for Lightweight Precast Concrete Application. Presented at 9th Symposium on High Performance Concrete. Rotoroa, New Zealand, August 9–11, 2011.

304. Presti S.L., Martines E., Mulone A. From Recycled Plastic Bottles to New Building Materials. Presented at International Seminar on Sustainable Development in Cement and Concrete Industries, Milan, Italy, October 17–18, 2003.

305. Ilic M., Colc N., Miletic D., Otovic S., Folic R. Recycling of Plastic, Chromium-Magnesite Bricks and Demolition Waste for Concrete Production. In *Third CANMET/ACI International Symposium on Sustainable Development of Cement and Concrete*, San Francisco, September 2001, pp. 451–457.

306. Trussoni M., Hays C.D., Zollo R.F. Comparing Lightweight Polystyrene Concrete Using Engineered or Waste Materials. *ACI Materials Journal*, 101–106, 2012.

307. Chew P., MacDougall C. Compressive Strength Testing of Hemp Masonry Mixture. Presented at International Conference on Sustainability in the Cement and Concrete Industry, Lillehammer, Norway, September 16–19, 2007.

308. Arnaud L., Samri D. Innovative Building Material Based on Lime and Hemp Particles: From Ecological to Technical Interests. Presented at International Conference on Sustainability in the Cement and Concrete Industry, Lillehammer, Norway, September 16–19, 2007.

309. Hou G.H., Zhu X., Shi S.Y. Hydration and Hardening of Cement Paste Mixed with Crop Straw. Presented at Proceedings of the 7th International Symposium on Cement and Concrete (ISCC 2010), Jinan, China, May 9–12, 2010.

310. Habermehl-Cwirzen K.M.E., Branco F.G., Penttala V., Cwirzen A. Papercrete: Mechanical Strength and Thermal Conductivity of a Lightweight and Cementitious Material. Presented at 9th Symposium on High Performance Concrete, Rotoroa, New Zealand, August 9–11, 2011.

311. Ng C.H., Ideris Z., Narayanan S.P., Mannan M.A., Kurian V.J. Engineering Properties of Oil Palm Shell (Ops) Hybrid Concrete for Lightweight Precast Floor Slab. In *International Conference on Sustainable Materials and Technologies*, Coventry, UK, June 11–13, 2007, pp. 41–44.

312. Naik T.R. Greener Concrete for Sustainable Construction. Presented at Sustainable Development in Cement and Concrete Industries, Milan, Italy, October 17–18, 2003.

313. Meyer C. Glass Concrete. *Concrete International*, June 2003.

314. Johansen K. *Blokkstein med lettilslag av glass fra ide til product*, 16–18. Report STF22 A05150. SINTEF, Trondheim, Norway, 2005.

315. Dahl P.A. Glassbetong–dokumentasjon og utførelse. Presented at Norwegian Concrete Day Kristiansand, Norway, 2000.

316. Uehara T., Umehara H., Kiriyama K., Hattori K., Morishima K. Properties of High-Fluidy Concrete with Siliceous Powdered Waste. In *Third CANMET/ ACI International Symposium on Sustainable Development of Cement and Concrete*, San Francisco, September 2001, pp. 139–153.

317. Corinaldesi V., Monosi S., Tittarelli F., Favoni O. Influence of Paper Mill Ashes on the Properties of Self Compacting Concrete. In *Ninth CANMET/ ACI International Conference on Recent Advances in Concrete Technology and Sustainability Issues*, Seville, Spain, October 13–16, 2008, pp. 587–596.

318. Anashkin N.S., Pavlenko S.I. Development of Technological Regulations in the Manufacture of Aggregate for Fine Concretes from the Dumped Open-Hearth Furnace Slags. In *Ninth CANMET/ACI International Conference on Fly Ash, Silica Fume Slag and Natural Pozzolans in Concrete*, Warsaw, Poland, May 20–25, 2007, pp. 189–199.

319. Lukhanin M.V., Pavlenko S.I. Raising the Efficiency of Heat-Resisting Concretes and Masses by Utilization of Secondary Mineral Resources. In *Ninth CANMET/ACI International Conference on Fly Ash, Silica Fume Slag and Natural Pozzolans in Concrete, Warsaw*, Poland, May 20–25, 2007, pp. 239–251.

320. Sakaia N., Cornelis G., Gerven T.V., Vandecasteele C. Utilization of Pb-Slag as Partial Substitution of Fine Aggregates: Strength Properties of Cement Mortar and Leaching Assessment. Presented at International Conference on Sustainability in the Cement and Concrete Industry, Lillehammer, Norway, September 16–19, 2007.

321. Tailings. http://en.wikipedia.org/wiki/Tailings (accessed January 30, 2012).

322. Cai J.-W., Wu J.-X., Lu Z.-H., Gao G.-L. Effects of Powdery Mill Tailings from Magnetite on Workability and Strength of Concretes. In *Proceedings of the International Symposium on Ecological Environment and Technology of Concrete*, Beijing, August 1–4, 2011, pp. 233–238.

323. Feng X., Xi X., Cai J., Chai H., Song Y. Investigation of Drying Shrinkage of Concrete Prepared with Iron Mine Tailings. In *Proceedings of the International Symposium on Ecological Environment and Technology of Concrete*, Beijing, August 1–4, 2011, pp. 35–48.

324. Davoodi M.G., Nikraz H., Jamieson E. Chemical and Physical Characterization of Coarse Bauxite Residue (Red Sand). In *International Conference on Sustainable Materials and Technologies, Coventry*, UK, June 11–13, 2007, pp. 41–44.

325. Hwang C.L., Laiw J.C. Properties of Concrete Using Copper Slag as a Substitute for Fine Aggregate. In *Third CANMET/ACI International Conference on Fly Ash, Silica Fume, Slag and Natural Pozzolans in Concrete*, Trondheim, Norway, 1989, vol. 2, pp. 1677–1695.

326. Haugen M., Skjølvold O. *Laboratorieundersøkelser av formsand fra Ulefoss NV*. Report 3D0592.02. SINTEF, Trondheim, Norway, 2009.

327. Mohammed A., Nehdi M., Adawi A. Recycling Waste Latex Paint in Concrete with Added Value. *ACI Materials Journal*, July–August 2008.

328. Lepech M.D., Li V.C., Robertson R.E., Keoleian G.A. Design of Green Engineered Cementitious Composites for Improved Sustainability. *ACI Materials Journal*, November–December 2008.

329. Zhou J., Ye G., Sui T. Performance of Engineered Cementitious Composites for Concrete Repairs Subjected to Different Shrinkage. In *Proceedings of Concrete Repair, Rehabilitation and Retrofitting III*, Cape Town, South Africa, September 3–5, 2012, pp. 387–389.

330. Brandstetr J., Havlica J. Properties and Some Possibilities of the Utilization of Solid Residue of Fluidized Bed Combustion of Coal and Lignite. In *Sixth CAMET/ACI International Conference on Fly Ash, Silica Fume, Slag and Natural Pozzolans in Concrete*, Bangkok, Thailand, 1998, pp. 133–160.

331. Jahren P. Do Not Forget the Other Chapters. *Concrete International*, July 2002.

332. Jahren P. Sustainable Development of Cement and Concrete: Two Practical Examples of Typical Implications for the Future. In *CANMET/ACI International Symposium on Sustainable Development of the Cement and Concrete Industry*, Ottawa, Canada, October 21–23, 1998, pp. 337–347.

333. Sorourshian P., Plasencia J., Ravanbakhsh S. Assessment of Reinforcing Effects of Recycled Plastic and Paper in Concrete. *ACI Materials Journal*, May/June 2003.

334. Silva F. de A., Zhu D., Mobasher B., Filho R.D.T. Impact Behavior of Sisal Fiber Cement Composites under Flexural Load. *ACI Materials Journal*, March/April 2011.

335. Shao Y., LeFort T., Moras S., Rodriguez D. Waste Glass: A Possible Pozzolanic material for Concrete. Presented at CANMENT/ACI International Symposium on Sustainable Development of the Cement and Concrete Industry, Ottawa, Canada, October 21–23, 1998.

336. de Rojas M.I.S., Frias M., de Lomas G., Mujika R. Waste Products from FCC as Pozzolanic Addition: kiNetic of Pozzolanic Reaction Study. In *Ninth CANMET/ ACI International Conference on Fly Ash, Silica Fume Slag and Natural Pozzolans in Concrete*, Warsaw, Poland, May 20–25, 2007, pp. 125–138.

337. Jacobsen S. Properties of Cement Kiln Dust (CKD) and Use in Cementitious Materials—An Overview. Presented at International Conference on Sustainability in the Cement and Concrete Industry, Lillehammer, Norway, September 2007.

338. Khanna O.S., Kerenides K., Footon R.D. Utilization of Cement Kiln Dust and High-Alkali Cements in Cementitious Materials. Presented at International Conference on Sustainability in the Cement and Concrete Industry, Lillehammer, Norway, September 2007.

339. Lachemi M., Hossain K.M.A., Lofty A., Shehata M., Sahmaran M. CLSM Containing Cement Kiln Dust, *Concrete International*, June 2009.

340. Sui T., Liu K., Fu S., Zhao Z., Huang T., Yang Y. High Belite Cement with Low Environmental Load and High Performance. In *CANMET/ACI International Symposium on Sustainable Development of the Cement and Concrete Industry*, Ottawa, Canada, October 21–23, 1998, pp. 502–510.

341. Norcem. *Rapport om bærekraftig utvikling 2007*. Norcem, Oslo, Norway, 2007. www.norcem.no.

342. Sui T. Current Situation of China Cement and Concrete Industry on Sustainability. Presented at Second ACF Sustainability Forum, Seoul, November 3–4, 2011.

343. Jahren P. Knapphet på grus? Konsekvenser for næringsliv og produksjon. Presented at Conference on Shortage of Natural Resources, Trondheim, Norway, 1977.

344. Grattan-Bellow P.E., Dolar-Mantuani L.M.M., Sereda P.J. The Aggregate Shortage and High Alkali Cement in Changing Energy Situation. *Canadian Journal of Civil Engineering*, 5(2), 250–261, 1978.

345. Danielsen S.W. Sustainability in the Production and Use of Concrete Aggregates. In *International Conference on Sustainability in the Cement and Concrete Industry*, Lillehammer, Norway, September 16–19, 2007, pp. 322–333.

346. Minnick B. Aggregate Shortage, Permitting Problems Concern Industry. *Seattle Daily Journal of Commerce*, May 15, 1998.

347. California Geological Survey. Aggregate Availability in California 2006. Map Sheet 52. State of California, Department of Conservation, 2006.

348. Mehta P.K., Burrows R.W. Building Durable Structures in the 21st Century. *Concrete International*, March 2001.

349. Niu J.G., Wang W.L., Niu D.T. Neutralization Mechanism of Concrete in CO_2 and Acid Rain. Presented at Proceedings of the 7th International Symposium on Cement and Concrete (ISCC 2010), Jinan, China, May 9–12, 2010.

350. Mehta P.K. Fundamental Principles for Radical Enhancement of Durability of Concrete. In *Second International Symposium on Concrete Technology for Sustainable Development*, Hyderabad, India, February 27–March 3, 2005, pp. 83–93.

351. Betongelementforeningen. *Betong for energieffective bygninger–fordelene med termisk masse.* Precast Concrete Association, BEF, Oslo, Norway. www.betonelement.no.
352. Cembureau. *The Potential for CO2 Savings from Buildings Should Be More Fully Exploited.* Eurobrief 123. Cembureau, Brussels, Belgium, April 2004.
353. Kleiven T. *Termisk masse–tunge byggematerialer kan redusere energibehov til kjøling. Mur+Betong*, 1-2009.
354. Biasioli F., Øberg M. Concrete for Energy Efficient and Comfortable Buildings. Presented at International Conference on Sustainability in the Cement and Concrete Industry, Lillehammer, Norway, September 16–19, 2007.
355. Hertwich E. Energispaing kan gi klimagassutslipp. *CICERONE, Norwegian Journal for Climate Research*, October 5, 2005.
356. U.S. Environment Protection Agency. Heat Island Effect. http://www.epa.gov/hiri/Internet (accessed March 8, 2012).
357. Wikipedia. Volatile Organic Compound. http://en.wikipwdia.org.wiki/volatile_organic_compound (accessed February 8, 2012).
358. Wikipedia. Urban Heat Island. http://en.wikipedia.org/wiki/Urban-heat-island (accessed March 8, 2012).
359. WBCSD. http://www.wbcsdcement.org/index.php?option?com (accessed February 2, 2012).
360. Fabeco. *Sikkerhetsdatablad fersk fabrikkblandet betong.* Norwegian Ready Mixed Concrete Association, Oslo, Norway, 2008.
361. Kjellsen K.O., Meijer C., Bremseth S.K. Cases of Skin Burn due to Prolonged Contact with Wet Concrete Paving Stones. Presented at International Conference on Sustainability in the Cement and Concrete Industry, Lillehammer, Norway, September 16–19, 2007.
362. Thomassen T.R., Sandweg S., Andorsen G.S. Reactive Airways Dysfunction Syndrome (RADS) after Exposure of Iron Sulphate Dust in Cement Plant. Presented at International Conference on Sustainability in the Cement and Concrete Industry, Lillehammer, Norway, September 16–19, 2007.
363. Thommassen Y. *Eksponering for sementstøv blant bygningsarbeidere og på sementfabriker.* Statens arbeidsmiljøinstitutt [Governmental Work Environment Institute], Norway. Accessed February 2, 2012.
364. Cementa. *Healthy Floors.* Information brochure. Cementa AB, Danderyd, Sweden. www.cementa.se.
365. Jahren P. NOx-reduserende betong. *Betong industrien.* Report 1-2001. Oslo, Norway, 2001.
366. Brochure Mitsubishi Materials Corp. *NOXER-NOx Removing Paving Blocks.* English ed. Tokyo, 2000.
367. Nilsson Å. TiOmix–Fotokatalytiske møyligheter. Cementa. asa.nilsson cementa.se.
368. Barbesta M., Schaffer D. Concrete That Cleans Itself and the Air. *Concrete International*, February 2009.
369. KLIF: Klimaendringer påvirker helsa. Norwegian Government Climate and Pollution Directorate. http://www.miljostatus.no/Toppmeny/Miljostatus_i-Europa-2010/miljoproblemer (accessed February 15, 2012).

370. Mahoutian M., Bakfshi M., Shekarchi S. Study on Gas Permeability of Air-Entrained Concrete. In *Ninth CANMET/ACI International Conference on Fly Ash, Silica Fume Slag and Natural Pozzolans in Concrete*, Warsaw, Poland, May 20–25, 2007, pp. 441–453.

371. Sui T. Review of Rising Energy Efficiency and Emission Reduction of China Cement Industry. Presented at Russia-China Cement Summit, Moscow, Russia, November 30-December 1, 2011.

372. EIPPCB (European Integrated Pollution Prevention and Control Bureau). *Reference Document on Best Available Techniques in the Cement, Lime and Magnesium Oxide Manufacturing Industries.* European Commission, Joint Research Centre, Institute for Prospective Technological Studies, May 2010. Seville. ftp://ftp.jrc.es/pub/eippcb/doc/clm_bref_0510.pdf (Accessed August 20, 2012).

373. EPA. *Materials Characterization Paper in Support of the Final Rulemaking: Identification of Non-Hazardous Secondary Materials That Are Solid Waste Cement Kiln Dust (CKD).* February 3, 2011. http://www.regulations.gov/#!documentDetail;D = EPA-HQ-RCRA-2008–0329–1814 (accessed August 19, 2012).

374. SBC (Secretariat of the Basel Convention). *Updated General Technical Guidelines for the Environmentally Sound Management of Wastes Consisting of, Containing or Contaminated with Persistent Organic Pollutants (POPs).* 2007. http://www.basel.int/pub/techguid/tg-POPs.pdf (accessed August 20, 2012).

375. Karstensen K.H. Formation, Release and Control of Dioxins in Cement Kilns—A Review. *Chemosphere*, 70, 543–560, 2008.

376. GTZ/Holcim. *Guidelines on Co-Processing Waste Materials in Cement Production.* GTZ-Holcim Public Private Partnership, 2006.

377. Achternbosch M., Bräutigam K.-R., Hartlieb N., Kupsch C., Richers U., Stemmermann P. *Heavy Metals in Cement and Concrete Resulting from the Co-Incineration of Wastes in Cement Kilns with Regard to the Legitimacy of Waste Utilization.* 2003. Forschungszentrum Karlsruhe in der Helmholtz-Gemeins, Forschungszentrum Karlsruhe chaft Wissenschaftliche Berichte GmbH, Karlsruhe.

378. Zhang J., Liu J., Li C., Nie Y., Jin Y. Comparison of the Fixation of Heavy Metals in Raw Material, Clinker and Mortar Using a BCR Sequential Extraction Procedure and NEN7341 Test. *Cement and Concrete Research*, 38, 675–680, 2008.

379. Achternbosch M., Brtigam K.R., Hartlieb N., Kupsch C., Richers U., Stemmermann P. Impact of the Use of Waste on Trace Element Concentrations in Cement and Concrete. *Waste Management Research*, 23, 328–337, 2005.

380. Murat M., Sorrentino F. Effect of Large Additions of Cd, Pb, Cr, Zn to Cement Raw Meal on the Composition and Properties of the Clinker and the Cement. *Cement and Concrete Research*, 26(3), 377–385, 1996.

381. Katyal N.K., Parkash R., Ahluwalia S.C., Samuel G. Influence of Titania on the Formation of Tricalcium Silicate. *Cement and Concrete Research*, 29, 355–359, 1999.

382. Katyal N.K., Ahluwalia S.C., Parkash R. Effect of Barium on the Formation of Tricalcium Silicate. *Cement and Concrete Research*, 29, 1857–1862, 1999.

383. Kolovos K., Tsivilis S., Kakali G. SEM Examination of Clinkers Containing Foreign Elements. *Cement and Concrete Composites*, 27, 163–170, 2005.

384. Cullinane M.J., Bricka R.M., Francingues N.R. *An Assessment of Materials That Interfere with Stabilization/Solidification Processes*. Report EPA/600/9/87/015. Environmental Protection Agency, WA, 1987.

385. Bhatty J.J., West P.B. Stabilization of Heavy Metals in Portland Cement Matrix: Effects on Paste Properties. In *Stabilization and Solidification of Hazardous, Radioactive and Mixed Wastes*, ed. T.M. Gilliam, C.C. Wiles. ASTM STP 1240, vol. 33. American Society for Testing Materials, West Conshohoken, PA, 1996.

386. Olmo I.F., Chacon E., Irabien A. Influence of Lead, Zinc, Iron (III) and Chromium (III) Oxides on the Setting Time and Strength Development of Portland Cement. *Cement and Concrete Research*, 31, 1213–1219, 2001.

387. Van der Sloot H.A., van Zomeren A., Stenger R., Schneider M., Spanka G., Stoltenberg-Hansson E., Dath P. *Environmental Criteria for Cement Based Products, ECRICEM*. ECN Report ECN-E—08-011. Energy Research Centre of the Netherlands (ECN), 2008.

388. Sui T., Fan L., Ni Z. *Control of Water Soluble Cr(VI) in Chinese Cement*. Research report. China Building Materials Academy, 2008.

389. CEN/TS 14405. *Characterization of Waste—Leaching Behavior Tests—Up-Flow Percolation Test (under Specified Conditions)*. 2004.

390. CEN/TS 14429. *Characterization of Waste—Leaching Behavior Tests—Influence of pH on Leaching with Initial Acid Base Addition*. 2005.

391. NEN 7371. *The Determination of the Availability of Inorganic Components for Leaching*. The Netherlands Normalization Institute Standard, 2004.

392. *Leaching Characteristics Determination of the Leaching of Inorganic Components from Granular Materials with a Column Test—Solid Earthy and Stony Materials*. NEN 7373:2004. The Netherlands Normalization Institute Standard.

393. *Leaching Characteristics: Determination of Leaching of Inorganic Components with the Diffusion Test*. NEN 7375:2004. The Netherlands Normalization Institute Standard.

394. DEV S4 DIN 38414 S4. *German Standard Procedure for Water, Waste Water and Sediment Testing—Group (Sludge and Sediment)*. Determination of Leachability (S4). Institüt für Normung, Berlin, 1984.

395. Sanchez F., Gervais C., Garrabrant A.C. Leaching of Inorganic Contaminants from Cement-Based Waste Materials as a Result of Carbonation during Intermittent Wetting. *Waste Management*, 22, 249, 2002.

396. Garrabrants A.C., Sanchez F., Kosson D.S. Changes in Constituent Equilibrium Leaching and Pore Water Characteristics of a Portland Cement Mortar as a Result of Carbonation. *Waste Management*, 24, 19, 2004.

397. Brameshuber W., Hohberg I., Uebachs S. Environmental Compatibility of Concrete in Contact with Soil and Ground Water-Testing and Evaluation. In *Third CANMET/ACI International Symposium on Sustainable Development of Cement and Concrete*, San Francisco, September 2001, pp. 339–353.

398. Sugiyama T., Takahash S., Honda M., Sakai E. Current State of the JSCE Standard on Test Method for Leaching of Trace Elements from Hardened Concrete. In *International Conference on Sustainable Materials and Technologies*, Coventry, UK, June 11–13, 2007, pp. 197–203.

399. Fava G., Ruello M., Sani D. Leaching Behavior and Environmental Impact of Concrete Manufactured with Biomass Ashes. In *International Conference on Sustainable Materials and Technologies*, Coventry, UK, June 11–13, 2007, pp. 295–301.

400. Marion A.-M., De Laneve M., De Graauw A. Study of the Leaching Behavior of Lean Concrete for Road Sub-Base: The Use of a Tank Test with Demineralized Water and Quantification of Heavy Metal Content in Leachate. In *Eight CANMET/ACI Conference on Fly Ash, Silica Fume, Slag and Natural Pozzolans in Concrete*, Las Vegas, May 23–29, 2004, pp. 327–341.

401. Li K.L., Jiang L.H., Cai Y.B. Investigation on Heavy Metal and Radioactive Metal Solidification by Geopolymer Materials. In *Proceedings of the First International Conference on Advances in Chemically-Activated Materials (CAM' 2010)*, Jinan, Shandong, China, May 9–12, 2010, pp. 233–248.

402. Yanai S., Ishikawa M., Miura K., Murata T. Material Selection and Construction Practices for Controlling Thermal Cracking in Construction of Underground Diaphragm wall. In *Sixth CANMET/ACI International Conference on Fly Ash, Silica Fume, Slag and Natural Pozzolans*, Bangkok, Thailand, 1998, pp. 621–640.

403. Serclerat I., Moszkowicz P., Pollet B. Retention Mechanisms in Mortars of the Trace Metals Contained in Portland Cement Clinkers. *Waste Management*, 20, 259, 2000.

404. http://en.wikipedia.org/wiki/Noise_pollution (accessed October 4, 2011).

405. http://www.klif.no/no/Tema/stoy/ (accessed October 4, 2011).

406. Brochure Cementa. *Cementa and Miljøn*. Cementa, Danderyd, Sweden, 1998.

407. Åkerløf L. Good Sound Environment (in Swedish). Cementa, Danderyd, Sweden, undated.

408. www.byggutengrenser.no.

409. http://www.sengpielaudio.com/calculator-RT60Coeff.htm.

410. Billington C. Reduction of Impact Sound Transmission from Precast Stairs. *Concrete*, August 2008. www.concrete.org.uk.

411. National Police Agency of Japan. Damage Situation and Police Counter-measures (from "deaths" template). July 11, 2012. http://www.npa.go.jp. Accessed July 15, 2012.

412. Shimbun Y. Japan. April 18, 2011. http://www.yomiuri.co.jp/science/news/20110319-OYT1T00743.htm (accessed March 19, 2011).

413. Hur J. Food Contamination Set to Rise as Japan Fights Radiation Crisis at Reactor. *Bloomberg*, March 27, 2011. Archived from the original on April 18, 2011.

414. Davis J.R., Johnson R., Stepanek J. *Fundamentals of Aerospace Medicine*, 221–230. Philadelphia, PA. Lippincott Williams & Wilkin, 2008.

415. UNSCEAR. *Sources and Effects of Ionizing Radiation. Report to General Assembly*. United Nations Scientific Committee on the Effect of Atomic Radiation (UNSCEAR), New York, 1993.

416. European Commission. *Radiological Protection Principles Concerning the Natural Radioactivity of Building Materials*. Radiation Report 112. 1999.

417. NRCP. Report 95. 1987. http://www.ncrppublications.org/reports/95 (accessed July 19, 2012).

418. NCRP. *Ionizing Radiation Exposure of the Population of the United States.* Report 160. http://www.ncrponline.org/Publications/Press_Releases/160press. html (accessed July 19, 2012).

419. Demecz D., et al. PPT on Radiation—Basic Principles and Effects on the Human Body. www.boslo.de (accessed July 19, 2012).

420. NRCP. Report 116. 1993. http://www.ncrppublications.org/reports/116 (accessed July 19, 2012).

421. ICRP. Publication 60. *Annals of ICRP*, 21(1–3), 1991.

422. International Congress Series. Cancer Risk due to Exposure to High Levels of Natural Radon in the Inhabitants of Ramsar, Iran. *Science Direct*, February 24, 2005. http://www.sciencedirect.com/science/article/pii/S0531513104018461. Accessed July 19, 2012.

423. Camphausen K.A., Lawrence R.C. Principles of Radiation Therapy. In *Cancer Management: A Multidisciplinary Approach*, ed. R. Pazdur, L.D. Wagman, K.A. Camphausen, W.J. Hoskins. pp. 1–9, Cancernetwork, PA. 11th ed. 2008.

424. Markkanen M. *Radiation Dose Assessments for Materials with Elevated Natural Radioactivity.* Report STUK-B-STO 32. Radiation and Nuclear Safety Authority—STUK, Finland, 1995.

425. Mustonen R., Pennanen M., Annanmäki M., Oksanen E. *Enhanced Radioactivity of Building Materials.* Final Report of Contract 96-ET-003 for the European Commission. Radiation and Nuclear Safety Authority–STUK, Finland, 1997; Radiation Protection 96, Luxembourg, 1999.

426. NRCP. Report 94. 1987, http://www.ncrppublications.org/reports/94 (accessed July 19, 2012).

427. European Commission. *Radiological Protection Principles Concerning the Natural Radioactivity of Building Materials.* Radiation Protection Report RP-112. European Commission, Luxembourg, 1999.

428. Executive Order of the Council of Ministers of the Republic of Poland. http:// www.paa.gov.pl/en/?frame = 1.2.

429. Khan K., Khan H.M. Natural Gamma-Emitting Radionuclides in Pakistani Portland Cement. *Applied Radiation and Isotopes*, 54, 861–865, 2001.

430. Zhang Y., Yu T., Mao Z., Zhang X., Li Z., Zhang S. The Investigation and Analysis on Natural Radioactivity Nuclide Specific Activity of Cement Product in China (in Chinese). *Examination and Accreditation*, 1003-8965, 6-0001-05, 2009.

431. Stranden E. Assessment of the Radiological Impact by Using Fly Ash in Cement. *Health Physics*, 45, 145, 1983.

431a. Beretka, J. and Matthew, P.J. *National Radioactivity of Australian Fly Ashes.* 2nd International Conference on Ash Technology and Marketing. London, UK. 1985.

431b. Lange, S.C.T. Juenger, M., and Siegel, J.A. Indoor Radon Exhalation Rates from Concrete with Fly Ash. Report of U.S. Environmental Protection Agency Star Fellowship Project No. F09l10140. 2010.

432. Johnson D. Radioactive Elements Present in Cementitious Grouts. Presented at RWIN IV Meeting, University of Sheffield, UK, July 19–20, 2005.

433. Somlai J., et al. Radiological Aspects of the Usability of Red Mud as Building Material Additive. *Journal of Hazardous Materials*, 150, 541, 2008.

434. Cozmuta A.I., Van der Graaf E.R., De Mejer R.J. Moisture Dependence of Radon Transport in Concrete: Measurements and Modeling. *Health Physics*, 85, 438, 2003.

435. Roelofs L.M.M., Scholten L.C. The Effect of Aging, Humidity, and Fly-Ash Additive on the Radon Exhalation from Concrete. *Health Physics*, 67, 266, 1994.

436. Kovler, K. et al. Radon Exhalation of Cementitious Materials Made with Coal Fly Ash. Part 1. Scientific Background and Testing of the Cement and Fly Ash Emanation. *Journal of Environmental Radioactivity*, 82, 321, 2005.

437. International Commission on Radiological Protection. *Protection against Radon-222 at Home and at Work*. Publication 65. Amsterdam, The Netherlands. Pergamon Elsevier, 1993.

438. Leung J.K.C., Tso M.Y.W., Ho C.W. Radon Action Level for High-Rise Buildings. *Health Physics*, 76, 537, 1999.

439. Wikipedia. http://en.wikipedia.org/wiki/Concrete_degradation.

440. Kan Kan Y.C., Pei K.C., Chang C.L. Strength and Fracture Toughness of Heavy Concrete with Various Iron Aggregate Inclusions. *Nuclear Engineering and Design*, 228, 119–127, 2004.

441. American Concrete Institute. Heavyweight and Radiation Shielding Concrete. ACI 304-3R. 2007, pp. 33–36.

442. Davis H.S. 1972a. Concrete for Radiation Shielding—In Perspective. In *Concrete for Nuclear Reactors*, ed. C.E. Kesler, 3–13. SP-34. American Concrete Institute, Farmington Hills, MI.

443. Jaeger R.G., Blizard E.P., Chilton A.B., Grotenhuis M., Höning A., Jaeger Th.A., Eisenlohr H.H. *Engineering Compendium on Radiation Shielding. Shielding Materials*. Vol. II. Springer, Berlin-Heidelberg, 1975.

444. Gencel O., Brostow W., Ozel C., Filiz M. An Investigation on the Concrete Properties Containing Colemanite. *International Journal of Physical Sciences*, 5(3), 216–225, 2010.

445. Volkman D.E. Concrete for Radiation Shielding. In *Significance of Tests and Properties of Concrete and Concrete-Making Materials*, ed. P. Klieger, J. Lamond, 540–546. ASTM STP 169C. ASTM, 1994.

446. Gencel O., Brostow W., Cozel C., Filiz M. Concretes Containing Hematite for Use as Shielding Barriers, *Materials Science*, 16(3), 2010.

447. El-Sayed A.A., Kansouh W.A., Megahid R.M. Investigation of Radiation Attenuation Properties for Barite Concrete. *Japanese Journal of Applied Physics*, 41, 7512–7517, 2002.

448. Macphee D.E., Glasser F.P. *Immobilization Science of Cement Systems*. Materials Research Society Bulletin XVIII. 1993, pp. 66–71.

449. Toutanji H., Delatte N., Aggoun S., Duval R., Danson A. Effect of Supplementary Cementitious Materials on the Compressive Strength and Durability of Short-Term Cured Concrete. *Cement and Concrete Research*, 34, 311–319, 2004.

450. Goñi S., Hernández M.S., Guerrero A. Cemented Matrices Used in the Storage of Low and Medium Radioactive Waste: Spanish Experience. Presented at 1st Spanish National Conference on Advances in Materials Recycling and Eco-Energy, Madrid, November 12–13, 2009.

451. El-Dakroury A., Gasser M.S. Effects of SF and Ilemenite on the Chemical, Mechanical and Radiation Behavior of Matrices Used as Solidification of Wastes. *Nature and Science*, 10(5), 2012.

452. Niwase K., Suguhashi N., Tsuji Y. Mortar Mix Design of Low Diffusion Layer for Sub-Base-Surface Radioactive Waste Disposal Facilities in Japan. Presented at Proceedings of the 7th International Symposium on Cement and Concrete (ISCC 2010), Jinan, China, May 9–12, 2010.

453. Scheetz B.E., Roy D.M., Duffy C. Physical and Mechanical Properties of Thermally Altered Cementitious Sealing Materials for a Nuclear Waste Repository in Tuff. In *Third CANMET/ACI International Conference on Fly Ash, Silica Fume, Slag and Natural Pozzolans in Concrete*, Trondheim, Norway, 1989, vol. 2, pp. 1597–1614.

454. Basu P.C. Blended-Cement Concrete for Structures of Nuclear Complex. In *Second International Symposium on Concrete Technology for Sustainable Development*, Hyderabad, India, February 27–March 3, 2005, pp. 409–423.

455. Langton C.A. Slag-Based Materials for Toxic Metal and Radioactive Waste Stabilization. In *Third CANMET/ACI International Conference on Fly Ash, Silica Fume, Slag and Natural Pozzolans in Concrete*, Trondheim, Norway, 1989, vol. 2, pp. 1697–1706.

456. PCA: Radioactive Waste. http://www.cement.org/waste/wt_apps_radioactive. asp (accessed February 16, 2012).

457. http://www.chemicalprocessing.com/industrynews/2004/45.html (accessed July 29, 2012).

458. http://no.wikipedia.org/wiki/Oosterscheldekering (accessed January 30, 2011).

459. Water. http://en.wikipedia.org/wiki/water (accessed February 23, 2012).

460. Mehta P.K. Reducing the Environmental Impact of Concrete. *Concrete International*, October 2001.

461. Muszynski L., Chini A., Bergin M. Re-Using Wash Water in Ready-Mixed Concrete Operations. *Concrete*, February 2002.

462. Malhotra V.M. Global Warming and Role of Supplementary Cementing Materials and Superplasticizrs in Reducing Greenhouse Gas Emissions from the Manufacturing of Concrete. In *Tenth ACI International Conference on Recent Advances in Concrete Technology and Sustainability Issues*, Seville, Spain, October 2009, pp. 421–440.

463. Holton I., Glass J., Clarke M. Case Studies Demonstrating Reductions in the Consumption of Natural Resources and Energy by the UK Precast Concrete Industry. Presented at International Conference on Sustainability in the Cement and Concrete Industry, Lillehammer, Norway, September 16–19, 2007.

464. Sorteberg A., Kvamstø N.G. Hvor mye vil nedbøren øke i et varmere klima? *KLIMA* (Norwegian journal for climate research), 4-2008.

465. Reiersen T. *Fordrøyning i betngrør-dimensjonering, løsninger og erfaringer*. Report 3. Norwegian Concrete Society, Environmental Committee, October 2011.

466. Muthanna T.M. *Det koster å ikke gjøre noe*. Report 3. Norwegian Concrete Society, Environmental Committee, October 2011.

467. Norwegian Concrete Society, Environmental Committee. *Økt focus på Miljø og Klimaendringer-nye muligheter for betong*. Report 2. Oslo, Norway, 2010.

468. Myhr K. *Veileder i dimensjonering og bruk av permeable dekker av belegningsstein*. Report 3. Norwegian Concrete Society, Environmental Committee, October 2011.

469. Justnes H. *Innvirkning av betong, vaskevann og–slam på miljøet ved deponi og muligheter for gjenbruk*. Report STF50 F05205. SINTEF, Trondheim, Norway, December 2005.

470. Donnaes P. Cutting Costs by Recycling Materials. *Concrete International*, July 2002.

471. Nakamoto J., Togawa K., Mitsuiwa Y. Effective Utilization of Sludge Water with a Setting Retarder. In *Seventh CANMET/ACI International Conference on Recent Advances in Concrete Technology*, Las Vegas, May 26–29, 2004, pp. 231–245.

472. Heimdal E. Resirkuleringsanlegg på betongstasjoner. Presented at Norwegian Concrete Day, Kristiansand, Norway, 2000.

473. Bartnes J., Serch-Hansen C. *Retningslinjer for sedimenteringsbasseng ved ferdigbetongfabrikker*. Norconsult Report 3179500 for the Norwegian Ready Mixed Association, Oslo, Norway, October 22, 1998.

474. Meininger R.C. *Recycling Mixed Wash Water—Its Effect on Ready-Mixed Concrete*. Technical Information Letter 298. National Ready-Mixed Concrete Association, Silver Spring, MD, March 1973.

475. Sæbø A., Hjermann M. Betongslamsom tilsetning til jord ved dirking av raigras og rødsvingel. *Bioforsk Rapport*, 3(114), 2008. www.bioforsk.no.

476. *FAO Yearbook Fishery Information: Capture Production*. Vol. 94/1. Rome, 2004.

477. Seaman W. Jr., Sprague L.M. *Artificial Habitats for Marine and Freshwater Fisheries*. Academic Press, San Diego, 1990.

478. Jahren P. *Kunstige Fiskerev*. State-of-the-Art Report. P.J. Consult AS for Norwegian Research Council (NFR), Oslo, Norway, October 1, 1998.

479. Jahren P. *Japansk Fiskehusteknologi–orienterende undersøkelser*. Preliminary Investigations. P.J. Consult AS/Fiskehus AS, April 10, 1990.

480. Rivenga, S. et al. European Artificial Reef Research. Presented at Proceedings of the 1st EARRN Conference, Ancona, Italy, March 1996.

481. Ceccaldi H.J. Figure in Private Letter. CERAM, Aix-Marceille, France, January 1998.

482. Charbonelle E. *Effects of Artificial Reef Design on Associated Fish Assemblage: A Review in France*. EARRN, Southampton University, UK, January 1998.

483. Edwards E., Clark S. Rehabilitation of Coral Reef Flats Using Precast Concrete. *Concrete*, January/February 1992.

484. Christie H., Mortensen C. *Feltrapport fra Fiskehusprosjekt i Lofoten ti år etter*. NIVA, Oslo, Norway, October 2004.

485. OECD. *Glossary of Statistical Terms*. 2003. Stats.oecd.org (http://stats.oecd.org/glossary/detail.asp?ID=2896) (accessed October 12, 2009).

486. European Commission. Environment: Waste. http://ec.europa.eu/environment/wasteindex.htm (accessed February 24, 2012).

487. Badanoiu A., Georgescu M., Zahanagiu A. Solidification/Stabilization of Hazardous Waste with Heavy Metal Content in Blended Cements. In *Ninth CANMET/ACI International Conference on Fly Ash, Silica Fume Slag and Natural Pozzolans in Concrete*, Warsaw, Poland, May 20–25, 2007, pp. 253–264.

488. Krol A. Immobilization of Heavy Metals in Composites Based on Binders with High Content of Mineral Additives. In *Ninth CANMET/ACI International Conference on Fly Ash, Silica Fume Slag and Natural Pozzolans in Concrete*, Warsaw, Poland, May 20–25, 2007, pp. 265–282.

489. Karstensen K.H., Justnes H. Environmentally Sound Management of Hazardous and Industrial Waste in Cement Kilns in China. In *International Conference on Sustainability in the Cement and Concrete Industry*, Lillehammer, Norway, September 16–19, 2007, pp. 335–342.

490. UNEP 2005, Stockholm Convention Expert Group on Best Available Techniques and Best Environmental Practices. *Guidelines on Best Available Techniques and Provisional Guidance on Best Environmental Practice Relevant to Article 5 and Annex C of the Stockholm Convention in Persistent Organic Pollutant.* http://www.pops.int/.

491. Basel Convention 2006. General Technical Guidelines for the Environmentally Sound Management of Wastes Consisting of, Containing or Contaminated with Persistent Organic Pollutants (POPs). Draft, unedited version. April 7, 2006. http://www.basel.int/techmatters/techguide/frsetmain.php?topicId=0.

492. Port of Helsinki. *TBT Tributtyltenn–Sanering av TBT-kontaminerade Sediment i Nordsjø hamn.* Port of Helsinki, Finland. www.vuosaarensatarna.fi/ymparisto.

493. Eriksson S. Hvordan lage nye boligtomter av forurenset slam I havner? Presented at Norwegian Concrete Day, Molde, Norway, October 26–27, 2006.

494. Laugesen J., Eriksson S.B. Recent Advances in Stabilization and Solidification of Contaminated Soil Sediments in the Nordic Countries. In *International Conference on Sustainability in the Cement and Concrete Industry*, Lillehammer, Norway, September 16–19, 2007, pp. 369–382.

495. *Bindemidler og metoder for stabilisering/solidifisering av forurensede massrer.* Project report. ST&SO, Oslo, Norway, 2006–2008. www.stabilgrunn.no.

496. Brendbekken G., Aabøe R., Dieseth H. *Flyveaske fra papirproduksjon som tilsetning i kalksementpeler.* Technology Report 2447. Norwegian Public Road Administration, Oslo, Norway, September 2007.

497. Zhan-guo L., Yin C., Xin H., Xia-Ming S. A Study of Stabilizer for Saline Sludge. In *Proceedings of the International Symposium on Ecological Environment and Technology of Concrete*, Beijing, August 1–4, 2011, pp. 185–189.

498. Dawadi S., Hansen M.R., Berdanier B.W. Encapsulation of Contaminated Soil in Concrete Mortar. *ACI Materials Journal*, 347–352, 2004.

499. Federal Interagency Stream Restoration Working Group (FISWG). *Stream Corridor Restoration: Principles, Processes, and Practices.* PB98–158348LUW. 1998.

500. Brekke I. Tyskland skal drives av sol og vind. *Aftenposten*, Oslo, Norway, November 13, 2011.

501. Markenes K.H. Norge griper ikke sjansen. *Teknisk Ukeblad*, Oslo, Norway, 43/2009.

502. Johansen S. Vindmøller i betong. Presented at Norwegian Concrete Day, Stavanger, Norway, October 21, 2010.

503. Inneo Torres. Precast Concrete Wind Towers. http://www.inneo.es (accessed January 5, 2010).

504. de Jaime R. Email to Per Jahren. June 20, 2011.

505. Demuynck A., Gunst N. Phase One of Wind Project Winds Down. *Concrete International*, October 2008.
506. Malhotra V.M. Sustainability Issues and Concrete Technology. In *11th International Conference on Advances in Concrete Technology and Sustainable Development*, Jinan, China, May 9–12, 2010, vol. 1, pp. 1–14.
507. Yan Z., Juan D. Government Looking at New Ways to Set Prices for Wind Power. *China Daily*, December 15, 2011.
508. Sui T., Liu K., et al. Study on the Properties of High Belite Cement. *Journal of Chinese Ceramic Society*, 4, 488–492, 1999.
509. Sui T., et al. Recent Progress in Special Cements in China. In *Proceedings of the 11th International Congress on the Chemistry of Cement (ICCC)*, Durban, South Africa, May 11–16, 2003, vol. 4, pp. 2028–2032.
510. Chatterjee A.K. High Belite Cement: Present Status and Future Technology Options. *Cement Concrete Research*, 26(8), 1213–1237, 1996.
511. Harada S., Masuda T., Takata M., Sasaki K. The Study on Mix Proportion and Fundamental Properties of High Performance Concrete Using Belite-Rich Portland Cement. *JCA Proceeding of Cement and Concrete*, 50, 630–635, 1996.
512. Sui T., Li J., Peng X., et al. A Comparison of HBC and MHC Massive Concrete for Three Gorges Project in China. In *Proceedings of Measuring, Monitoring and Modeling Concrete Properties, Greece*, March 2006, pp. 341–346.
513. Sui T., Guo S., Liu K., et al. Research on High Belite Cement. Parts I and II. Presented at 4th Beijing International Symposium on Cement and Concrete, October 26–29, 1998.
514. Sui T., Wang J., Wen Z., et al. Strength and Pore Structure of High Belite Cement. In *Proceedings of the 5th International Symposium on Cement and Concrete*, Shanghai, October 2002, vol. 1, pp. 261–265.
515. Sui T., Fan L., Wen Z., et al. Study on the Properties of High Strength Concrete Using High Belite Cement. *Journal of Advanced Concrete Technology*, 2(2), 201–206, 2004.
516. Taylor H.F.W. *Cement Chemistry*. 2nd ed. Thomas Telford, London, 1997.
517. Sui T., et al. Materials Innovation on Cement Based Materials—China's Effort. Presented at Proceedings of the 17th International Conference on Building Materials, Weimar, Germany, 2009.
518. Jahren P. Examples of High Strength Concrete Abroad. Presented at Norwegian Concrete Day, Oslo, Norway, October 31–November 1, 1986.
519. http://en.wikipedia.org7wiki/World_population (accessed October 11, 2011).

Index